Recent Results
in Cancer Research

99

Founding Editor
P. Rentchnick, Geneva

Managing Editors
Ch. Herfarth, Heidelberg · H. J. Senn, St. Gallen

Associate Editors
M. Baum, London · V. Diehl, Köln
C. von Essen, Villigen · E. Grundmann, Münster
W. Hitzig, Zürich · M. F. Rajewsky, Essen

Recent Results in Cancer Research

Peptide Hormones in Lung Cancer

Edited by
K. Havemann, G. Sorenson, and C. Gropp

With 100 Figures and 63 Tables

Springer-Verlag Berlin Heidelberg GmbH

Professor Dr. Klaus Havemann
Klinikum der Philipps-Universität Marburg
Zentrum für Innere Medizin
Medizinische Klinik und Poliklinik
Schwerpunkt Hämatologie, Onkologie, Immunologie
Baldingerstrasse, 3550 Marburg, FRG

George Sorenson, MD
Dartmouth Medical School
Department of Pathology
Hanover, NH 03756, USA

Priv. Doz. Dr. Claus Gropp
Klinik Bergisch-Land
Im Saalscheid 5, 5600 Wuppertal-Ronsdorf 21, FRG

ISBN 978-3-540-15504-1   ISBN 978-3-642-82533-0 (eBook)
DOI 10.1007/978-3-642-82533-0

© Springer-Verlag Berlin Heidelberg 1985
Originally published by Springer-Verlag Berlin Heidelberg New York Tokyo in 1985
Softcover reprint of the hardcover 1st edition 1985

Typesetting, : Appl, Wemding
2125/3140-543210

Preface

The incidence of lung cancer has reached epidemic proportions throughout the civilized world. One indication of the dimensions of this problem is that in the United States lung cancer has become the leading cause of cancer death in women as well as men. In 1912 there was a "nearly complete consensus of opinion that primary malignant neoplasms of the lung (were one) of the rarest forms of disease," according to Adler. By 1937, however, it had become clear that the incidence of lung cancer was increasing significantly; this increase has been progressive ever since.

It is now well known that some lung cancers give rise to a variety of hormones which, at times, produce clinical manifestations. The association of hormone production with a "nonendocrine" tumor raises many questions, the answers to which may shed some light on the etiology of this prevalent form of cancer. This fascinating problem has stimulated a wide variety of studies in both the clinical and the basic sciences. A number of the more recent studies in this field were discussed at the International Symposium on Peptide Hormones and Lung Cancer held in Marburg, West Germany, on June 18–20, 1984. This volume contains the papers that were presented on this occasion.

Do the cell of origin for lung cancer produce hormones even before the cancer develops? Or is this hormone production a phenotypic alteration associated with malignant transformation? Cells producing hormones such as calcitonin, bombesin, leu-enkephalin, and neurotensin have been described in the bronchial mucosa. If these cells are the cells of origin for lung cancer, then it follows that these tumors can be considered comparable to other tumors derived from endocrine cells already producing hormones. If, on the other hand, a phenotypic alteration occurs, a different question arises: Since hormone production occurs in all forms of lung tumors, including adenocarcinomas, why does it occur only rarely, if at all, in adenocarcinomas arising in other tissues, such as the breast or colon?

Another issue discussed at the conference was the comparison of hormone production in lung tumors and in normal cells. Is there a difference between these two processes? If so, what does it consist in? And if not, what can be learned about normal hormone production by studying hormone production in tumor cells? Much of the current knowledge in this area has been gleaned from studies of long-term cultures that have been established from lung tumor cells. Several papers at the symposium dealt with this rewarding work.

Hopes were raised earlier that hormone production would prove useful as a tumor marker for diagnostic purposes and also as a tool for measuring the amount of tumor present and following the course of tumor growth. These hopes have not been completely realized, though there have been some notable successes, which were described at the symposium.

A major question in this area is: Can studies of hormones production lead to better methods of treatment? The answer to this question appears to be yes, insofar as the hormones produced by these tumors may act as autocrine growth factors in many cases, so that intervention in this process may lead to the inhibition of cell growth.

The introduction of molecular biological techniques has caused an explosive increase in the range of laboratory techniques available. These new approaches have shifted the emphasis of studies in this area from the peptide hormone to the nucleic acid level, where they focus either on mRNA or on the genome. This significant shift in the scientific approach has prepared the way for exciting new discoveries in this field, many of which are presented in the papers contained in this volume.

K. Havemann
G. Sorenson
C. Gropp

Contents

List of Contributors*

Abe, K. *107*[1]
Adachi, I. *107*
Baylin, S. B. *237*
Becker, K. L. *17, 167*
Bennett, H. P. J. *34*
Bepler, G. *157*
Bister, K. *221*
Bleehen, N. M. *175*
Bloom, S. R. *1*
Bonner, T. I. *221*
Bork, E. *180*
Carney, D. N. *157, 167*
Cate, C. C. *130, 143*
Chahinian, A. P. *187*
Craig, R. K. *71*
Deftos, L. J. *167*
Edbrooke, M. R. *71*
Emson, P. C. *175*
Gazdar, A. F. *157, 167*
Go, W. *167*
Goedert, M. *175*
Griffin, C. A. *237*
Gropp, C. *79, 117, 194*
Hansen, M. *180*
Havemann, K. *79, 117, 194*
Hernandez, O. *56*
Hesch, R.-D. *88*
Holle, R. *194*
Ihle, J. *221*
Jansen, H. W. *221*

Kameya, T. *107*
Keenan, K.-P. *94*
Kern, H. F. *117*
Kimura, S. *107*
Kodama, T. *107*
Krauss, S. *177*
Lazarus, L. H. *56*
LeRoith, D. *209*
Liang, V. *167*
Luster, W. *79, 117*
Marangos, P. J. *167*
Maurer, L. H. *187*
Mayer, H. *88*
McDowell, E. M. *94*
McVey, J. H. *71*
Milhaud, G. *67*
Moelling, K. *221*
Moody, T. W. *167*
North, W. G. *187*
O'Donnell, J. *187*
Parker, D. *71*
Perry, M. *187*
Pettengill, O. S. *130, 143*
Polak, J. M. *1*
Rapp, U. R. *221*
Ratcliffe, J. G. *46*
Reeve, J. G. *175*
Richardson, J. *29*
Riley, J. H. *71*
Röher, H. D. *117*

---

* The address of the principal author is given on the first page of each contribution
1 Page on which contribution begins

Occurrence and Distribution of Regulatory Peptides in the Respiratory Tract

J. M. Polak and S. R. Bloom

Departments of Histochemistry and Medicine, Royal Postgraduate Medical School, Hammersmith Hospital, Du Cane Road, London W12 OHS, Great Britain

Introduction

It has been recognised for some time that regulatory peptides are distributed throughout the body in a diffuse neuroendocrine system (Polak and Bloom 1983).

The respiratory tract is well provided with a wide variety of regulatory peptides (Table 1) which are found in specialised mucosal endocrine cells (previously known as Feyrter [clear] cells [Feyrter 1938], Kultchitzky cells or APUD cells [Pearse 1969] and/or in the lung innervation (Polak and Bloom 1982a, b, 1984a, b).

Table 1. Main established regulatory peptides found in the respiratory tract (known molecular variants are omitted for simplicity)

Peptide	Found in[a, b]	Main known actions[b]
Vasoactive intestinal polypeptide (VIP)	N	Muscle relaxation, vasodilatation, secretion
Peptide with histidine and isoleucine (pig) (PHI)[1] or methionine (man) (PHM)	N	Muscle relaxation, secretion
Cholecystokinin (CCK)	NK	NK
Gastrin-releasing peptide (GRP)	C	Trophic
(Bombesin)	C	Trophic C
Substance P	N	Sensory, vasodilation, muscle relaxation
Neuropeptide with tyrosine (NPY)[7]	N	Vasoconstriction
Somatostatin	NK	Release and action of many peptides; may be antitrophic
Calcitonin	C	NK
Calcitonin gene-related peptide (CGRP)[9]	N and C	May be sensory
Enkephalin (met- and leu-)	C	NK

[a] *N*, nerve cells; *C*, endocrine cells
[b] *NK*, not known

Means of Investigating the Regulatory Peptide System of the Lung

Many modern techniques have been used to study the regulatory peptides of the lung. The neurobiological approach includes surgical and pharmacological treatment to determine the nature of the peptide-containing nerves; physiological studies of the mode of action of the peptides have been carried out in experimental animals and in isolated tissues or cultured cells; and biochemistry has been used in receptor assays.

We ourselves have been using radiommunoassay combined with chromatography (Bloom and Long 1982) and immunocytochemistry (Polak and Van Noorden 1983) to analyse the occurrence and distribution of regulatory peptides in the respiratory tract of man and other animals. Radioimmunoassay and chromatography provide information about the chemical nature of an extractable peptide, and immunocytochemistry is used to determine its precise localisation in endocrine cells and nerves.

Immunocytochemistry using antibodies to neuron-specific enolase, which is a glycolytic enzyme originally extracted from the brain, permits the complete depiction of this diffuse neuroendocrine system comprising all the specialised endocrine cells containing regulatory peptides and/or amines and all classes of nerve containing both classical and novel neurotransmitters, including regulatory peptides (Polak and Marangos 1984) (Fig. 1).

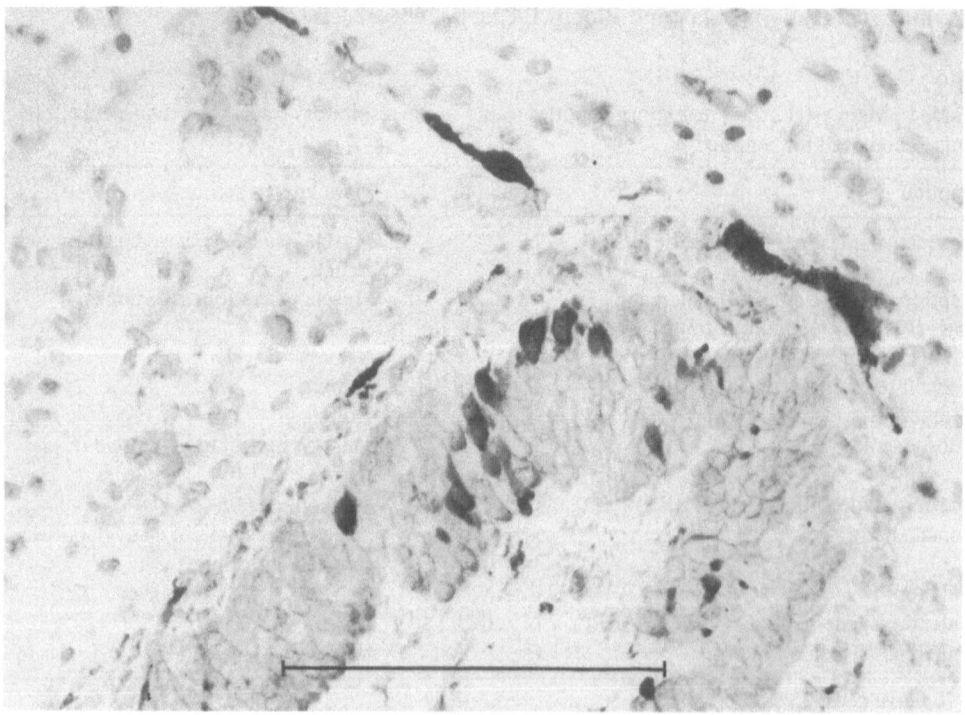

Fig. 1. Neuron-specific enolase (NSE) immunoreactivity in endocrine cells, neuronal cells and nerves in the human fetal lung. Benzoquinone vapour fixation; 5 μm section; peroxidase anti-peroxidase method. *Scale bar* 100 μm

Individual Regulatory Peptides

Substance P and Other Peptides of the Tachykinin Family

Substance P is an 11 amino acid peptide originally discovered by Von Euler and Gaddum (1931); its amino acid sequence was later disclosed by Leeman and Hammerschlag (1967). Substance P belongs to a large family of related peptides called the tachykinins (Fig. 2), many of which were originally found in amphibian skin. Substance P is present in the respiratory tract of man and other animals. It is present in the lung in a network of nerve fi-

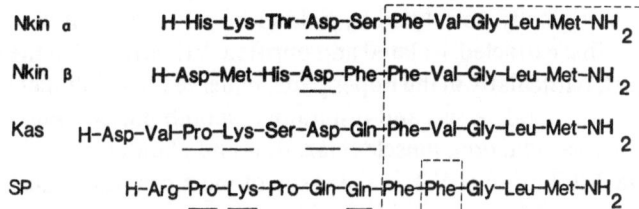

Fig. 2. Schematic representation of the amino acid sequences of various peptides of the tachykinin family

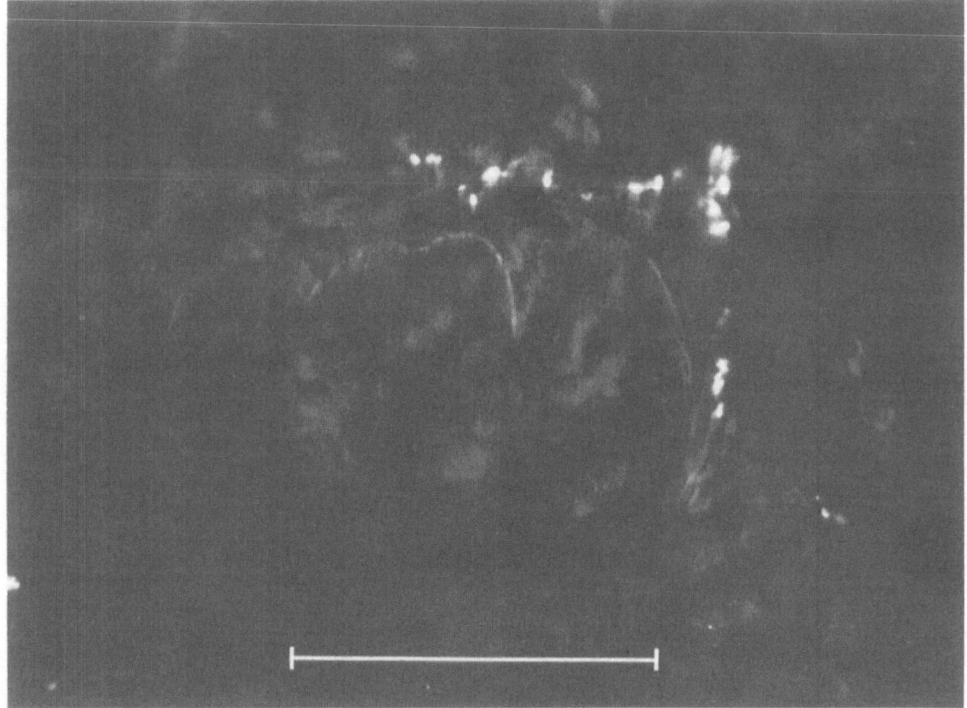

Fig. 3. Substance P-immunoreactive nerve fibre penetrating the respiratory epithelium in the trachea. Benzoquinone fixation; 20-µm section; immunofluroescence method. *Scale bar* 100 µm

bres which innervate the bronchial epithelium and the vascular and non-vascular smooth muscle (Fig. 3). Substance P nerve fibres originate from primary sensory neurons in the nodose ganglion. This finding was obtained in experimental animals by use of neurosurgical and pharmacological procedures including vagal ligation and treatment with capsaicin, an agent extracted from red pepper that induces a marked release of neurotransmitter from the terminals of sensory nerve fibres (Fitzgerald 1983; Lundberg and Saria 1983; Sheppard et al. 1983; Terenghi et al. 1983).

Vasoactive Intestinal Polypeptide/Peptide with Histidine and Isoleucine

The existence of vasoactive intestinal polypeptide (VIP), a 28 amino acid peptide (Fig. 4), was originally postulated by Said (1967). It was from the gastrointestinal tract that VIP was first extracted, isolated and purified. VIP is found in the innervation of the respiratory tract, particularly in the upper parts, in nerve fibres intimately associated with seromucous glands, blood vessels and respiratory smooth muscle (Fig. 5). The main effects of VIP include vasodilation, muscle relaxation and glandular secretion, and match very well with the distribution of VIP in the tissue of the respiratory tract.

A novel 27 amino acid peptide, found some time after its discovery to be associated with VIP, has recently been described. Its discoverers termed this peptide PHI (peptide with histidine and isoleucine) (Fig. 4) (Tatemoto and Mutt 1981). When specific antibodies to PHI were raised its presence in the respiratory tract was noted (Christofides et al. 1985) and, as in many other organs, it was observed that its distribution was very similar to that of VIP. Studies on the concentrations of the two peptides showed an almost equimolar distribution of VIP and PHI in the various areas of the respiratory tract.

Because of their closeness of localisation it was suspected that these two quite separate peptides originated from the same precursor (Christofides et al. 1982a). This hypothesis has now been shown to be correct by two separate groups using recombinant DNA technology and mRNA extracted from two different tumours – one pancreatic and one neural – producing VIP. Both Itoh et al. (1983) and Bloom et al. (1983) have shown that the complete sequence of PHI (in man PHM, as the last C-terminal acid is methionine) is included in the sequence of the pre-pro VIP molecule (Fig. 6).

PHI, like VIP, is especially prevalent in nerves of the upper regions of the respiratory tract around seromucous glands, blood vessels and smooth muscle (Fig. 6).

The main source of both classes of nerves is local, from cell bodies present, in particular, in the tracheal wall. The sphenopalatine ganglion (Lundberg et al. 1981) is the origin of VIP nerve fibres found in the nasal mucosa.

	1	2	3	4	5	6	7	8	9	10
VIP	His-	Ser-	Asp-	Ala-	Val-	Phe-	Thr-	Asp-	Asn-	Tyr
PHI	His-	Ala-	Asp-	Gly-	Val-	Phe-	Thr-	Ser-	Asp-	Phe

	11	12	13	14	15	16	17	18	19	20
VIP	Thr-	Arg-	Leu-	Arg-	Lys-	Gln-	Met-	Ala-	Val-	Lys
PHI	Ser-	Arg-	Leu-	Leu-	Gly-	Gln-	Leu-	Ser-	Ala-	Lys

	21	22	23	23	25	26	27	28
VIP	Lys-	Tyr-	Leu-	Asn-	Ser-	Ile -	Leu-	Asn-NH₂
PHI	Lys-	Tyr-	Leu-	Glu-	Ser-	Leu-	Ile-NH₂	

Fig. 4. Amino acid sequences making up VIP and PHI

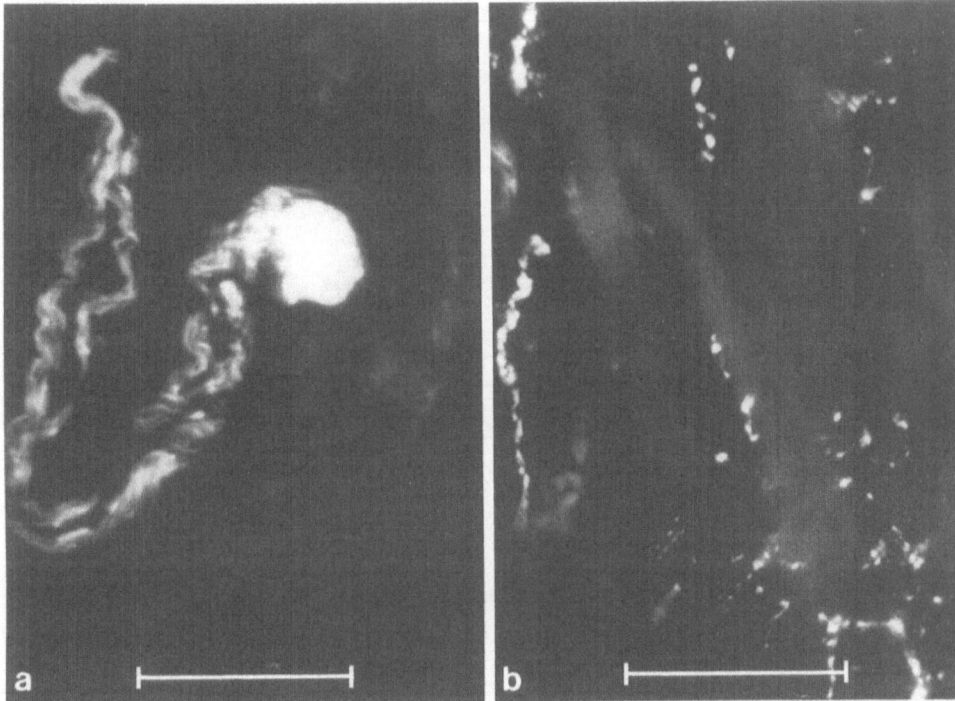

Fig. 5a, b. VIP-immunoreactive ganglion cell (**a**) and nerve fibres (**b**) in the smooth muscle of the respiratory tract. Benzoquinone fixation; 20-μm section; immunofluorescence method. *Scale bar* 5 μm

While VIP is a potent vasodilator and PHI is not, they both relax smooth muscle (Christofides at al. 1982b) and are secretomotor (Ghiglione et al. 1982), and they are probably equipotent for these two actions. Their physiological role in controlling pulmonary blood flow, on both a whole organ and a local tissue basis, and their influence on airway diameter and epithelial secretion are still unclear, but are likely to be important. This has led to the proposal that VIP may be useful in the treatment of asthma, relaxing bronchial smooth muscle and increasing watery secretions to loosen mucosal plugs. Unfortunately, "drug delivery" is a problem. Aerosols of VIP with the requisite droplet diameter are difficult to obtain, and the local tissue penetration is poor. The rapid destruction rate of VIP, which has two intrinsic double basic bonds, rendering it very susceptible to trypsin-like enzyme proteolysis, makes it particularly difficult to achieve adequate tissue levels by airway administration. Whether PHI would be better in this regard is untested. An alternative approach, i. v. infusion, has proved unacceptable, at least for VIP, because of the severe systemic vasodilation and tachycardia (Barnes et al. 1984).

Neuropeptide Y

Neuropeptide Y (NPY) is a 36 amino acid peptide also discovered by Tatemoto and Mutt of the Karolinska Institutet (Tatemoto et al. 1982). It belongs to the PP family of peptides (Fig. 7) and the structure of pre-pro NPY has recently been disclosed by recombinant

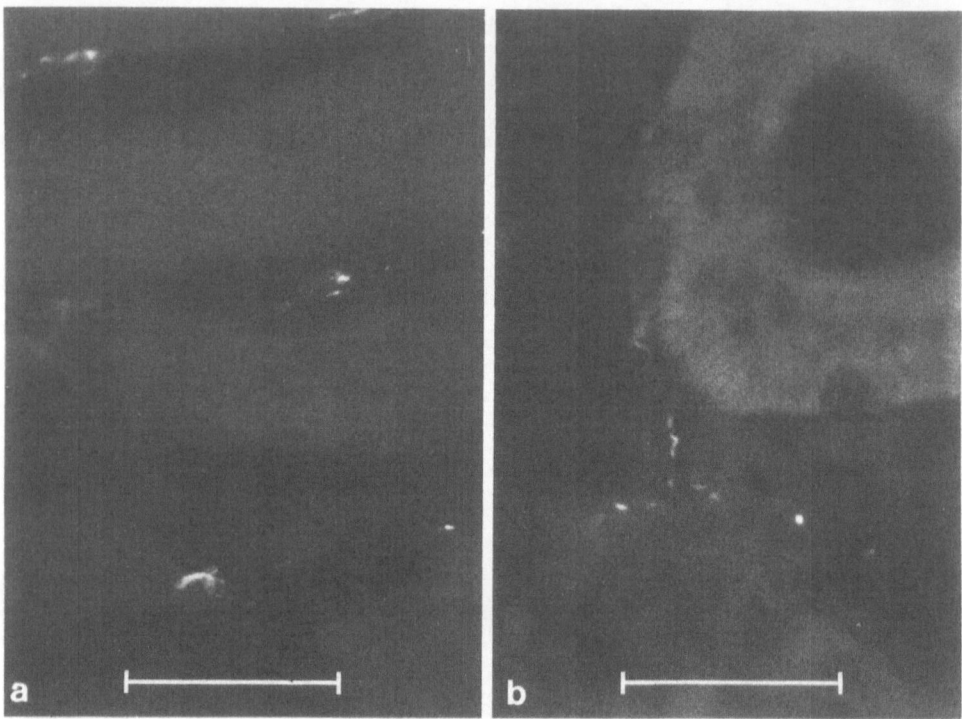

Fig. 6. a PHI immunoreactivity in the smooth muscle of pig trachea. Benzoquinone solution fixation; 10-μm cryostat section; immunofluorescence method. *Scale bar* 50 μm. **b** PHI nerve fibre near the ducts of seromucous glands of pig nasal mucosa. Benzoquinone solution fixation; 10-μm cryostat section; immunofluorescence method. *Scale bar* 50 μm

PP: (APP)   Gly Pro Ser Gln Pro Thr Tyr Pro Gly Asp Asp Ala

             Pro Val Glu Asp Leu Ile Arg Phe Tyr Asp Asn Leu

             Gln Gln Tyr Leu Asn Val Val Thr Arg His Arg Tyr

             amide

NPY: Tyr Pro Ser Lys Pro Asp Asn Pro Gly Glu Asp Ala

             Pro Ala Glu Asp Leu Ala Arg Tyr Tyr Ser Ala Leu

             Arg His Tyr Ile Asn Leu Ile Thr Arg Gln Arg Tyr

             amide

PYY: Tyr Pro Ala Lys Pro Glu Ala Pro Gly Glu Asp Ala

             Ser Pro Glu Glu Leu Ser Arg Tyr Tyr Ala Ser Leu

             Arg His Tyr Leu Asn Leu Val Thr Arg Gln Arg Tyr

             amide

Fig. 7. Amino acid sequence of pancreatic polypeptide and related peptides

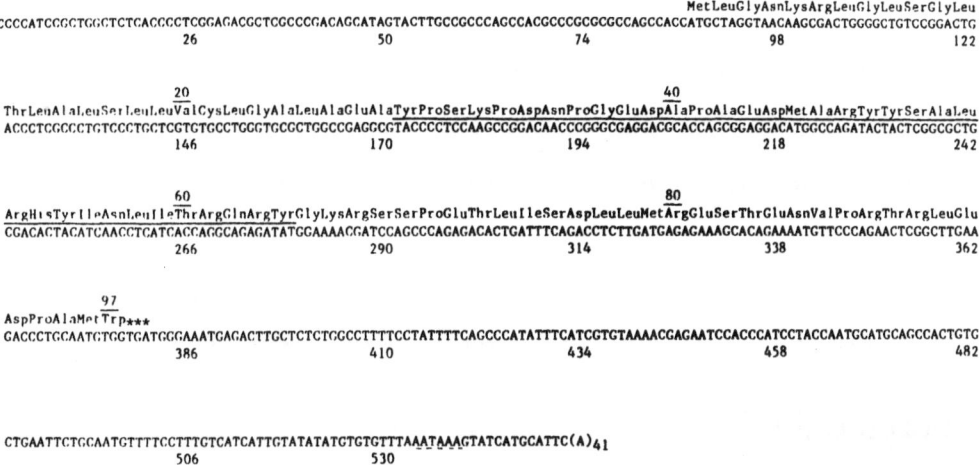

```
                                                                           MetLeuGlyAsnLysArgLeuGlyLeuSerGlyLeu
ACCCCATCCGCTGGCTCTCACCCCTCGGAGACGCTCGCCCGACAGCATAGTACTTGCCGCCCAGCCACGCCCGCGCGGCCAGCCACCATGCTAGGTAACAAGCGACTGGGGCTGTCCGGACTG
              26                    50                    74                    98                   122
```

```
                      20                                                  40
            ThrLeuAlaLeuSerLeuLeuValCysLeuGlyAlaLeuAlaGluAlaTyrProSerLysProAspAsnProGlyGluAspAlaProAlaGluAspMetAlaArgTyrTyrSerAlaLeu
ACCCTCGCCCTGTCCCTGCTCGTGTGCCTGGGTGCGCTGGCCGAGGCGTACCCCTCCAAGCCGGACAACCCGGGCGAGGACGCACCAGCGGAGGACATGGCCAGATACTACTCGGCGGCTG
              146                   170                   194                   218                   242
```

```
                      60                                                  80
            ArgHisTyrIleAsnLeuIleThrArgGlnArgTyrGlyLysArgSerSerProGluThrLeuIleSerAspLeuLeuMetArgGluSerThrGluAsnValProArgThrArgLeuGlu
CGACACTACATCAACCTCATCACCAGGCAGAGATATGCAAAACGATCCAGCCCAGAGACACTGATTTCAGACCTCTTGATGAGAGAAAGCACAGAAAATGTTCCCAGAACTCGGCTTGAA
              266                   290                   314                   338                   362
```

```
             97
AspProAlaMetTrp***
GACCCTGCAATGTGGTGATGGGAAATGAGACTTGCTCTCTGGCCCTTTTCCTATTTTCAGCCCATATTTCATCGTGTAAAACGAGAATCCACCCATCCTACCAATGCATGCAGCCACTGTG
              386                   410                   434                   458                   482
```

```
CTGAATTCTGCAATGTTTTCCTTTGTCATCATTGTATATATGTGTGTTTAAATAAAGTATCATGCATTC(A)41
              506                   530
```

Fig. 8. The nucleotide sequence encoding the precursor to human NPY. The amino acid sequence of pre-pro NPY is numbered 1 through 97. The termination codon is denoted by asterisks. The mature hormone is underlined. The putative signal for polyA addition is underlined with a broken line. (Minth et al 1984)

DNA technology (Minth et al. 1984) (Fig. 8). In the respiratory tract NPY is found mostly co-stored with noradrenergic nerves of the respiratory tract (Sheppard et al. 1984), as the immunocytochemical location of NPY is indistinguishable from that of the catecholamine-synthesising enzyme which delineates noradrenergic nerves.

Bombesin and Gastrin-Releasing Peptide

Bombesin was first extracted from the skin of the discoglossid frog, *Bombina bombina,* which gave its name to the peptide. The counterpart of bombesin in mammals is known as mammalian bombesin or gastrin-releasing peptide (GRP) (McDonald 1981). GRP is composed of 27 amino acids, 7 of which (at the C-terminal end) are shared with bombesin. Lately another peptide, neuromedin C, has been extracted from the mammalian spinal cord (Roth et al. 1983) and has been found to correspond to a large form of GRP (Fig. 9).

Bombesin/GRP is localised to mucosal endocrine cells of the respiratory tract, particularly in man. Antibodies reacting with the C-terminal of bombesin were used for the first demonstration of this peptide in a subclass of mucosal endocrine cells of the lung (Wharton et al. 1978). This finding was later confirmed by several authors.

There are high concentrations of bombesin/GRP immunoreactivity in the fetal and neonatal lung (Ghatei et al. 1983). Bombesin-immunoreactive cells are particularly numerous at this stage and drop in number soon after birth (Wharton et al. 1978). These findings led to the postulate that bombesin/CGRP may play a trophic role, which has been supported recently by observations in human pathology and tissue culture (Rozengurt and Sinnett-Smith 1983).

It is interesting that bombesin has been found to trigger the release of a number of other regulatory peptides (Ghatei et al. 1983). It may be that bombesin acts in part as a regulator of regulators.

GRP

Ala Leu Val Thr Gly Gly Gly Val Ser Val Pro Ala

Lys

Met Tyr Pro Arg Gly Asn His Trp Ala Val Gly His Leu Met amide

Bombesin

pGlu Gln Arg Leu Gly Asn Gln Trp Ala Val Gly His Leu Met amide

Neuromedin B

Gly Asn Leu Trp Ala Thr Gly His Phe Met amide

Neuromedin C

Gly Asn His Trp Ala Val Gly His Leu Met amide

Fig.9. Amino acid sequence similarities of bombesin and related peptides

Calcitonin Gene-Related Peptide

Calcitonin gene-related peptide (CGRP) is a 37 amino acid peptide discovered by Rosen-feld et al. (1983) during their investigation of the calcitonin gene. The calcitonin gene en-codes this novel peptide, which originates after transcription of the heterogeneous nuclear RNA into two separate messenger RNAs, one of which, expressed mostly in neural tissue, directs its synthesis (Figs. 10 and 11).

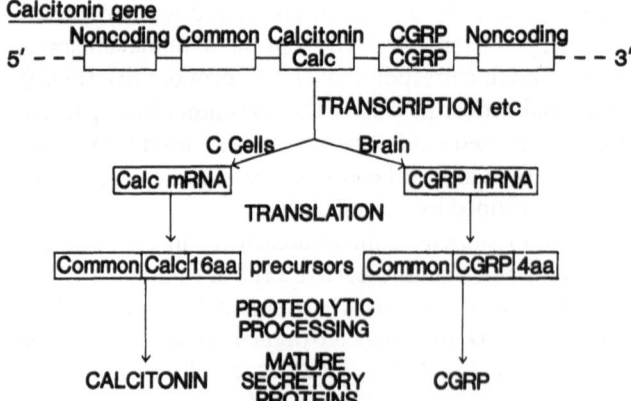

Fig.10. Schematic representation of the common gene encoding the sequence for both calcitonin and CGRP and of the possible translational mechanism for the two peptides in separate types of tis-sue

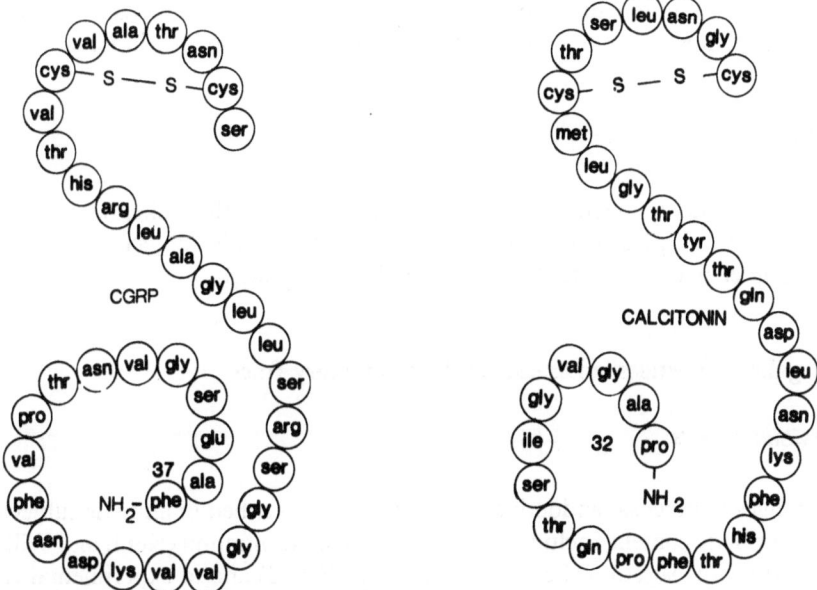

Fig. 11. Schematic representation of the aminoacid sequence of calcitonin gene-related peptide (CGRP) and calcitonin

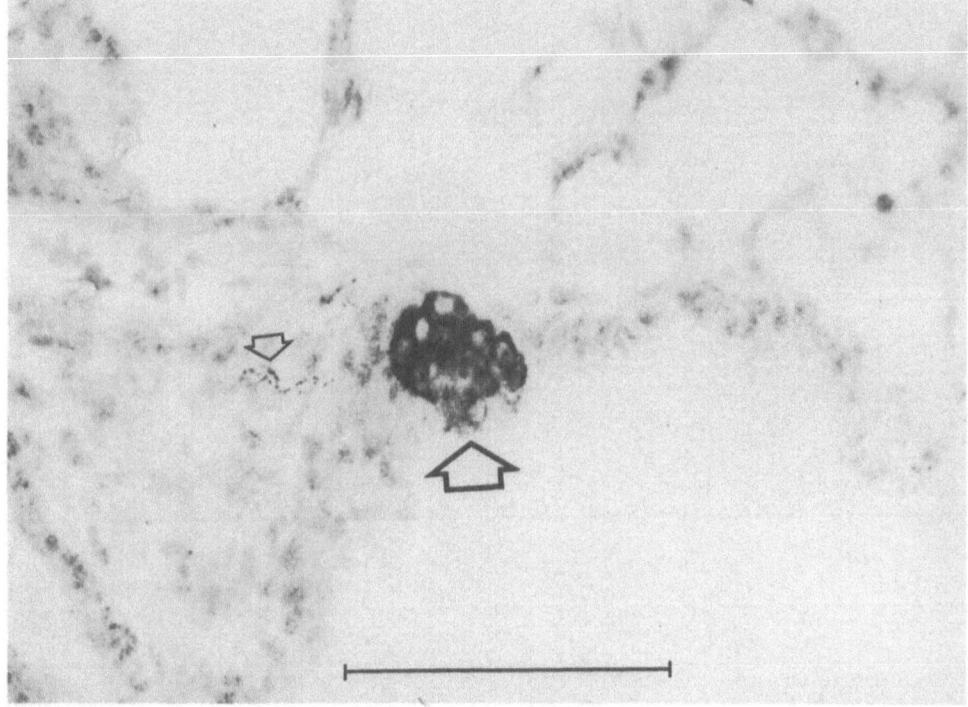

Fig. 12. Group of neuroendocrine cells *(large arrow)* and nerve fibres *(small arrow)* with CGRP-like immunoreactivity in bronchiole of 20-day-old rat. Bouin's fluid fixation; immunogold silver technique. *Scale bar* 100 μm

CGRP is found in the respiratory tract and is localised in mucosal endocrine cells separate from those immunoreactive for bombesin, and in the lung innervation (Fig. 12).

Others

Other regulatory peptides, including calcitonin and the enkephalins, have been discovered to be present in mucosal endocrine cells. This subject is discussed elsewhere (Gropp et al. this volume; Schmelzer et al. this volume).

Regulatory Peptides in Diseases of the Respiratory Tract

Animal Models

Changes in mucosal endocrine cells have been described in animals subjected to altered atmospheric content, including varied oxygen concentration (Keith and Will 1982), to saturation with asbestos (Sheppard et al. 1982) (Fig. 13) and to experimental carcinogenesis induced by nitrosamine (Linnoila et al. 1984; Reznik-Schuller 1984).

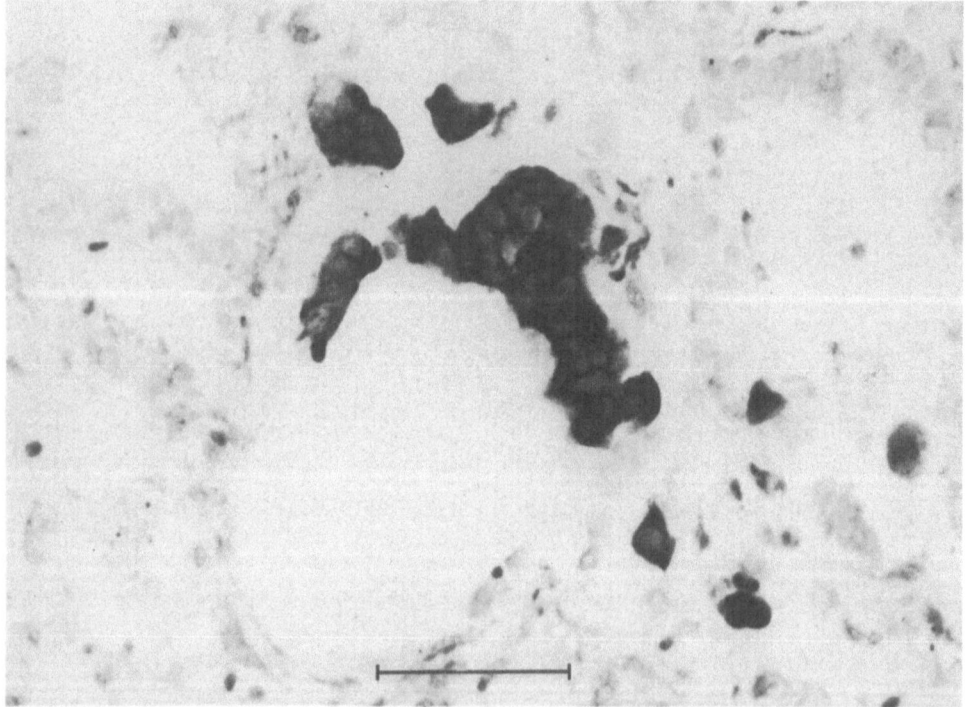

Fig. 13. Neuron-specific enolase (NSE) immunoreactivity in numerous hyperplastic cells in the lung of rat following prolonged exposure to asbestos fibres. Benzoquinone vapour fixation; 5-μm section; peroxidase anti-peroxidase method. *Scale bar* 50 μm

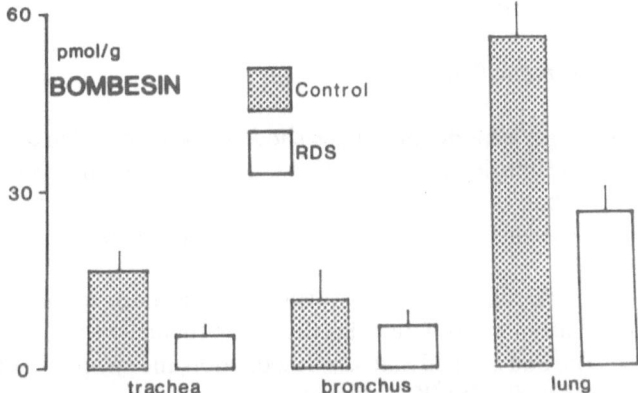

Fig. 14. Bombesin levels in various areas of respiratory tract of newborns: *shaded bar* control; *white bar* newborns with respiratory distress syndrome (RDS). Reproduced, with permission, from *Endocrinology*

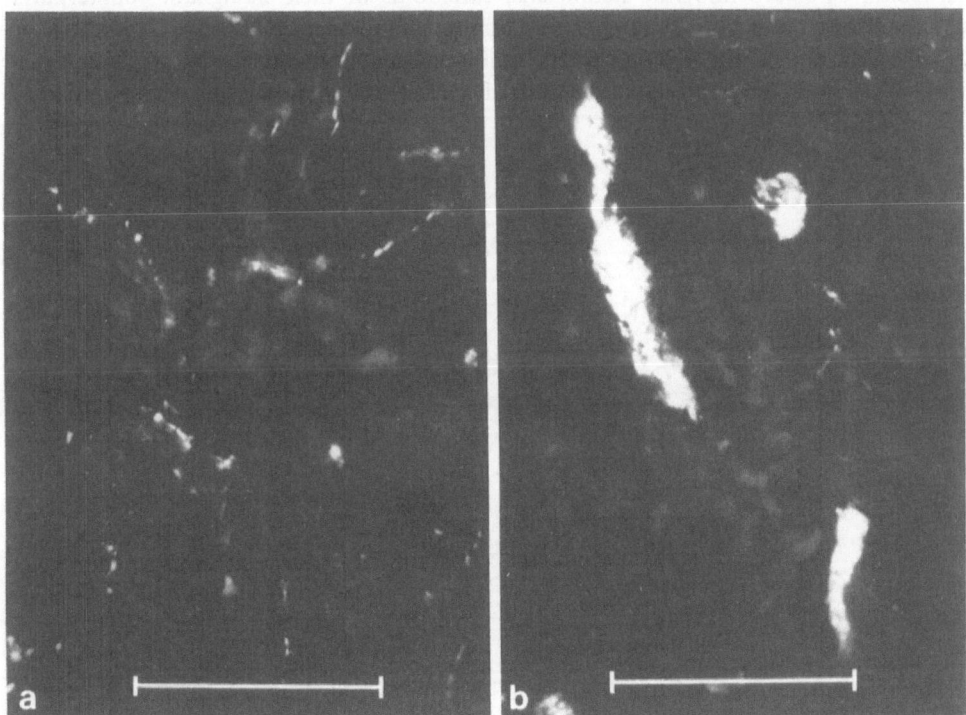

Fig. 15. VIP-immunoreactive nerve fibres in **a** normal nasal mucosa and **b** nasal mucosa of a patient suffering from vasomotor rhinitis. Benzoquinone solution fixation; 20-μm section; immunofluorescence method. *Scale bar* 50 μm

Studies in Man

Non-Tumour Pathology

Regulatory peptides have been shown to be involved in diseases of the lung. For instance, a marked depletion of the bombesinergic system has been observed in the hyperplastic lung of babies dying from respiratory distress syndrome (Ghatei et al. 1983) (Fig. 14). This finding is interesting in view of the postulated growth-related role for bombesin/GRP (see below).

A marked hyperplasia of VIP nerves has been found in a group of patients suffering from vasomotor rhinitis. This disease is characterised by excessive watery secretion from the nose and marked mucosal vasodilation, the symptoms often being resistant to atropine treatment (Fig. 15) (Kurian et al. 1983).

Tumour Pathology

It has long been recognised that hormones may be produced by tumours of the lung, in particular, by neuroendocrine neoplasms, carcinoids or their malignant counterparts, small cell carcinomas of the lung. These tumours show neuroendocrine differentiation which can be demonstrated by the use of antibodies to neuron-specific enolase and is particularly useful in the poorly granulated small cell carcinoma of the lung (Sheppard et al.

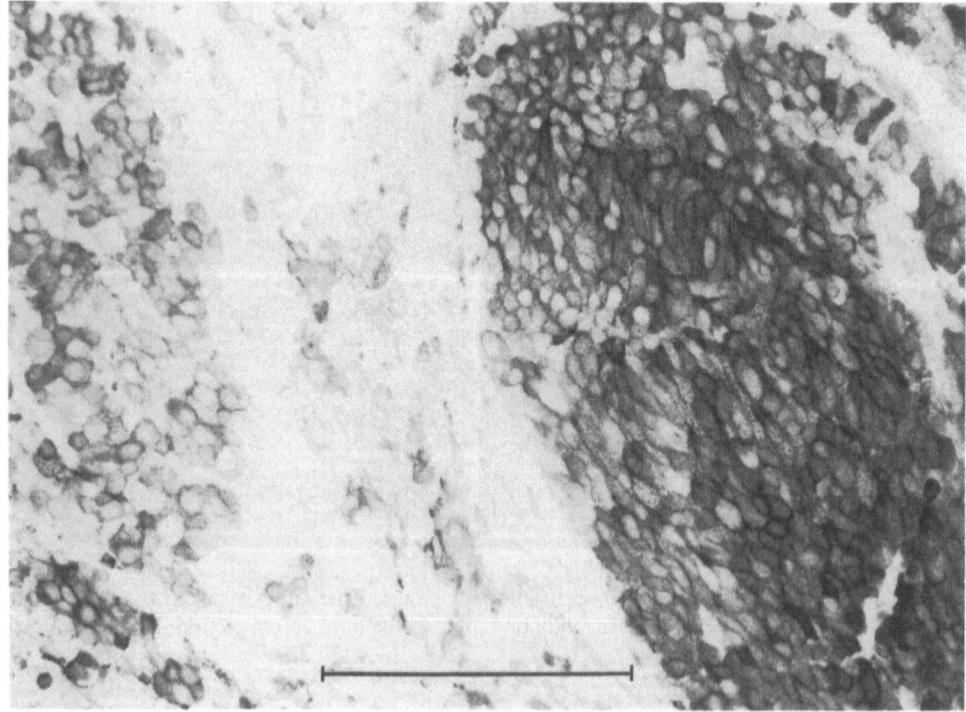

Fig. 16. NSE-immunoreactivity in small cell carcinoma of the lung. Benzoquinone vapour fixed; 5-μm section, peroxidase anti-peroxidase method. *Scale bar* 100 μm

1984a) (Fig. 16). Bombesin/GRP is commonly produced by neuroendocrine neoplasms of the lung (Wood et al. 1981; Erisman et al. 1982; Moody et al. 1983; Sorenson et al. 1982), and its production by small cell carcinomas has been used as a model to support the hypothesis that bombesin is a putative growth factor. Tumour cell lines from small cell carcinoma of the lung are better maintained in culture after the addition of bombesin/GRP to the culture medium (Oie et al. 1984), and small cell carcinomas injected into nude mice fail to grow if the animal is also treated with monoclonal antibodies to bombesin (Moody et al. 1981). Details of bombesin/GRP production in small cell carcinomas are given by Gazdar et al. (this volume) (Fig. 17). Other peptides may also be produced (Wood et al. 1981), and we have recently found both growth hormone releasing factor (GRF) and CGRP production by small cell carcinomas.

Conclusion

A multimembered system of active regulatory peptides has been found in the respiratory tract. Studies on the neuroendocrine marker, neuron-specific enolase, indicate that there are still further types of mucosal endocrine cells and nerves whose products remain to be elucidated. This is not surprising in view of the rate of discovery of novel regulatory peptides by new biochemical methods and recombinant DNA technology. It is well known that the lung possesses functions in addition to mere respiration, and these functions may well be regulated by the various regulatory peptides present in the respiratory tract. Many

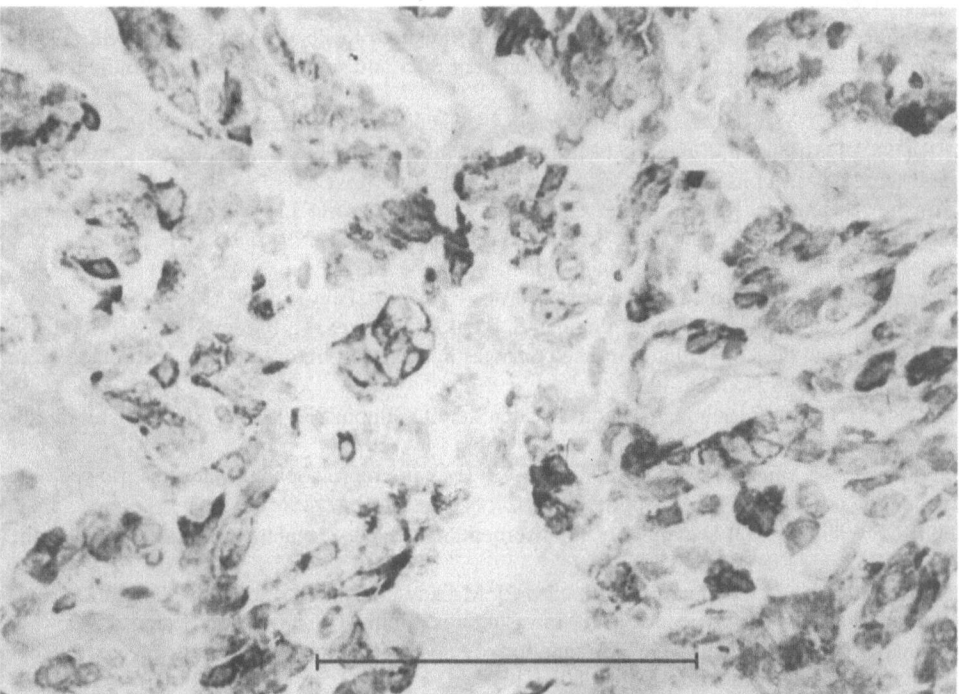

Fig. 17. Bombesin immunoreactivity in a neuroendocrine neoplasm of the lung. Benzoquinone vapour fixation; 5-μm section; peroxidase anti-peroxidase method. *Scale bar* 100 μm

diseases of the lung are still poorly understood. The discovery of regulatory peptides involved in regulating mucosecretion, blood flow and muscle tone will undoubtedly contribute to the further understanding of these diseases.

The basic mechanism of tumour growth and the biology of tumour cells will also be better understood after further study of the active regulatory peptides, many of which play an important role as autocoids in the maintenance of tumour growth.

It is clear that regulatory peptides are also becoming a province of the pharmaceutical industry. Analogues are continually being discovered which are capable of blocking or enhancing the release of regulatory peptides, or of reducing their effect at the receptor site.

Many years of very exciting multidisciplinary research clearly lie ahead for all of us.

Acknowledgements. We would like to express our thanks to the Council for Tobacco Research, USA, who kindly provided the financial support for the investigations which led to this paper.

References

Barnes PJ, Bloom SR, Dixon CMS (1984) VIP and asthma. Lancet I: 112

Bloom SR, Long RG (eds) (1982) Radioimmunoassay of gut regulatory peptides. Saunders, London

Bloom SR, Christofides ND, Delamarter J, Buell G, Kawashima E, Polak JM (1983) Diarrhoea in VIPoma patients associated with cosecretion of a second active peptide (peptide histidine isoleucine) explained by single coding gene. Lancet II: 1163-1165

Christofides ND, Yiangou Y, Blank MA, Tatemoto K, Polak JM, Bloom SR (1982a) Are peptide histidine isoleucine and vasoactive intestinal peptide co-synthesised in the same pro-hormone. Lancet II: 1398

Christofides ND, Yiangou Y, Piper PJ, Ghatei MA, Tatemoto K, Polak JM and Bloom SR (1982b) Distribution of PHI in the mammalian respiratory tract and some aspects of its pharmacology. Regul Pep 4: 359

Christofides ND, Yiangou Y, Piper PJ, Ghatei MA, Sheppard MN, Tatemoto K, Polak JM, Bloom SR (1985) Distribution of peptide histidine isoleucine (PHI) in the mammalian respiratory tract and some aspects of its pharmacology. Endocrinology (in press)

Erisman MD, Linnoila RI, Hernandez O, di Augustine RP, Lazarus LH (1982) Human lung small cell carcinoma contains bombesin. Proc Natl Acad Sci USA 79: 2379-2381

Feyrter F (1938) Uber diffuse endokrine epitheliale Organe. Barth, Leipzig, pp 6-17

Fitzgerald M (1983) Capsaicin and sensory neurons - a review. Pain 15: 109-130

Ghatei MA, Sheppard MN, Henzen-Logman S, Blank MA, Polak JM, Bloom SR (1983) Bombesin and VIP in the developing lung: marked changes in acute respiratory distress syndrome. J Clin Endocrinol Metab 57: 1226-1232

Ghiglione M, Christofides ND, Yiangou Y, Uttenthal LO, Bloom SR (1982) PHI stimulates intestinal fluid secretion. Neuropeptides 3: 79-82

Itoh N, Obata K, Yanaihara N, Okamoto H (1983) Human preprovasoactive intestinal polypeptide contains a novel PHI-27-like peptide, PMM-27. Nature 304: 547-459

Keith IM, Will JA (1982) Dynamics of the neuroendocrine cell - regulatory peptide system in the lung. Exp Lung Res 3: 387-402

Kurian SS, Blank MA, Sheppard MN, Stanley PJ, Mackay IS, Cole PJ, Bloom SR, Polak JM (1983) Vasoactive intestinal polypeptide (VIP) in vasomotor rhinitis. IRCS Medical Science 2: 425-426

Leeman SE, Hammerschlag R (1967) Stimulation of salivary secretion by a factor extracted from hypothalamic tissue. Endocrinology 81: 803-810

Linnoila RI, Becker KL, Silver OL, Snider RH, Moore CF (1984) Pulmonary endocrine cells in the normal and nitrosamine-treated hamster. In: Becker KL, Gazdar AF (eds) The endocrine lung in health and disease. Saunders, Philadelphia, pp 352-362

Lundberg JM, Saria A (1983) Capsaicin-induced desensitization of airway mucosa to cigarette smoke, mechanical and chemical irritants. Nature 302: 251-253

Lundberg JM, Angegard A, Emson P, Fahrenkrug J, Hökfelt T (1981) Vasoactive intestinal polypeptide and cholinergic mechanisms in cat nasal mucsa: studies on choline acetyl-transferase and release of vasoactive intestinal polypeptide. Proc Natl Acad Sci USA 78: 5255-5259

McDonald TJ (1981) Non-amphibian bombesin-like peptides. In: Polak JM, Bloom SR (eds) Gut hormones, 2nd edn Churchill Livingstone, Edinburgh, pp 407-412

Minth C, Bloom SR, Polak JM, Dixon JE (1984) Cloning characterisation and DNA sequence of a human cDNA encoding neuropeptide tyrosine (NPY). Proc Natl Acad Sci USA 81: 4577-4581

Moody TW, Pert CB, Gazdar AF, Carney DN, Minna JD (1981) High levels of intracellular bombesin characterize human small cell lung carcinoma. Science 214: 1246-1248

Moody TW, Russel EK, O'Donohue TL, Linden CD, Gazdar AF (1983) Bombesin-like peptides in small cell lung cancer: biochemical characterization and secretion from a cell line. Life Sci 32: 487-493

Oie HK, Brower M, Carney DN (1984) Growth factor requirements for in vitro growth of endocrine and nonendocrine lung cancers in serum-free defined media. In: Becker KL, Gazdar AF (eds) The endocrine lung in health and disease. Saunders, Philadelphia, pp 469-475

Pearse AGE (1969) The cytochemistry and ultrastructure of polypeptide hormone-producing cells of the APUD series, and the embryologic, physiologic and pathologic implications of the concept. J Histochem Cytochem 17: 303-313

Polak JM, Bloom SR (1982a) Regulatory peptides and neuron-specific enolase in the respiratory tract of man and other mammals. Exp Lung Res 3: 313-328

Polak JM, Bloom SR (1982b) Distribution of regulatory peptides in the respiratory tract of man and mammals. In: Bloom SR, Polak JM, Lindenlaub E (eds) Systemic role of regulatory peptides - Symposion Oosterbeek, Netherlands, 2.-6. May 1982. Schattauer, Stuttgart, pp 241-269

Polak JM, Bloom SR (1983) Regulatory peptides: key factors in the control of bodily functions. Br Med J 286: 1461-1466

Polak JM, Bloom SR (1984a) Regulatory peptides of the respiratory tract: a newly discovered control system. In: Markini L, Ganong WE (eds) Frontiers in neuroendocrinology vol 8. Raven, New York, pp 199-221

Polak JM, Bloom SR (1984b) Regulatory peptides. In: Becker KL, Gazdar AF (eds) The endocrine lung in health and disease. Saunders, Philadelphia

Polak JM, Marangos PJ (1984) Neuron-specific enolase: A marker for neuroendocrine cells. In: Falkmer S, Hakanson R, Sundler F (eds) Evolution and tumor pathology of the neuroendocrine system: The Proceedings of the Erik K Fernström Symposium. Elsevier Biomedical, Amsterdam, pp 433-480

Polak JM, Van Noorden S (eds) (1983) Immunocytochemistry - practical applications in pathology and biology. Wright, Bristol

Reznik-Schüller HM (1984) An overview of experimental carcinogenesis and endocrine tumours of the lung. In: Becker KL, Gazdar AF (eds) The endocrine lung in health and disease. Saunders, Philadelphia, pp 338-344

Rosenfeld MG, Mermod JJ, Amara SG, Swanson LW, Sawchenko PE, Rivier J, Vale WW, Evans RM (1983) Production of a novel neuropeptide encoded by the calcitonin gene via tissue-specific RNA processing. Nature 304: 129-135

Roth KA, Evans CJ, Lorenz RG, Weber E, Barches JD and Chang J-K (1983) Identification of gastrin releasing peptide-related substances in guinea pig rat and brain. Biochem Biophys Res Commun 112: 528-536

Rozengurt E, Sinnett-Smith J (1983) Bombesin stimulation of DNA synthesis and cell division in cultures of Swiss 3T3 cells. Proc Natl Acad Sci USA 80: 2936-40

Said SI (1967) Vasoactive substances in the lung. In: Proceedings of the Tenth Aspen Emphysema Conference, Aspen Colorado, USA. US Public Health Service Publication no 1787, 223

Sheppard MN, Johnson NF, Cole GA, Bloom SR, Marangos PJ, Polak JM (1982) Neuron specific enolase (NSE) immunostaining: a useful tool for the light microscopical detection of endocrine cell hyperplasia in adult rats exposed to asbestos. Histochemistry 74: 505-513

Sheppard MN, McGregor GP, Kurian SS, Hayes AG, Tyers MB, Morrison SR, Bloom SR, Polak JM (1983) Systemic administration of capsaicin to neonatal rats induces dramatic depletion of substance P containing nerve fibres from the lung. In: Skrabanek P, Powell D (eds) Substance P – Dublin 1983. Boole, Dublin, pp 221–222

Sheppard MN, Corrin B, Bennett MH, Marangos PJ, Bloom SR, Polak JM (1984) Immunocytochemical localisation of neuron specific enolase in small cell carcinomas and carcinoid tumours of the lung. Histopathology 8: 171–181

Sheppard MN, Polak JM, Allen JM, Bloom SR (1984) Neuropeptide tyrosine (NPY): a newly discovered peptide is present in the mammalian respiratory tract. Thorax 39: 326–330

Sorenson GD, Bloom SR, Ghatei MA, Del Prete SA, Cate CC, Pettengill OS (1982) Bombesin production by human small cell carcinoma of the lung. Regul Pept 4: 59–66

Tatemoto K, Mutt V (1981) Isolation and characterisation of the intestinal peptide porcine PHI (PHI-27), a new member of the glucagon secretin family. Proc Natl Acad Sci USA 78: 6603–6607

Tatemoto K, Carlquist M, Mutt V (1982) Neuropeptide Y – a novel brain peptide with structural similarities to peptide YY and pancreatic polypeptide. Nature 296: 659–662

Terenghi G, McGregor GP, Gu J, Huang W-M, Morrison JFB, Bloom SR, Polak JM (1983) Substance P-containing sensory fibres in the rat bladder originate from pelvic and hypogastric nerves. In: Skrabanek P, Powell D (eds) Substance P – Dublin 1983. Boole, Dublin, pp 161–162

von Euler US, Gaddum JH (1931) An unidentified depressor substance in certain tissue extracts. J Physiol (Lond) 192: 74–87

Wharton J, Polak JM, Bloom SR, Ghatei MA, Solcia E, Brown MR, Pearse AGE (1978) Bombesin-like immunoreactivity in the lung. Nature 273: 769–770

Wood SM, Wood JR, Ghatei MA, Lee YC, O'Shaughnessy D and Bloom SR (1981) Bombesin, somatostatin and neurotensin-like immunoreactivity in bronchial carcinoma. J Clin Endocrinol Metab 53: 1310–1312

Peptide Hormones and Their Possible Functions in the Normal and Abnormal Lung

K. L. Becker

Endocrinology, Veterans Administration Medical Center and George Washington University, Medicine and Physiology, George Washington University, Washington, DC, USA

Introduction

The normal or abnormal lung contains a host of hormones: nonpolypeptide and polypeptide. The *nonpolypeptide* hormones produced by the lung include the *arachidonic acid metabolites* and the *biogenic amines*. The arachidonic acid metabolites which are produced in the lungs include protaglandin E_2, protaglandin $F_{2\alpha}$, protaglandin D_2, prostaglandin I_2 (prostacyclin), thromboxane, and the leukotrienes. The term biogenic amines has been somewhat broadly applied to a group of amines which are contained within and act upon many tissues, including the lung. Often, these substances have neurotransmitter roles, and some are found in the dense core granules of neurons and endocrine cells. They include dopamine, norepinephrine, epinephrine, serotonin, and histamine. Two somewhat related compounds are the neurotransmitter, acetylcholine, and the amino acid, gammaaminobutyric acid. All these aforementioned substances play important roles in both the normal and the diseased lung, and are being extensively studied.

The purpose of this chapter is to discuss the *polypeptide* hormones produced by the lung, to outline the conditions in which these hormones are increased within the lung or in the blood, and to speculate on the possible roles of these substances in health and disease. The pulmonary polypeptides to be discussed include adrenocorticotropin (ACTH), angiotensin II, bombesin-like immunoreactivity/gastrin-releasing peptide (BLI/GRP), bradykinin, calcitonin, neurotensin, the opioid peptides, somatostatin, substance P, and vasoactive intestinal peptide (VIP). Table 1 lists these polypeptides and indicates their location within the lung, their major locations elsewhere in the body, and non-neoplastic and neoplastic pulmonary conditions in which blood levels of the hormone are increased (for further details, see Becker 1984).

The elucidation of the action of an intrapulmonary polypeptide is difficult. Until a few years ago, most of the known polypeptide hormones were thought to be produced exclusively by anatomically discrete endocrine glands. Often, these glands lent themselves relatively easily to extirpation without causing the immediate loss of life of the experimental animal. In addition, in the human, there often were pathologic conditions in which the endocrine gland was destroyed, or surgical conditions in which the gland was removed. Thus, one could obtain important information concerning the action of a hormone by observing the physiologic effects of its absence.

Subsequently, one could treat the animal or person with the hormone in question and document the extent to which normalization occurred. However, as we have become suc-

Table 1. Pulmonary polypeptide hormones, their location in the lung and elsewhere, and non-neoplastic or neoplastic pulmonary conditions in which their blood levels can be increased

Polypeptide	Location within lung	Found by RIA of lung tissue	Major locations elsewhere in body	Radioimmunoassay available	Increased blood levels in pulmonary disease or other conditions in which lung might be involved	Increased blood levels in lung cancer
ACTH	Pulmonary endocrine cell of fibrotic bronchioles of man	Normal pig; diseased human and dog lung	Anterior pituitary, brain, gastric mucosa, pancreatic islets, gut wall nerves	Yes	Chronic obstructive pulmonary disease	All cell types, but mostly SCCL
Angiotensin II	Produced by angiotensin-converting enzyme in endothelium	–	Ubiquitous	Yes	Increased angiotensin-converting enzyme: thiourea lung damage in rats, paraquat lung damage in mice, humans with pulmonary sarcoidosis, humans with talc granulomatosis. Decreased angiotensin-converting enzyme: hypoxic rabbits, asthma in humans, bleomycin lung damage in rabbits, pulmonary hypertension in rats, human respiratory distress syndrome of sepsis	– –
BLI/GRP	Pulmonary endocrine cell	Human, hamster, rat, guinea pig, cat	Brain, gut nerves, gut endocrine cells	Yes	–	SCCL
Bradykinin	Kininogens in pulmonary blood	–	Ubiquitous	Yes	Hypoxia, hyperoxia, hypercapnic acidosis, oxygenated fetal lamb, pulmonary embolism, anaphylaxis, respiratory distress syndrome in human neonate, endotoxic shock	Carcinoid

Calcitonin	Pulmonary endocrine cell	Human, monkey hamster, rat, pigeon, lizard	Thyroid C cells, thymus, hypothalamus, gut	Yes	Acute pneumonitis, pulmonary tuberculosis, chronic obstructive pulmonary disease, cystic fibrosis, endotoxic shock	All cell types, but SCCL and bronchial carcinoid in particular
Neurotensin	Pulmonary nerves? Pulmonary endocrine cells?	Chicken, rat	Hypothalamus, gut endocrine cells, gut nerve fibers, adrenal medulla	Yes	–	–
Opioid peptides (β-endorphin, met-enkephalin, leu-enkephalin, met[5]-enkephalin-arg[6]-phe[7])	Pulmonary endocrine cells (leu-enkephalin); pulmonary endocrine cell of fibrotic human lung (β-endorphin)	Rat, pig, guinea pig	Brain, gut nerves, gut endocrine cells, sympathetic ganglia, adrenal medulla	Yes (β-endorphin, met-enkephalin)	Endorphins: hypoxic sheep, human with high altitude pulmonary edema, endotoxic shock	–
Somatostatin	Pulmonary endocrine cell	Rat, cat, diseased human lung	Hypothalamus, gut nerves and gut endocrine cells, pancreatic islets, thyroid C cells	Yes	–	SCCL
Substance P	Pulmonary nerves	Rat, cat, guinea pig	CNS and peripheral nerves, gut nerves and gut endocrine cells	Yes	–	Bronchial carcinoid?
VIP	Pulmonary nerves and ganglia	Rat, cat, guinea pig	CNS and peripheral nerves, pulmonary mast cells	Yes	Anaphylaxis, acute hypoxia and acute respiratory acidosis in dogs, asbestos exposure in rats (increased VIP content of lung tissue and lavage fluid)	Non-SCCL

cessively aware of the endocrine gut, the endocrine brain, and now the endocrine lung, the study of the pathophysiology of polypeptide hormones has become more difficult.

The polypeptides of the lung are diffusely distributed, and it is likely that the principal tissue upon which these hormones act is the lung itself. The total extirpation of the pulmonary hormone under study would involve the total extirpation of the lungs. Obviously, this drastically alters the nature of the experimental preparation. It also removes the very tissue whose effects one wants to study. Furthermore, even if one wished to investigate the impact of total pulmonary extirpation on a distal organ or tissue, one would still have to contend with the large amounts of the hormone which is produced elsewhere in the body.

As a result, most experiments which study the effects of a pulmonary polypeptide involve the administration of pharmacologic dosages of the hormone to an experimental animal (i. v. or directly into the pulmonary arterial circulation) or in vitro studies of pulmonary tissues, e. g., contractile effects on tracheal spiral preparations, mucus production by short-term epithelial cultures, secretion of histamine by enriched preparations of pulmonary mast cells, etc. (Becker 1984).

Pulmonary Peptides

Adrenocorticotropin

Adrenocorticotropin (ACTH) of extrapulmonary origin is known to influence the maturation of the lung; perhaps intrapulmonary ACTH plays a similar role. When fetal sheep are hypophysectomized they demonstrate poor pulmonary maturation, even if they have been treated with cortisol. However, when ACTH is adminstered to such animals, pulmonary distensibility and alveolar surfactant levels return to normal (Liggins et al. 1981). Within the canine lung, the endothelial enzyme angiotensin-converting enzyme, which activates angiotensin I and degrades bradykinin, is inhibited noncompetitively by ACTH (Verma et al. 1982). Thus, ACTH might also modulate the intrapulmonary levels of these two hormones.

Angiotensin

The pulmonary hormone, angiotensin II, is an octapeptide which is produced within the endothelium by the above-mentioned enzymatic cleavage of the inactive plasma decapeptide, angiotensin I. Angiotensin II produces pulmonary vasoconstriction in several species (Chand an Altura 1980). It is known that angiotensin II is released from hypoxic lung, and it has been invoked as a mediator of the hypoxic pulmonary vasoconstrictive reflex (Allison and Clay 1976; Berkov 1974). Intrapulmonary arachidonic acid metabolites, including prostacyclin, are released by angiotensin II (Gryglewski et al. 1980). Interestingly, in the rabbit, the administration of angiotensin II into the right ventricle caused ultrastructural lesions of the alveolar epithelium, changes which could eventuate in alveolar flooding (Gil and McNiff 1983).

Bombesin-Like Immunoreactivity/Gastrin-Releasing Peptide

It seems likely that the mammalian form of bombesin is gastrin-releasing peptide. However, until this question is resolved by chemical analysis, it will be referred to as BLI/GRP. Experimentally, when this hormone is administered intracerebrally to cats there is an increased respiratory tidal volume (Gillis et al. 1983), and in rats there is an increase of minute ventilation due to an increase of both respiratory rate and tidal volume, as well as periodic sighing respirations (Niewoehner et al. 1983). Experiments in which rat lung primordia were transplanted into chick embryos and allowed to develop have demonstrated that fetal hormonal factors influence lung maturation (Marin and Dameron 1981). Perhaps the large amounts of BLI/GRP which occur in fetal lung indicate a role in fetal pulmonary development (Track and Cutz 1982). Indeed, Willey et al. (1984) found that bombesin and GRP enhance the colony-forming efficiency and the clonal growth rate of cultured normal human bronchial epithelial cells. In pharmacologic experiments, the administration of bombesin or GRP into the pulmonary artery of normoxic and hypoxic newborn lambs did not affect pulmonary vascular tone (Kulik et al. 1983). BLI/GRP produces bronchoconstriction in the guinea pig lung which does not appear to be mediated by acetylcholine, histamine, or serotonin (Impicciatore and Bertaccini 1973).

Bradykinin

The nonapeptide bradykinin is made throughout the body by the action of the plasma enzyme kallikrein, which acts upon plasma kininogen. Bradykinin is also generated within the lung; it is inactivated within the lungs by the angiotensin-converting enzyme of the pulmonary endothelium. The effect of bradykinin on the pulmonary vasculature varies with the species and the experimental design. In the dog, bradykinin exerts a vasodilatory effect on isolated intrapulmonary arteries. However, if the endothelium is destroyed, bradykinin causes vasoconstriction (Chand and Altura 1981). This suggests that some of the pulmonary vascular effects of the hormone are mediated by prostacyclin. Indeed, the pulmonary release of prostacyclin, as well as of thromboxane and prostaglandins, can be induced by bradykinin (Vargaftig and Dao Hai 1972; Barst et al. 1983). Also, it is possible that the release of arachidonic acid metabolites might be responsible for the bronchoconstriction caused by bradykinin (Simonsson et al. 1973; Said et al. 1980b; Parrat and Sturgess 1975). In the case of the dog, bronchoconstriction appears to occur via the stimulation of afferent vagal C fibers (Kaufman et al. 1980). In the intact dog, bradykinin was found to increase both pulmonary artery and tracheal pressure (Sugiyama et al. 1982). The substance is a known releaser of histamine from mast cells (Johnson and Erdös 1973). Bradykinin increases the permeability of the pulmonary endothelium of hypoxic sheep (O'Brodovich et al. 1981). In the rat, the hormone competitively inhibits the conversion of angiotensin I to angiotensin II (Ercan et al. 1982). In addition, it is possible that the release of bradykinin, which in the case of the fetal lamb has been shown to be induced by oxygenation (Heymann et al. 1969), assists in causing the pulmonary vasodilation following aeration at birth (Addonizio and Harken 1980; Regoli and Barabé 1980). It has been suggested that it also may play a role in the constriction of the ductus arteriosus in the newborn.

Calcitonin

The total pulmonary content of calcitonin in the lungs of man exceeds that found within the thyroid gland (Becker et al. 1984). Outside the lung, when calcitonin is administered intracerebrally in rats, there is an increased respiratory minute volume due to increases of both respiratory rate and tidal volume (Niewoehner et al. 1983). Within the lung, the high levels of this hormone in the fetus may indicate an influence on the growth of this organ. In this regard, the hormone is known to stimulate the growth of cartilage (Burch 1982). Calcitonin increases the generation of prostacyclin in the pig (Clopath and Sinzinger 1980). Within guinea pig lung, it inhibits the synthesis of prostaglandins and thromboxane (Cesarini et al. 1979).

Many actions within the lung are mediated by the cytoplasmic location and concentration of calcium, e. g., secretion from pulmonary endocrine cells of the fetal rabbit (Sonstegard et al. 1979), release of histamine from pulmonary mast cells (Church et al. 1982), pulmonary vasoconstriction (Voelkel et al. 1980), and the synthesis and pulmonary effects of arachidonic acid metabolites (Wong and Cheung 1979; Weichman et al. 1983). In this regard, calcitonin influences transcellular and intracellular movements of calcium (Yamaguchi et al. 1975; Parkinson and Radde 1969), and the intrapulmonary role of this hormone merits further investigation.

Neurotensin

Neurotensin, which has been found in normal chicken lung (Carraway et al. 1983), also occurs in the lungs of mammals. Perhaps it acts as a neurotransmitter within the lung, as it does elsewhere (Rökaeus 1981). Also, it enhances phagocytosis (Goldman et al. 1982), increases vascular permeability (Carraway and Leeman 1975), and can degranulate mast cells and release histamine (Rökaeus 1981; Sydbom 1982). Interestingly, in experiments in the sheep, the lung did not appreciably clear infused neurotensin (Shulkes et al. 1983).

Opioid Peptides

In the dog, the opioid peptides diminish the synthesis of angiotensin II (Sander et al. 1980). In pharmacologic experiments with the intact dog, leu-enkephalin increased the respiratory rate (Sander et al. 1982). While the hormone increased pulmonary artery pressure if the dog was conscious, it was decreased following anesthesia. None of these reactions occurred following the administration of naloxone, an opioid antagonist. In another in vivo study in anesthetized dogs, the intracoronary injection of met-enkephalin analogs caused no changes in pulmonary circulation (Fitzal et al. 1980). In the rat, an enkephalin analog stimulated J receptors, producing apnea. In another study, the administration of an enkephalin analog into the right atrium of spontaneously breathing decerebrate rats caused an increase in lung resistance and a decrease in dynamic compliance (Willette et al. 1983). These responses, which were not altered by pretreatment with antihistaminic, antiserotinergic, or antimuscarinic agents, appeared to be due to a vagally mediated pulmonary chemoreflex which was triggered by an interaction with pulmonary opiate receptors.

Studies of isolated canine tracheal smooth muscle strips demonstrated that met- and leu-enkephalin depress the contractile response induced by field electrical stimulation, but only minimally blunted the contractile effect of exogenous acetylcholine. The investigators suggested that the enkephalins inhibit the release of endogenous acetylcholine from postganglionic, parasympathetic neurons (Russell and Simons 1984). An analog of enkephalin has induced asthmatic wheezing in man. However, in human asthmatics, the blockage of endogenous opioids with naloxone does not decrease their bronchoconstrictive response to the inhalation of cold air (Weinberger et al. 1982). In a study of chronic obstructive pulmonary disease, endogenous opiates were found to prevent an increase in respiratory effort when the patients were presented with added airway resistance (Santiago et al. 1981).

The i. v. administration of met-enkephalin in awake dogs increased inspiratory minute volume, due to an increase of both tidal volume and respiratory rate. Mean inspiratory flow also increased. The investigators concluded that met-enkephalin may function as a stress hormone (Evanich et al. 1984). Studies in the isolated perfused rat lung suggest that the pulmonary degradation of met-enkephalin may be due, in part, to angiotensin-converting enzyme (Manwaring and Mullane 1984).

Somatostatin

In rats, somatostatin can protect against the hemorrhagic pulmonary lesions which are caused by the oral administration of cysteamine (Schwedes et al. 1978). In humans with hepatic cirrhosis, the i. v. administration of the hormone increased mean pulmonary artery pressure (Hallemans et al. 1981).

Substance P

Substance P, which has a peptidergic function (Nicoll et al. 1980), probably plays the same role within the lung. In this regard, it may function as a sensory neurotransmitter (Terenghi et al. 1983). On the other hand, the hormone can cause tracheobronchoconstriction in the cat and guinea pig (Nilsson et al. 1977; Andersson and Persson 1977). While in one study of isolated specimens of human bronchi there was no response to substance P (Cerrina et al. 1983), another study demonstrated a dose-dependent contraction (Lundberg et al. 1983a). Experiments in guinea pigs, using a substance P antagonist, suggest that substance P nerves cause local bronchial smooth muscle contraction and increase vascular permeability (Lundberg et al. 1983b). It is likely that substance P causes pulmonary vasodilation (Hallberg and Pernow 1975; Pernow and Rosell 1975). In addition, studies of canine tracheal explants have revealed that substance P stimulates the synthesis and release of mucus (Baker et al. 1977). Coles et al. (1984) suggested that the potent induction of glycoprotein release from in vitro canine tracheal mucosa indicates that this hormone might function as a mediator of irritant-induced mucus hypersecretion in the airway. Studies of rat mast cells in vitro show that substance P releases histamine (Johnson and Erdös 1973; Sydbom 1982). It has been postulated that the flushed skin and the pulmonary bronchoconstriction that are associated with the carcinoid tumor might be due to substance P (Skrabanek et al. 1978). Interestingly, the hydrolysis of substance P is catalyzed by rabbit angiotensin-converting enzyme (Cascieri et al. 1984).

Vasoactive Intestinal Peptide

VIP probably plays a peptidergic role in the lungs (Said et al. 1980a). The hormone increases ventilation and causes bronchodilation (Said et al. 1980b; Kitamura et al. 1980; Frandsen et al. 1978). In the guinea pig, the relaxant effect on the trachea is potentiated by corticosteroids and by alpha receptor blockade (Kitamura et al. 1980). An aerosol of VIP protects the dog against bronchoconstriction induced either by histamine or by prostaglandin $F_{2\alpha}$ (Hara et al. 1975). It also counteracts the constrictor effects of leukotriene D_4 on the isolated guinea pig trachea, pulmonary artery, and lung parenchyma (Hamasaki et al. 1982). In cats, VIP dilates pulmonary vessels following the production of increased pulmonary artery pressure with a prostaglandin analog (Mojarad et al. 1980). In the intact dog, the administration of VIP into the left pulmonary artery increased pulmonary artery pressure, increased blood flow to the left lower lobe of the lung, and decreased pulmonary vascular resistance in that lobe (Sugiyama et al. 1982). Studies with in vitro segments of both human bronchi and pulmonary arteries suggest that VIP induces a prolonged relaxation of these structures which is not mediated by cyclo-oxygenase metabolites (Saga and Said 1984). In other in vitro experiments, VIP decreased basal human bronchial mucus production (Coles et al. 1981).

Studies in dogs and in rats suggest that VIP might ameliorate acute hypoxic vasoconstriction. In addition, VIP inhibits antigen-induced histamine release from minced lung of the guinea pig (Undem et al. 1983). Morice et al. (1983) found that the i.v. administration of VIP caused bronchodilation in asthmatic humans and also ameliorated histamine-induced bronchoconstriction.

Further Discussion and Conclusions

The foregoing listing of hormonal polypeptides within the lung is not intended to be complete. There are other polypeptides which may be of physiologic importance within the lung. Also, most biologically active polypeptides are synthesized from larger precursors, which then undergo further intracellular or extracellular processing – perhaps the most marked example of this phenomenon is pro-opiomelanocortin, which contains β-lipotropin, ACTH, β-endorphin, met-enkephalin, melanocyte-stimulating hormone, CLIP (corticotropin-like intermediate lobe peptide), and an N-terminal glycopeptide. Thus, some or all of these substances may occur in lung. Furthermore, it is possible that alternative post-translational processing steps within the diseased lung might produce different polypeptide fragments or different proportions of polypeptides.

It should be emphasized that the synthesis of polypeptide hormones might be a characteristic of all tissues. In studies of extracted tissues, all the tissues we studied contained at least some detectable insulin (Rishi and Becker 1967) or calcitonin (Becker et al. 1979). The difference between a polypeptide-secreting cell and one which does not secrete polypeptides may be quantitative and not qualitative. Some cells, such as the pulmonary endocrine cells of the lung, the islet cells of the pancreas, or the cells of the anterior pituitary, both contain the biochemical machinery (e.g., high levels of dopa decarboxylase and of neuron-specific enolase) and have the requisite structural integration (e.g., the appropriate endoplasmic reticulum and secretion granules) to synthesize large amounts of one or more polypeptide hormones and to secrete them rapidly in response to the appropriate stimulus. Other cells which lack such extensive and finely tuned synthesizing and processing equipment probably can do the same, although at greatly reduced capabilities.

Another important aspect of polypeptide production by the normal lung is discussed at length by Sorenson et al. and Gazdar et al. (this volume), namely, the endocrine characteristics of the tumors which arise from pulmonary cells. In particular, studies of small cell cancer of the lung (SCCL) provide a wealth of information concerning the pulmonary endocrine cell which it so closely resembles (Becker and Gazdar 1983). However, one should interject a note of caution: increased blood levels of a hormone do not necessarily prove that the hormone in question originates from a patients's tumor. For example, a patient with epidermoid cancer of the lung is usually a heavy smoker with chronic obstructive pulmonary disease and/or smoking-related bronchitis. Patients with these last complications often have increased blood levels of calcitonin, ACTH, and β-lipotropin, which are not necessarily produced by the tumor (Becker et al. 1984). Also, interpretations of studies which involve immunoperoxidase staining of a tumor or radioimmunoassay of extracted tumor tissue must take into consideration the fact that pulmonary cancers often are not homogeneous; they may include different cell types (Gazdar 1984) and also may include areas of non-neoplastic adjacent pulmonary parenchyma which contain areas of hyperplastic pulmonary endocrine cells (Becker et al. 1981).

The discovery of polypeptide hormones in the normal lung and the finding that the intrapulmonary and/or blood levels of many of these hormones are increased in acute and chronic benign lung diseases and in neoplasms of the lung have opened up a new field of medicine. To date, several pharmacologic studies have demonstrated that these hormones exert important effects on pulmonary function and pulmonary metabolic activities. One of the pressing tasks of the next several years is the elucidation of the pathophysiologic role of these hormones in the normal human and in the patient with pulmonary disease.

References

Addonizio VP Jr, Harken AH (1980) The surgical implications of nonrespiratory lung function. J Surg Res 28: 86–95

Allison DJ, Clay T (1976) Angiotensin II release from the hypoxic lobe of the intact dog lung. J Physiol (Lond) 260: 32P–33P

Andersson P, Persson H (1977) Effect of substance P on pulmonary resistance and dynamic pulmonary compliance in the anesthetized cat and guinea-pig. Acta Pharmacol Toxicol 41: 444–448

Baker AP, Hillegass LM, Holden DA, Smith WJ (1977) Effect of kallidin, substance P, and other basic polypeptides on the production of respiratory macromolecules. Am Rev Respir Dis 115: 811–817

Barst RJ, Stalcup SA, Mellins RB (1983) Bradykinin-induced changes in circulating prostanoids in unanesthetized sheep. Fed Proc 42: 302

Becker KL (1984) The endocrine lung. In: Becker KL, Gazdar AF (eds) The endocrine lung in health and disease. Saunders, Philadelphia, pp 3–46

Becker KL, Gazdar AF (1983) The pulmonary endocrine cell and the tumors to which it gives rise. In: Reznick-Schuller HM (ed) Comparative respiratory tract carcinogenesis, vol II. CRC Press, Boca Raton, pp 161–187

Becker KL, Snider RH, Moore CF, et al (1979) Calcitonin in extrathyroidal tissues of man. Acta Endocrinol (Copenh) 92: 746–751

Becker KL, Nash DR, Silva OL, et al (1981) Increased serum and urinary calcitonin in pulmonary disease. Chest 79: 211–216

Becker KL, Silva OL, Snider RH, et al (1984) The pathophysiology of pulmonary calcitonin. In: Becker KL, Gazdar AF (eds) The endocrine lung in health and disease. Saunders, Philadelphia, pp 227–299

Berkov S (1974) Hypoxic pulmonary vasoconstriction in the rat. The necessary role of angiotensin II. Circ Res 35: 256–261

Burch W (1982) Calcitonin: a growth factor avian embryonic cartilage in vitro. Clin Res 30: 854A

Carraway R, Leeman S (1975) Structural requirements for the biological activity of neurotensin, a new vasoactive peptide. In: Walter R, Meinenhofer J (eds) Peptides: chemistry, structure and biology. Ann Arbor Science Publishers, Ann Arbor, pp 679–685

Carraway RE, Ruane SE, Ritsema RS (1983) Radioimmunoassay for Lys[8], Asn[9], neurotensin 8–13: Tissue and subcellular distribution of immunoreactivity in chickens. Peptides 4: 111

Cascieri MI, Bull HG, Mumford RA et al (1984) Carboxyl-terminal tripeptidyl hydrolysis of sub-
    stance P by purified rabbit lung angiotensin-converting enzyme and the potentiation of substance
    P activity in vivo by captopril and MK-422. Mol Pharmacol 25: 287–293
Cerrina J, Boullet C, Labat C, et al (1983) Pharmacology of isolated human bronchial muscle. Fed
    Proc 42: 908
Ceserani R, Colombo M, Olgiatri VR, Pecile A (1979) Calcitonin and prostaglandin system. Life Sci
    25: 1851–1855
Chand N, Altura BM (1980) Reactivity and contractility of rat main pulmonary artery to vasoactive
    agents. J Appl Physiol 49: 1016–1021
Chand N, Altura BM (1981) Acetylcholine and bradykinin relax intrapulmonary arteries by acting
    on endothelial cells: Role in lung vascular diseases. Science 213: 1376–1379
Church MK, Pao, GJ-K, Holgate ST (1982) Characterization of histamine secretion from mechani-
    cally dispersed human lung mast cells: effects of anti-IgE, calcium ionophore A23187, compound
    48/80, and basic polypeptides. J Immunol 129: 2116–2121
Clopath P, Sinzinger H (1980) Calcitonin increases porcine vascular prostacyclin formation. Pros-
    taglandins 19: 1
Coles SJ, Said SI, Reid LM (1981) Inhibition by vasoactive intestinal peptide of glycoconjugate and
    lysozyme secretion by human airways in vitro. Am Rev Respir Dis 124: 531–536
Coles SJ, Neill KH, Reid L (1984) Substance P and related peptide induce glycoprotein release by
    canine airway mucosa. Fed Proc 43: 826
Ercan ZS, Bor NM, Torunoglu M, Türker RK (1982) Alteration by kallikrein and bradykinin of the
    conversion of angiotensin I to angiotensin II in the isolated perfused rat lung. Arzneimittelforsch
    32: 30–31
Evanich MJ, Sander G, Giles T (1984) Ventilatory response to intravenous enkephalin in awake
    dogs. Fed Proc 43: 1006
Fitzal S, Gilly H, Pauser G, et al (1980) Cardiovascular effects of two synthetic enkephalin analogues
    following intracoronary administration in dogs. Anaesthesist 29: 579–85
Frandsen EK, Krishna GA, Said SI (1978) Vasoactive intestinal polypeptide promotes cyclic adeno-
    sine 3', 5'-monophosphate accumulation in guinea-pig trachea. Br J Pharmacol 62: 367–369
Gazdar AF (1984) The pathology of endocrine tumors of the lung: an overview. In: Becker KL, Gaz-
    dar AF (eds) The endocrine lung in health and disease. Saunders, Philadelphia, pp 364–372
Gil J, McNiff JM (1983) Alveolar epithelial lesions induced by angiotensin in rabbit lungs. Am J Pa-
    thol 113: 331–340
Gillis RA, Jensen RI, Buller A, et al (1983) Central respiratory and cardiovascular effects of bombe-
    sin. Fed Proc 42: 358
Goldman R, Bar-Shavit Z, Shezen E, et al (1982) Enhancement of phagocytosis by neurotensin, a
    newly found biological activity of the neuropeptide. Adv Exp Med Biol 155: 133–141
Gryglewski RJ, Splawinski J, Korbut R (1980) Endogenous mechanisms that regulate prostacyclin
    release. Adv Prostaglandin Thromboxane Res 7: 777–787
Hallberg D, Pernow B (1975) Effect of substance P on various vascular beds in the dog. Acta Physiol
    Scand 93: 277–285
Hallemans R, Naeije R, Melot C, et al (1981) Systemic and pulmonary haemodynamic effects of
    somatostatin. Lancet I: 1270
Hamasaki Y, Saga T, Mojarad M, Said SI (1982) VIP reduces or abolishes the contractile effects of
    leukotriene D_4 on guinea pig trachea, pulmonary artery and lung parenchyma. Clin Res 30:
    879A
Hara N, Geumei A, Chijimatsu Y, Said SI (1975) Vasoactive intestinal peptide aerosol protects
    against histamine and prostaglandin $F_{2\alpha}$-induced bronchoconstriction. Clin Res 23: 347A
Heymann MA, Rudolph AM, Nies AS, Melmon KL (1969) Bradykinin production associated with
    oxygenation of the fetal lamb. Circ Res 25: 521–534
Impicciatore M, Bertaccini G (1973) The bronchoconstrictor action of the tetradecapeptide bombe-
    sin in the guinea-pig. J Pharm Pharmacol 25: 872–875
Johnson AR, Erdös EG (1973) Release of histamine from mast cells by vasoactive peptides. Proc Soc
    Exp Biol Med 142: 1252–1256

Kaufman MP, Coleridge HM, Coleridge JCG, Baker DG (1980) Bradykinin stimulates afferent vagal C-fibers in intrapulmonary airways of dogs. J Appl Physiol 48: 511-517

Kitamura S, Ishihara Y, Said SI (1980) Effect of VIP, phenoxybenzamine and prednisolone on cyclic nucleotide content of isolated guinea-pig lung and trachea. Eur J Pharmacol 67: 219-223

Kulik TJ, Johnson DE, Elde RP, Lock JE (1983) Pulmonary vascular effects of bombesin and gastrin-releasing peptide in conscious newborn lambs. J Appl Physiol 55: 1093-1097

Liggins GC, Kitterman JA, Campos GA, et al (1981) Pulmonary maturation in the hypophysectomised ovine fetus. Differential responses to adrenocorticotrophin and cortisol. J Dev Physiol 3: 1-14

Lundberg JM, Martling CR, Saria A (1983a) Substance P and capsaicin-induced contraction of human bronchi. Acta Physiol Scand 119: 49-53

Lundberg JM, Saria A, Brodin E, et al (1983b) A substance P antagonist inhibits vagally induced increase in vascular permeability and bronchial smooth muscle contraction in the guinea pig. Proc Natl Acad Sci USA 80: 1120-1124

Manwaring D, Mullane K (1984) Disappearance of enkephalins in the lung. Fed Proc 43: 831

Marin L, Dameron F (1981) Role of the fetal hormonal factors in lung maturation. In: Levine SZ (ed) Physiological and biochemical basis for perinatal medicine. Karger, Basel, pp 59-66

Mojarad M, Said SI, Hyman AL (1980) Vasoactive intestinal peptide dilates pulmonary vessels in anesthetized cats. Clin Res 28: 894A

Morice A, Unwin RJ, Sever PS (1983) Vasoactive intestinal peptide protects against histamine-induced bronchoconstriction in asthmatic subjects. Lancet II: 1225-1227

Nicoll RA, Schenker C, Leeman SE (1980) Substance P as a transmitter candidate. Annu Rev Neurosci 3: 277-268

Niewoehner DE, Levine AS, Morley JE (1983) Central effects of neuropeptides on ventilation in the rat. Peptides 4: 277-281

Nilsson G, Dahlberg K, Brodin E, et al (1977) Distribution and constrictor effect of substance P in guinea pig tracheobronchial tissue. In: Von Euler US, Pernow B (eds) Subtance P. Raven, New York, p 75

O'Brodovich HM, Stalcup SA, Pang LM, et al (1981) Bradykinin production and increased pulmonary endothelial permeability during acute respiratory failure in unanesthetized sheep. J Clin Invest 67: 514-522

Parkinson DK, Radde IC (1969) Calcitonin action on membrane ATPase-an hypothesis. In: Calcitonin 1969. Heinemann, London, p 466

Parrat JR, Sturgess RM (1975) Evidence that prostaglandin release mediates pulmonary vasoconstriction induced by E. coli endotoxin. J Physiol (Lond) 246: 79P-80P

Pernow B, Rosell S (1975) Effect of substance P on blood flow in canine adipose tissue and skeletal muscle. Acta Physiol Scand 93: 139-141

Regoli D, Barabé J (1980) Pharmacology of bradykinin and related kinins. Pharmacol Rev 32: 1-46

Rishi S, Becker KL (1967) Insulin content of mammalian tissues. Diabetes 16: 516

Rökaeus A (1981) Studies on neurotensin as a hormone. Acta Physiol Scand [Suppl 501]: 1-62

Russell JA, Simons EJ (1984) Inhibition of acetylcholine release in canine tracheal smooth muscle by enkephalin. Fed Proc 43: 435

Saga T, Said SI (1984) Vasoactive intestinal peptide (VIP) relaxes human bronchial and pulmonary arterial smooth muscle. Clin Res 32: 563A

Said SI, Giachetti A, Nicosia S (1980a) VIP: Possible functions as a neural peptide. In: Costa E, Trabucchi M (eds) Neural peptides and neuronal communication. Raven, New York, pp 75-82

Said SI, Mutt V, Erdös EG (1980b) The lung in relation to vasoactive polypeptides. Ciba Found Symp 78: 217-237

Sander GE, Lorenz PE, Verma PS (1980) Inhibition of the partially purified canine lung angiotensin I converting enzyme by opioid peptides. Biochem Pharmacol 29: 3115-3118

Sander GE, Giles T, Kastin A, et al (1982) Leucine-enkephalin: Reversal of intrinsic cardiovascular stimulation by pentobarbital. Eur J Pharmacol 78: 467-470

Santiago TV, Remolina C, Scoles V III, Edelman NH (1981) Endorphins and the control of breathing. Ability of naloxone to restore flow-resistive load compensation in chronic obstructive pulmonary disease. N Engl J Med 304: 1190–1195

Schwedes U, Szabo S, Usadel KH (1978) Prevention of chemically induced adrenal hemorrhage and
lung injury by somatostatin in the rat. Metabolism [9 Suppl 1] 27: 1377–1380

Shulkes A, Fletcher DR, Hardy KJ (1983) Organ and plasma metabolism of neurotensin in sheep.
Am J Physiol 245: E457–E462

Simonsson BG, Skoogh B-E, Bergh NP, et al (1973) In vivo and in vitro effect of bradykinin on bronchial motor tone in normal subjects and patients with airways obstruction. Respiration 30:
378–388

Skrabanek P, Cannon D, Kirrane J, Powell D (1978) Substance P secretion by carcinoid tumours. Ir J
Med Sci 147: 47

Sonstegard K, Wong V, Cutz E (1979) Neuro-epithelial bodies in organ cultures of fetal rabbit lungs.
Cell Tissue Res 199: 159–170

Sugiyama Y, Hsu L-H, Ishihara Y, et al (1982) Effect of acetylcholine, bradykinin, and vasoactive intestinal polypeptide on the canine airways and pulmonary vascular bed. Kokyu To Junkan 30:
387–391

Sydbom A (1982) Histamine release from isolated rat mast cells by neurotensin and other peptides.
Agents Actions 12: 90–93

Terenghi G, McGregor GP, Bhuttacharji S, et al (1983) Vagal origin of substance P-containing nerves
in the guinea pig lung. Neurosci Lett 36: 229–236

Track NS, Cutz E (1982) Bombesin-like immunoreactivity in developing human lung. Life Sci 30:
1553–1556

Undem BJ, Dick EC, Buckner CK (1983) Inhibition by vasoactive intestinal peptide of antigen-induced histamine release from guinea-pig minced lung. Eur J Pharmacol 88: 247–250

Vargaftig BB, Dao Hai N (1972) Selective inhibition by mepacrine of the release of "rabbit aorta con-
tracting substance" evoked by the administration of bradykinin. J Pharm Pharmacol 24: 159–161

Verma PS, Miller RL, Taylor RE, et al (1982) Inhibition of canine lung angiotensin converting enzyme by ACTH and structurally related peptides. Biochem Biophys Res Commun 104: 1484–1488

Voelkel NF, Morris KG, McMurtry IF, Reeves JT (1980) Calcium augments hypoxic vasoconstriction in lungs from high-altitude rats. J Appl Physiol 49: 450–455

Weichman BM, Tucker SS, Muccitelli RM, Wasserman MA (1983) Dependency of the LTD_4-in-
duced contraction of the guinea pig trachea on both intracellular and extracellular calcium. Fed
Proc 42: 442

Weinberger SE, Weiss ST, Johnson TS, et al (1982) Naloxone does not affect bronchoconstriction induced by isocapnic hyperpnea of subfreezing air. Am Rev Respir Dis 126: 468–471

Willette RN, Barcas PB, Krieger AJ, Sapru HN (1983) Pulmonary resistance and compliance
changes evoked by pulmonary opiate receptor stimulation. Eur J Pharmacol 91: 181–183

Willey JC, Lechner JF, LaVeck M, Harris CC (1984) Bombesin and the C-terminal portion of gastrin-releasing peptide stimulate the clonal growth of normal human bronchial epithelial cells. Clin
Res 32: 424A

Wong PY-K, Cheung WY (1979) Calmodulin stimulates human platelet phospholipase A_2. Biochem
Biophys Res Commun 90: 473–480

Yamaguchi M, Takei Y, Yamamoto T (1975) Effect of thyrocalcitonin on calcium concentration in
liver of intact and thyroparathyroidectomized rats. Endocrinology 96: 1004–1008

Neurotransmitters and Their Role in Pulmonary Physiology

J. Richardson

Montreal General Hospital, Research Institute, Room 3818, Livingston Hall, 1650 Cedar Avenue, Montreal, Quebec H3G 1A4, Canada

The lung contains a variety of biologically active peptides, which are located in specialized epithelial cells, neurons, and other tissues (Table 1) (Ghatei et al. 1982; Said 1982a; Hakanson et al. 1983). These peptides have diverse actions on many physiological functions and several of these peptides have been proposed as important modulators of pulmonary physiology (Said 1982a, b; Lundberg et al. 1984). The principal areas of physiological interest for these peptides in the lung are gland secretion, epithelial water and electrolyte transport, vascular and airways smooth muscle control, and modulation of the neural pathways in the lung. Some of the actions of these peptides are direct on the effector organs, while others are indirect, either through an intermediary substance or through their effect on the neural pathways. The exact role of these peptides in the lung is, at the present time, unknown but two peptides, vasoactive intestinal peptide (VIP) and substance P (SP), have been studied in some detail for possible effects on smooth muscle and on mucosal inflammatory changes secondary to irritation (Said 1982a; Lundberg et al. 1984).

Vasoactive intestinal peptide has been shown to have an effect on water and electrolyte secretion in intestinal mucosa where excess amounts of this peptide are associated with a syndrome of watery diarrhea (Said 1980), and it has also been shown to have a stimulatory effect on tracheal submucosal gland secretion (Peatfield et al. 1983), which may be through stimulation of cyclic AMP in the glands (Lazarus et al. 1985). It also stimulates

Table 1. Peptides

Vasoactive intestinal peptide
Substance P
Bombesin or gastrin-releasing peptide
Bradykinin
Cholecystokinin
Somatostatin
Spasmogenic lung peptide
β-Endorphin
Neurotensin
Neuropeptide Y
Met- and leu-enkephalins
Angiotensin II

Recent Results in Cancer Research. Vol 99
© Springer-Verlag Berlin · Heidelberg 1985

ion and water transport across airway epithelium (Nathanson et al. 1983). The lung arises from the foregut and shares many of the anatomical features of the gastrointestinal tract, and it is thus not surprising that VIP has been shown to have direct actions on the pulmonary glands and epithelium. Other studies, however, indicate that VIP may actually inhibit the discharge of macromolecules from the glands (Coles et al. 1981). The role of the other peptides in secretion and epithelial function is unknown at the present time but it is likely that others will be shown to have effects on both secretion from glands and water and electrolyte transport across the airway epithelium.

Several of the pulmonary peptides are localized in the neuroendocrine cells in the airways (Cutz et al. 1984) and it has been suggested that during hypoxia these cells might release these peptides, which would then have an action on the pulmonary circulation, epithelial secretion, or airways smooth muscle (Lauweryns et al. 1978). Decreased levels of bombesin and VIP have been demonstrated in acute respiratory distress in infants (Ghatei et al. 1983). Bombesin or gastrin-releasing peptide is in high concentration in these cells and it is a likely candidate for this function (Polak and Bloom 1982). Recent studies with fetal lambs, however, have shown that this peptide has no significant effect on the pulmonary vasculature under normal or hypoxic conditions (Kulik et al. 1983a), even though hypoxia has been shown to produce degranulation of these cells (Lauweryns et al. 1978). Vasoactive intestinal peptide, on the other hand, is a very potent vasodilator of the pulmonary vasculature, and its presence in nerves adjacent to the vessels may indicate that it has such a role (Kulik et al. 1982; Polak and Bloom 1982; Said 1982a, b; Hakanson et al. 1983). Substance P also dilates the pulmonary vasculature, but its effects are weak and variable in the fetal lung (Kulik et al. 1982). The effect of VIP on the vasculature, while thought to be a direct action, has been shown to occur through mediation by prostaglandins, whose synthesis is blocked by indomethacin in the newborn lamb, which is used as a model for the hypoxic human infant. When the synthesis is blocked the VIP-induced vasodilation is abolished (Kulik et al. 1983b). The mechanisms of action of these peptides are therefore more complex than originally thought. Although peptides such as bradykinin have an effect on the microcirculation in inflammatory states there is no evidence at the present time that such peptides produce pulmonary damage by increasing the permeability of the microvasculature, although they may play a role in hypoxic damage (Said 1982b).

Substance P and VIP are both plentiful in the lung and both are mainly localized in neurons and nerve trunks (Polak and Bloom 1982; Hakanson et al. 1983). Substance P is particularly associated with sensory nerve endings and VIP with both neurons and nerve trunks which are most probably motor in function (Hakanson et al. 1983). The current interest in SP concerns its possible role in neurogenic inflammation of the airways mucosa following irritation to the mucosa, and thus may be of great importance in chronic bronchitis and asthma as well as other respiratory diseases (Lundberg et al. 1983a, 1984). Vasoactive intestinal peptide is of interest because there is the possibility that it might be the neurotransmitter of the third autonomic neural pathway in the lung, which in the human is the only demonstrable inhibitory neural pathway (Nadel and Barnes 1984).

Substance P - immunoreactive nerves have been demonstrated in the respiratory tract of many species (Polak and Bloom 1982; Hakanson et al. 1983) and are probably of vagal sensory origin, as similar immunoreactivity has been demonstrated in the nodose ganglion (Polak and Bloom 1982). Vagal stimulation results in vasodilation and edema of the tracheal mucosa, and these effects are blocked by prior treatment with capsaicin or an antagonist of SP (Lundberg et al. 1984). Capsaicin treatment results in the liberation of SP from these nerves (Jessel et al. 1978), and the treatment of neonatal animals with this compound results in the degeneration of the chemosensitive neurons (Jancso et al. 1977). In the normal untreated animal SP is thought to be released following stimulation of the irritant receptors in the airways. This release of SP following stimulation of sensory fibers, which may also have a central SP-mediated connection (Jessel 1982), is similar to the response in the skin where SP plays a role in the development of the inflammatory response (Lundberg et al. 1983a). In the airways the released SP produces increased vascular permeability and edema in the mucosa (Lundberg et al. 1984), and bronchoconstriction may also be produced by the SP, which has both a direct action on the smooth muscle (Lundberg et al. 1983a) and an indirect effect through the augmentation of the release of acetylcholine from the cholinergic nerves in the airways (Tanaka and Grunstein 1984). In animals in which the SP-containing sensory nerves have been destroyed by treatment during the fetal stage with capsaicin, irritation of the airways with cigarette smoke does not result in the local inflammatory reaction in the mucosa (Lundberg and Saria 1983). This and other studies indicate that SP probably plays an important role in airways defense and also disease (Lundberg et al. 1983a, b). Inflammation of the airways is a constant finding in the asthmatic patient, and it is thought that the inflammation is important in the production of the smooth muscle hyperreactivity which is characteristic of this disease. The possibility that irritation of the mucosa with subsequent relase of SP and resulting inflammation of the mucosa may contribute to the smooth muscle hyperreactivity either through a direct action on the muscle or through augmentation of the cholinergic pathway is currently being investigated.

The demonstration of VIP immunoreactivity in nerve trunks and in neurons in pulmonary and enteric ganglia (Dey et al. 1981; Ghatei et al. 1982; Polak and Bloom 1982; Said 1982a; Hakanson et al. 1983) plus the fact VIP has a relaxant effect on the airways smooth muscle and on the smooth muscle of the gastrointestinal tract made this peptide a likely candidate for the neurotransmitter of the nonadrenergic noncholinergic neural pathway (Said et al. 1980; Fahrenkrug and Emson 1982; Said 1982a). This neural pathway, which is also referred to as the purinergic (Burnstock 1972) and as the peptidergic (Baumgarten et al. 1970) pathway, arises from ganglion cells in the enteric and the pulmonary connective tissues, and in both the airways and the gastrointestinal tract it has an inhibitory effect on the smooth muscle. In some species this pathway is found in conjunction with inhibitory adrenergic nerves, while in others it is the only inhibitory pathway to the smooth muscle (Richardson 1979). Its actions on glands and the vasculature in the lung are as yet undefined, but it is likely that it does play a role in the control of these systems. This autonomic component has been more thoroughly studied in the gastrointestinal tract, where it has been found to be exceedingly complex with many modulators and neural connections involved in its control; it is too soon to assign definite functions to the peptides found there (Costa and Furness 1982). There is evidence to support an inhibitory effect on airway smooth muscle, as addition of VIP to guinea pig tracheal smooth muscle preparations results in relaxation of these preparations (Matsuzaki et al. 1980). Furthermore, stimulation of the tracheal nerves by field stimulation of the tissue resulted in the relaxation of the tissue and the release of VIP from the tissue (Matsuzaki et al. 1980). There is however evidence against such a role for VIP, both in the gastrointestinal tract (MacKenzie and Burnstock 1980) and from airways (Karlsson and Persson 1983), where in some species nonadrenergic inhibition can be demonstrated with field stimulation of the nerves but VIP has no effect. This includes human airways, where VIP has only a slight inhibitory effect on the smooth muscle (Davis et al. 1982). There is also evidence that the action of VIP on guinea pig tracheal smooth muscle is indirect, probably through the action of prostaglandins as the effect of VIP can be abolished by prior treatment with indomethacin (Regal et al. 1983). Attempts to implicate other peptides as the neurotransmitter of this autonomic pathway have been unsuccessful and it would be premature to term this system peptidergic.

The peptides present in the lung could participate in many physiological functions, from gland secretion to the control of vascular and airways smooth muscle. Evidence for such functions is, however, just being accumulated and at the present time no single peptide, with the possible exception of substance P, can be ascribed to one function in particular as being the most important if not the only mediator of that function. There is great interest in SP, as it has been shown to be needed for the irritant-induced inflammation of the airways mucosa and this inflammation may be of great importance in common respiratory diseases such as chronic bronchitis and asthma. Work to date clearly shows that SP can cause an inflammatory response as well as other effects on airways smooth muscle in both experimental animals and human tissue, and extension of this work to diseased human tissue can be expected. The role of the other peptides in neurotransmission is not clear but it is likely that some, such as VIP and somatostatin, may play an important action in the modulation of the pulmonary ganglia and thus have an effect on the smooth muscle of the vessels and the airways, but more work is needed to define their functions.

References

Baumgarten HG, Holstein AF, Owman CH (1970) Auerbach's plexus of mammals and man. Electron microscopical identification of three different types of neuronal processes in myenteric ganglia of the large intestine from rhesus monkeys, guinea-pigs and man. Z Zellforsch 106: 376–397

Burnstock G (1972) Purinergic nerves. Pharmacol Rev 24: 509–581

Coles SJ, Said SI, Reid LM (1981) Inhibition by vasoactive intestinal peptide of glycoconjugate and lysozyme secretion by human airway in vitro. Am Rev Respir Dis 124: 531–536

Costa M, Furness JB (1982) Neuronal peptides in the intestine. Br Med Bull 38: 247–252

Cutz E, Gillan JE, Track NS (1984) Pulmonary endocrine cells in the developing human lung and during neonatal adaptation. In: Becker KL, Gazdar AF (eds) Endocrine lung in health and disease. Saunders, Philadelphia, pp 210–231

Davis C, Kannan MS, Jones TR, Daniel EE (1982) Control of human airway smooth muscle: in vitro studies. J Appl Physiol 53: 1080–1087

Dey DA, Shannon WA, Said SI (1981) Localization of VIP-immunoreactive nerves in airways and pulmonary vessels of dogs, cats and human subjects. Cell Tissue Res 220: 231–238

Fahrenkrug J, Emson PC (1982) Vasoactive intestinal polypeptide: functional aspects. Br Med Bull 38: 265–270

Ghatei MA, Sheppard N, O'Shaughnessy DJ, Adrian TE, McGregor GP, Polak JM, Bloom SR (1982) Regulatory peptides in the mammalian respiratory tract. Endocrinology 111: 1248–1254

Ghatei MA, Sheppard N, Henzen-Logman S, Blank MA, Polak JM, Bloom SR (1983) Bombesin and vasoactive intestinal polypeptide in the developing lung: marked changes in acute respiratory distress syndrome. J Clin Endocrinol Metab 57: 1226–1232

Hakanson R, Sundler F, Moghimzadeh E, Leander S (1983) Peptide-containing nerve fibres in the airways: distribution and functional implications. Eur J Respir Dis 64: 115–140

Jancso N, Kiraly E, Jancso-Gabor A (1977) Pharmacologically induced selective degeneration of chemosensitive primary sensory neurons. Nature 270: 741–743

Jessell TM (1982) Fifty years of substance P. Nature 295: 551–553

Jessell TM, Iversen LL, Cuello AC (1978) Capsaicin induced depletion of substance P from primary sensory neurons. Brain Res 152: 183–188

Karlsson JA, Persson CGA (1983) Evidence against vasoactive intestinal peptide (VIP) as a dilator and in favour of substance P as a constrictor in airway neurogenic responses. Br J Pharmacol 79: 634–636

Kulik TJ, Johnson DE, Niemi T, Fuhrman BP, Lock JE (1982) Effect of putative peptide neurotransmitters on the neonatal pulmonary circulation. Pediatr Res 16: 102 A

Kulik TJ, Johnson DE, Elde RP, Lock JE (1983 a) Pulmonary vascular effects of bombesin and gastrin-realising peptide in conscious newborn lambs. J Appl Physiol 55: 1093–1097

Kulik TJ, Johnson DE, Fuhrman BP, Lock JE (1983 b) Indomethacin blocks pulmonary vasodilation caused by vasoactive intestinal peptide. Am Heart Assoc: (abstract)

Lauweryns JM, Cokelaere M, Lerut T, Theunynck P (1978) Cross circulation studies on the influence of hypoxia and hypoxaemia on neuroepithelial bodies in young rabbits. Cell Tissue Res 193: 373–386

Lazarus SC, Basbaum CB, Barnes PJ, Gold WM (1985) Mapping of VIP receptors by use of an immunocytochemical probe for the intracellular mediator cyclic AMP. (in press)

Lundberg JM, Saria A (1983) Capsaicin-induced desensitization of the airway mucosa to cigarette smoke, mechanical and chemical irritants. Nature 302: 251–253

Lundberg JM, Martling CR, Saria A (1983 a) Substance P and capsaicin-induced contraction of human bronchi. Acta Physiol Scand 119: 49–53

Lundberg JM, Brodin E, Saria A (1983 b) Effects and distribution of vagal capsaicin-sensitive substance P neurons with special reference to the trachea and lungs. Acta Physiol Scand 119: 243–252

Lundberg JM, Brodin E, Hua X, Saria A (1984) Vascular permeability changes and smooth muscle contraction in relation to capsaicin-sensitive substance P afferents in the guinea-pig. Acta Physiol Scand 120: 217–227

MacKenzie I, Burnstock G (1980) Evidence against vasoactive intestinal polypeptide being the non-adrenergic, non-cholinergic inhibitory transmitter released from nerves supplying the smooth muscle of the guinea-pig taenia coli. Eur J Pharmacol 67: 255–264

Matsuzaki Y, Hamasaki Y, Said SI (1980) Vasoactive intestinal peptide: a possible transmitter of non-adrenergic relaxation of guinea pig airways. Science 210: 1252–1253

Nadel JA, Barnes PJ (1984) Autonomic regulation of the airways. Ann Rev Med 35: 451–467

Nathanson I, Widdicombe JH, Barnes PJ (1983) Effect of vasoactive intestinal peptide across dog tracheal epithelium. J Appl Physiol 55: 1844–1848

Peatfield AC, Barnes PJ, Bratcher C, Nadel JA, Davis B (1983) Vasoactive intestinal peptide stimulates tracheal submucosal gland secretion in ferret. Am Rev Respir Dis 128: 89–93

Polak JM, Bloom SR (1982) Regulatory peptides and neuron-specific enolase in the respiratory tract of man and other mammals. Exp Lung Res 3: 313–328

Regal JF, Johnson DE (1983) Indomethacin alters the effects of substance P and VIP on isolated airway smooth muscle. Peptides 4: 581–584

Richardson JB (1979) Nerve supply to the lungs. Am Rev Respir Dis 119: 785–802

Richardson JB (1981) Nonadrenergic inhibitory innervation of the lung. Lung 159: 315–322

Said SI (1980) Vasoactive intestinal peptide (VIP): isolation, distribution, biological actions, structure-function relationships and possible functions. In: Glass GBJ (ed) Gastrointestinal hormones. Raven, New York, pp 245–273

Said SI (1982 a) Vasoactive peptides in the lung, with special reference to vasoactive intestinal peptide. Exp Lung Res 3: 343–348

Said SI (1982 b) Vasoactive peptides and the pulmonary circulation. Ann NY Acad Sci 384: 207–212

Said I, Giachetti A, Nicosia S (1980) VIP: possible functions as a neural peptide. In: Costa E, Trabucchi M (eds) Neural peptides and neuronal communications. Raven, New York, pp 75–82

Tanaka DT, Grunstein MM (1984) Mechanisms of substance -P induced contraction of rabbit airway smooth muscle. J Appl Physiol 57: 1551–1557

Biochemistry of Peptide Hormones and Neurotransmitters Produced by Lung Cancer

Peptide Hormone Biosynthesis – Recent Developments

H. P. J. Bennett

Endocrine Laboratory, Royal Victoria Hospital, 687 Pine Avenue, West Montreal, Quebec H3A 1A1, Canada

Peptide Hormone Biosynthesis – From Nucleus to Secretory Granule

The biosynthesis of peptide hormones is now recognized to be a highly ordered series of events. These biosynthetic events can frequently be localized to a particular subcellular organelle (Fig. 1). Each organelle will be discussed in turn, to emphasize the sequential nature of the maturation process. There will be frequent references to the synthesis and processing of the corticotropin/endorphin precursor within the pituitary. This multihormone precursor is often referred to as pro-opiomelanocortin (POMC). POMC is an interesting model to study, since many post-translational modifications are manifest within its maturation products. More intriguing is the fact that processing of this multihormone precursor is tissue specific. In the pars distalis or anterior lobe of the pituitary, ACTH, β-lipotropin (β-LPH), and an N-terminal or 16 K fragment are the major biosynthetic products. In the pars intermedia or intermediate lobe processing is more complete, and α-melanotropin (α-MSH), corticotropin-like intermediate lobe peptide (CLIP), γ-LPH, various forms of acetylated endorphin, γ-MSH, and a shortened N-terminal ($16\,K_{1-49}$) fragment are the major products (Fig. 2). More is known of this system than of many others, primarily because pituitary tissue is relatively easy to maintain in culture and lends itself well to classical pulse/chase biosynthetic experiments. Also the AtT-20 mouse pituitary tumour cell line has been an extremely useful and accurate model for anterior pituitary processing (for extensive reviews on this subject see Eipper and Mains 1980; Chrétien and Seidah 1981).

The study of peptide hormone biosynthesis, whether at the transcriptional level or at the level of post-translational processing, is progressing rapidly on several fronts. It will not be possible to discuss all aspects of this multistep process in detail. Instead, some of the recent developments which are of particular interest in our own laboratory are discussed.

The Nucleus

In eukaryotic organisms gene transcription is a nuclear event. The resulting RNA transcript must be transported to the cytoplasm for translation to take place. Recently it has been shown that conditions that acutely alter ACTH release from the anterior pituitary also have longer term effects upon the levels of POMC mRNA levels in the cytoplasm. For

DIAGRAMMATIC REPRESENTATION OF PEPTIDE HORMONE PROCESSING

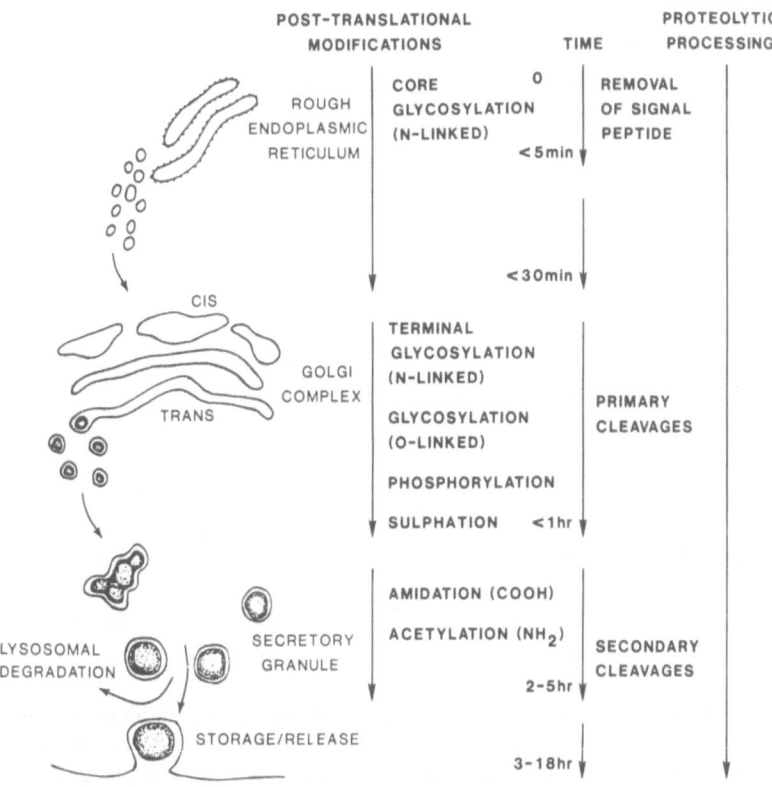

Fig. 1. Peptide hormone biosynthesis. The time scale emphasizes the sequential nature of the process and is approximate

instance, POMC mRNA levels are decreased by glucocorticoids and markedly increased by adrenalectomy (Eberwine and Roberts 1984). Intermediate pituitary secretion is essentially unaffected by glucocorticoids, and POMC mRNA levels are refractory to alterations in steroid output by the adrenal cortex. Secretion of POMC-derived peptides from the intermediate pituitary is under dopaminergic inhibitory control. POMC mRNA levels are decreased by ergocryptine, a dopamine agonist, and increased by haloperidol, a dopamine antagonist (Chen et al. 1983). These drugs have no apparent effect on ACTH secretion by the anterior pituitary or upon POMC mRNA levels.

While these examples represent endocrine control of absolute mRNA levels, details of other control mechanisms have emerged recently. It has been appreciated for some time that human growth hormone exists as two structurally distinct forms of 20 000 and 22 000 molecular weight. The former variant is brought about by an internal deletion of 15 amino acids and constitutes about 10% of the pituitary growth hormone. The two forms of growth hormone show subtle differences in biological activity. Analysis of growth hormone mRNA suggests that the message for the 20 000 molecular weight form is identical with that for the 22 000 molecular weight form except for a deletion of 45 nucleotides occurring exactly where the deleted amino acid sequence would predict (Moore et al. 1982). It seems that alternative RNA splicing is one way of altering hormonal expres-

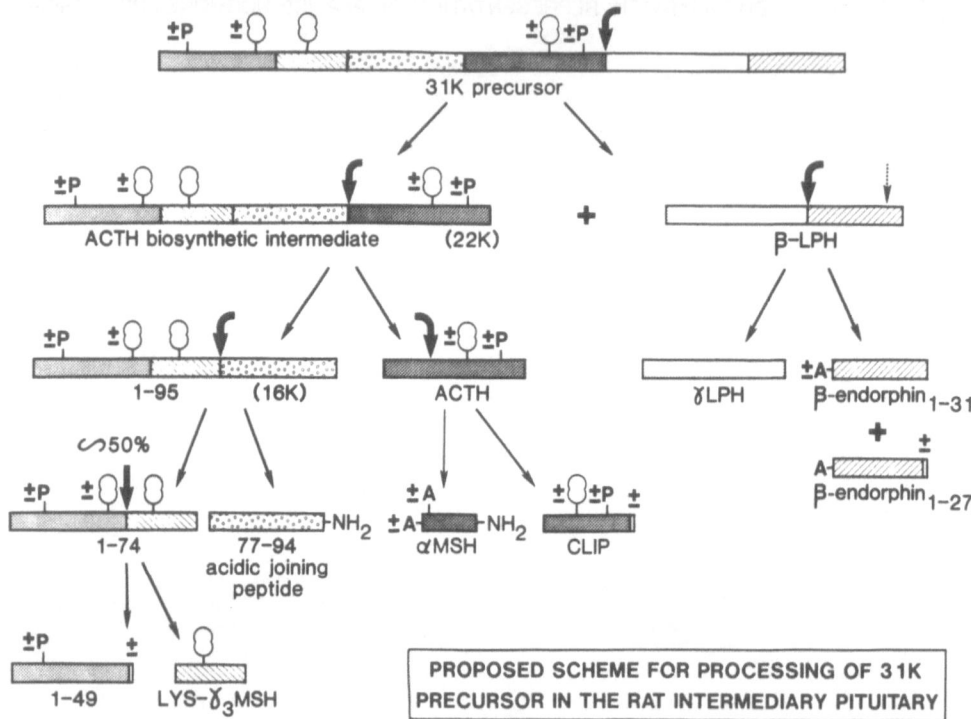

Fig. 2. Pro-opiomelanocortin processing in the rat intermediate pituitary. Intermediate lobe process-ing involves cleavage or partial cleavage at all the available dibasic signal sequences. Following re-moval of the signal peptide in the endoplasmic reticulum, the 31K precursor is cleaved to form the 23K ACTH biosynthetic intermediate and β-LPH. This is thought to take place mainly within the Golgi apparatus. Subsequent cleavages are thought to take place mainly within the secretory granule fraction in two phases. The 16K N-terminal fragment, joining peptide, α-MSH, CLIP, γ-LPH and β-endorphin are formed rapidly. $16K_{1-49}$, $Lys^1\gamma_3MSH$ and the C-terminally shortened forms of β-en-dorphin are formed more slowly. There are three sites of glycosylation (O). Two are N-linked through asparagine residues (i.e. within γ_3MSH and CLIP). While glycosylation of MSH is com-plete, CLIP is only partially glycosylated. The N-linked glycosylation is initiated in the endoplasmic reticulum and completed in the Golgi apparatus. The glycosylation site within the $16K_{1-49}$ sequence is thought to be O-linked, probably through a threonine residue at position 45. This takes place in the Golgi apparatus with approximately 50% efficiency. Glycosylation at this position appears to influ-ence the potential for 16K cleavage at the $Arg_{49}Lys_{50}$ sequence and only the subpopulation of 16K without the O-glycosylation site is successfully cleaved to $16K_{1-49}$ and $Lys^1\gamma_3MSH$. The 16K and CLIP sequences are partially phosphorylated (P) by a casein kinase probably located in the Golgi apparatus. Acetylation (A) of α-MSH and β-endorphin and amidation (NH_2-) of α-MSH and joining peptide are known to take place in the secretory granule

sion. Another intriguing example of alternative RNA processing is provided by the calci-tonin gene. Transcription of the calcitonin gene produces a primary transcript encoding a pre-pro sequence, calcitonin, and a 37 amino acid calcitonin-related peptide (CGRP, cal-citonin gene-related peptide). The primary transcript is spliced to form one of two possi-ble forms of mature mRNA. One spliced mRNA encodes a calcitonin precursor while the other encodes a CGRP precursor. Both precursors have nearly identical pre-pro se-quences. This process is very carefully controlled, so that calcitonin mRNA predominates in the thyroid while CGRP mRNA appears to predominate in the hypothalamus (Amara et al. 1982). In medullary thyroid carcinoma tumours are known to switch spontaneously from high to low states of calcitonin secretion. Alternative RNA splicing appears to be the mechanism behind this phenomenon. There are now several examples of prohormones containing several closely related sequences. It is probable that alternative RNA processing will be demonstrated in other multihormone precursors.

Endoplasmic Reticulum

It seems that almost all secretory proteins are synthesized as high-molecular-weight precursors containing N-terminal extensions called signal or pre sequences. These sequences play an important role in the co-translational binding of ribosomes to the endoplasmic reticulum (ER) membrane and the insertion of the growing peptide chain into the lumen of the ER, where posttranslational modifications begin. The signal peptide first binds to a soluble cytoplasmic signal recognition particle (Walter and Blobel 1982), which in turn brings about binding of the ribosome to a receptor protein on the cytoplasmic side of the ER membrane (Meyer et al. 1982). It has been shown recently that about 60 amino acids must be polymerized before binding of the ribosome to the ER can take place (Eskridge and Shields 1983). This implies that polypeptides must be synthesized as part of a precursor of at least 60–80 amino acid residues in order to be successfully inserted into the ER. This is a fundamental mechanism common to the synthesis of secretory proteins in both prokaryotic and eukaryotic organisms. The signal or pro sequence is characterically comprised of 20–30 amino acids, the majority of which are hydrophobic in nature. This property is thought to assist in the translocation of the polypeptide chain through the hydrophobic environment of the ER membrane. Before translation is completed the signal peptide is removed by a signal peptidase and the nascent protein enters the intracisternal space of the ER. The point of signal peptidase cleavage is generally after an amino acid residue with a small neutral side chain (e. g. alanine, glycine or serine).

Modification of the nascent protein begins with N-glycosylation of asparagine residues. This takes place through a mechanism common to all eukaryotic organisms. Studies of the biosynthesis of plasma membrane, viral and secretory glycoproteins have revealed that N-linked glycosylation begins with assembly of a high mannose precursor oligosaccharide attached to a dolichol lipid carrier (see Lennarz 1980 for extensive reviews). The oligosaccharide has the generalized structure $glucose_2$-$mannose_9$-N-acetylglucosamine$_2$ with linkage to the lipid carrier via N-acetylglucosamine. The primary amino acid sequence Asn-X-Thr/Ser is obligatory for N-glycosylation of asparagine, where X can be any amino acid except perhaps aspartic acid. The Thr/Ser hydroxyl group is important in the transfer of the oligosaccharide chain from the lipid carrier (Bause and Legler 1981). This results in a core-glycosylated protein with the inner N-acetylglucosamine now attached to the amide nitrogen of asparagine. N-Linked glycosylation is frequently, but not always, a co-translational event. Terminal glucose and mannose residues are rapidly removed. Capping sugars are not added until the protein reaches the Golgi apparatus. Not all asparagine residues present within Asn-X-Thr/Ser sequences become glycosylated. Secondary and tertiary structure undoubtedly influence the site and extent of glycosylation. For instance, the precursor of the adrenal medulla enkephalin precursor (proenkephalin A) has two possible N-linked glycosylation sites but the precursor has been shown not to be glycosylated (Kilpatrick et al. 1983). The rat ACTH precursor has been shown to have two possible N-glycosylation sites (Fig. 2). While glycosylation of the asparagine residue within the γ-MSH sequence is complete, the asparagine found within ACTH itself is only partially glycosylated. mRNA translation, ER insertion, cleavage of the signal peptide, and core glycosylation are thought to be completed within 5 min of the initiation of translation. Certainly all these processes are complete by the time the maturing protein is transferred via small transport vesicles to the Golgi apparatus where the N-linked oligosaccharides are further modified (< 20 min). Other than signal peptide cleavage no proteolytic modifications take place until the protein reaches the Golgi apparatus. This is the main reason why 20 min is chosen as the pulse period in many peptide biosynthesis studies.

Golgi Apparatus

The Golgi apparatus is best recognized for its role in the elongation and capping of complex oligosaccharides (see Farquhar and Patade 1981 for an extensive review). Processing of the glycogrotein in the ER has reduced the oligosaccharide to the generalized structure mannose$_5$-N-acetylglucosamine$_2$. The terminal sugars are added in a sequential manner by appropriate transferase enzymes. The maturation is thought by some to occur in a highly ordered manner as the protein passes from the *cis* to the *trans* face of the Golgi stacks. Some oligosaccharides become phosphorylated or sulphated in the Golgi apparatus. Our recent findings and those of others suggest that all three oligosaccharides in rat POMC are at least partially sulphated (Hoshina et al. 1982; Seger and Bennett to be published). The exact location of the sulphate in the sugar chain is unknown. Up to this point the biosynthesis of peptide hormone precursors and membrane proteins is very similar. However, membrane proteins are rapidly transported from the Golgi apparatus to the cell surface, while secretory proteins are packaged into granules and are not capable of being released for at least another 2 h. Moore et al. (1983) have suggested that sulphation may be part of the mechanism responsible for targeting POMC for packaging into secretory granules. Lysosomal glycoprotein enzymes are segregated in the Golgi apparatus based upon a mechanism that recognizes mannose-6-phosphate (Sly and Fischer 1982).

O-Linked glycosylation of threonine or serine takes place in the Golgi apparatus by the sequential addition of sugars without the participation of a lipid carrier (see Lennarz 1980). Linkage to the hydroxyl group of threonine or serine is normally though N-acetylgalactosamine. O-Linked oligosaccharides are generally much simpler in structure than the N-linked variety. Our recent findings suggest that the O-glycosylation site within the 1–49 sequence of the N-terminal fragment of rat POMC contains two residues of N-acetylgalactosamine and one residue of N-acetylglucosamine and galactose. Interestingly, this oligosaccharide also contains sulphate (Seger and Bennett to be published). O-Linked glycosylation is very rare amongst the peptide and protein hormones. The best-known example is the C-terminal sequence of chorionic gonadotropin, which has several O-Linked glycosylation sites. Within POMC the O-Linked glycosylation appears to influence the extent of processing of the 16 K N-terminal fragment (see Fig. 2).

It is clear that for some peptide hormones the Golgi apparatus is an important site of proteolytic processing. The initial cleavage of POMC to the 23 K ACTH biosynthetic intermediate and β-LPH is generally thought to be a Golgi event. It is possible that the Golgi apparatus may also be responsible for other post-translational events, such as sulphation of tyrosine residues and phosphorylation of serine residues.

We have recently discovered that rat and human ACTH are phosphorylated at serine residue 31 (Bennett et al. 1981, 1983). The serine is present within the tripeptide sequence – Ser – Ala – Glu–, which conforms to the recognition sequence for casein kinase. Casein kinase in mammary tissue is responsible for the multiple phosphorylation of casein. We have prepared a membrane fraction from lactating bovine mammary tissue which is capable of phosphorylating synthetic human CLIP in vitro using ATP as phosphate donor (Seger and Bennett 1983).

Secretory Granules

It is only recently that the secretory granule itself has been shown to participate in peptide hormone biosynthesis. The study of vasopressin and oxytocin biosynthesis has clearly indicated that cleavage of these hypothalamic peptides from their respective neurophysins takes place within the secretory granules as they pass down the axons of the pituitary stalk to the posterior pituitary (Brownstein et al. 1981). Cleavage of β-LPH to γ-LPH and β-endorphin in the rat intermediate pituitary is thought to take place, at least in part, within the secretory granule fraction (Glembotski 1981). β-Endorphin is subject to further cleavage at its carboxyl terminus to form β-endorphin 1-27 and 1-26. These events take place over a prolonged period (3-18 h post-translation), which clearly implicates the secretory granule as the site of processing (Eipper and Mains 1981). Studies in our own laboratory have shown that cleavage of the N-terminal fragment of POMC (16 K) to $16 K_{1-49}$ and $\gamma_3 MSH$ occurs 2-5 h post-translation, again indicating secretary granule processing (Seger and Bennett to be published).

Lysates of secretory granules isolated from the rat intermediate pituitary seem to be capable of accurately processing toad POMC to products characteristic of the intermediate lobe (see Fig. 1). The processing enzyme is a thiol protease with a pH optimum of about 5 (Chang and Loh 1984). It apparently cleaves on the carboxyl terminal side of the dibasic processing signal sequences. Removal of the basic amino acids is thought to be brought about by the action of an enzyme with properties similar to carboxypeptidase B. Similar enzymes have been implicated in the processing of pro-insulin to insulin by secretory granules prepared from pancreatic islets (Docherty et al. 1982). Very recently a protease capable of cleavage between paired basic residues has been isolated from yeast. This enzyme has a pH optimum of 7.5 (Mizuno and Matsuo 1984). For an extensive review of this subject see Lazure et al. 1983.

In recent elegant studies by Glembotski's group, C-terminal amidation of α-MSH and N-terminal acetylation of β-endorphin and α-MSH have clearly been demonstrated as secretory granule events. It seems that the N-acetylation of $tyrosine_2$ of endorphin and the N- and O-acetylation of $serine_1$ of α-MSH are brought about by the same acetyltransferase. This enzyme uses acetyl-CoA as co-substrate and has a neutral pH optimum (Glembotski 1982). The enzyme which brings about C-terminal amidation also has a neutral pH optimum. It uses C-terminally glycine-extended peptides as substrates. The amidating enzyme catalyzes a reaction in which the glycine residue donates the amide nitrogen. For optimal activity the enzyme requires molecular oxygen, ascorbate and copper ions (Eipper et al. 1983). Interestingly, these are the co-factor requirements displayed by dopamine-β-hydroxylase. This is another secretory granule enzyme involved in norepinephrine biosynthesis.

The role of the secretory granule in peptide hormone biosynthesis is quite controversial. The difference in pH optimum for the cleavage enzyme on the one hand (pH 5) and for the amidating and acetylating enzymes on the other (pH 7) is hard to explain. At present we have little idea how the hormone precursor and the processing enzymes, which in themselves are products of protein biosynthesis, are packaged together in such a carefully controlled manner. Moore et al. (1983) have shown that secretory granules prepared from AtT-20 cells contain several high-molecular-weight proteins unrelated to POMC. Together they make up about one half the protein content of the granule. These authors suggest that these proteins may participate in the sorting mechanisms involved in granule formation. For a review of approaches to studying the processing enzymes see Mains et al. (1983).

General Structural Characteristics of Peptide Prohormones

In the past 5 years numerous prohormones have been sequenced through cDNA techniques. The list includes POMC (Uhler et al. 1983) and the enkephalin precursors, proenkephalin A and B (Noda et al. 1982; Horikawa et al. 1983).

This mass of information has become available long before the structures of these precursors can be determined by protein chemistry. Indeed, it is unlikely that many will ever

Table 1. Post-translational processing of prohormones

Post-translational event	Recognition sequence for enzyme	Some known functions
Signal peptide cleavage	Cleavage C-terminal to small neutral amino acid (e.g. Ser, Ala, Gly)	Removal of hydrophobic signal peptide after ER insertion
Internal cleavage	Cleavage at (between?) pairs of basic amino acids (– Lys – Arg –, – Lys – Lys –, – Arg – Arg, rarely – Arg – Lys –)	Formation of active peptides
N-Glycosylation (asparagine)	– Asn – X – Thr/Ser (x: any amino acid except Asp)	Important for bioactivity of glycoprotein hormones of anterior pituitary
O-Glycosylation (serine/threonine)	Not clear but generally in proline-rich sequences	Important for bioactivity of hCG; may direct processing of N-terminal fragment of POMC
Sulphation (carbohydrate)	Not known	May direct processing
Sulphation (tyrosine)	Not known	Important for optimal activity of cholecystikinin
Phosphorylation (carbohydrate)	Not known	May direct processing
Phosphorylation (serine/threonine)	Thr/Ser – X – acidic residue	Not known
C-Terminal amidation	C-Terminal glycine residue	Important for optimal activity of many peptide hormones. Metabolic stabilization. Important for optimal activity of α-MSH
N-Terminal acetylation	Not known	Abolishes analgesic activity of β-endorphin. Metabolic stabilization

be completely sequenced in this way. To date only the prohormones for parathyroid hormone, insulin and ACTH/endorphin have been sequenced by classical means. Examination of the predicted primary structures has revealed the presence of numerous possible biologically important peptides (e.g. the γ-MSH sequence in POMC). Thus, putative hormones are being discovered while "side-stepping" many years of methods development (i.e. bioassays, protein isolation and sequencing).

How can we predict which sequences are likely to be biologically important and which are merely intervening sequences? The finding of evolutionally conserved sequences is a strong indication of an important biological function. For instance, in POMC there are four regions which are conserved, namely the N-terminal fragment, ACTH β-MSH, and β-endorphin. The enzyme recognition sequences listed in Table 1 indicate where proteolytic cleavage and post-translational modification may occur. Again these may indicate sequences of biological importance. Other clues come from inspection of the organization of the genes encoding prohormones. Examination of genomic sequences reveals the presence of intervening nucleotide sequences (introns). The introns are spliced out of the initial RNA transcript to form the mature mRNA. Introns do not appear to be found within regions encoding the biologically important amino acid sequences. For instance, in the genomic sequences for POMC proenkephalins A and B one intron is found in the extreme amino terminal region of each precursor. The biologically important sequences are toward the carboxyl terminus. It seems that splicing errors within biologically important sequences are an undesirable liability to the organism (cf. alternative splicing mechanism discussed earlier).

Evolution and Prohormones and Biosynthetic Mechanisms

Numerous studies over the past 10 years have indicated that the ACTH family of peptides are present in the pituitaries of lower vertebrates. ACTH from dogfish, salmon and ostrich pituitaries have been shown to be closely related in structure to that isolated from mammalian pituitaries. The steroidogenic 1–20 sequence in particular has been highly conserved throughout the vertebrates (Lowry and Scott 1975) (this may not be true of the cyclostomes; see below). Isolation of α-MSH and CLIP from dogfish intermediate pituitaries indicated that many of the mechanisms involved in α-MSH biosynthesis were established early in vertebrate evolution (Bennett et al. 1974; Lowry et al. 1974). More recently, peptides corresponding to all the major components of POMC have been isolated from salmon pituitaries (see Kawauchi 1983). While α-MSH is partially amidated in both dogfish and salmon, only the salmon appears to have the capacity for N-terminal acetylation of α-MSH. The N-terminal acetylation mechanism discussed earlier may have evolved after the teleost line diverged from the more primitive elasmobranchs. The amidation reaction, as will be discussed later, appears to be the more ancient mechanism. Other curiosities have emerged from these studies. The salmon peptide corresponding to the N-terminal fragment of POMC does not appear to contain a γ-MSH sequence. However, a γ-MSH-related peptide has been isolated from the dogfish pituitary (McLean and Lowry 1981). Further studies are required to determine the phylogenetic origins of this part of POMC. Kawauchi and co-workers have also made the discovery that salmon pituitaries synthesize two structurally distinct forms of each of the POMC-related peptides (i.e. the N-terminal fragment, α- and β-MSH, CLIP and endorphin). While these investigators conclude that each individual salmon pituitary contains two distinct forms of POMC, it is possible that there are two subpopulations of this species of fish, each with its own prohormone. It is natural to assume that POMC constitutes the precursor for ACTH and related peptides throughout the vertebrates. However, it should be pointed out that among the lower vertebrates, the appropriate precursor-product relationships have been clearly demonstrated only in the reptile pituitary (Dores 1982).

There have been some preliminary studies of the nature of ACTH and related peptides in the most primitive vertebrates available for study, the cyclostomes. The lamprey pituitary appears to contain biologically active ACTH. However, its lack of cross-reactivity with ACTH antisera indicates that the structure of this peptide must be quite distinct from that of mammalian ACTH (Nozaki and Gorbman 1984). Recently, it has been shown that hormones previously associated only with vertebrates may be found in extremely simple organisms, including protozoa and bacteria (LeRoith et al. 1982). Material resembling mammalian ACTH has been identified in extracts of the unicellular organism tetrahymena by the fairly rigorous criteria of biological activity and behaviour in various chromatographic systems. If ACTH from tetrahymena and lamprey have evolved from a common ancestral hormone it seems unlikely that tetrahymena ACTH should have evolved to resemble mammalian ACTH more closely than lamprey ACTH. Indeed, the concept of convergent evolution must be invoked to accommodate these findings. Careful structural studies are needed to resolve this apparent paradox.

Two recent studies of the structures of precursors for biologically active peptides from yeast and aplysia have shown that many of the features described for vertebrate peptide hormone biosynthesis are also found in much more primitive organisms. Mating in yeast is brought about by small pheromones or mating factors. α-Mating factor is a 13 amino acid peptide whose precursor structure has been determined recently by cDNA techniques (Kurjan and Herskowitz 1982). The α-mating factor precursor contains four tandem copies of mature α-mating factor. The precursor contains an N-terminal hydrophobic signal sequence, a pro sequence with three N-glycosylation sites, and each copy of α-mating factor is flanked by dibasic amino acid groupings. This establishes both of these processing signals as universal and evolutionarily ancient mechanisms. A recent study has shown that the signal or pre sequence is not removed during biosynthesis (Julius et al. 1984). Interestingly, each copy of the α-factor is preceded by a six-amino-acid N-terminal extension which, if not removed, renders the α-factor inactive. The N-terminally extended α-mating factor must first be processed by a dipeptidyl aminopeptidase to generate the active peptide (Julius et al. 1983). A similar mechanism is thought to be involved in the biosynthesis of mellitin, a bee venom peptide. There is no evidence as yet to suggest that such an aminopeptidase is important in the biosynthesis of peptide hormones from higher organisms. However, the α-mating factor N-terminal extensions, which have the generalized structure glu-ala-glu/asp-ala-glu-ala, are reminiscent of some intervening sequences found within mammalian prohormones. One notable example is the repeated glu-ala sequence found in the N-terminal portion of bovine β-LPH. This type of primary structure is optimal for α-helix formation (Chou and Fasman 1978) and may function as spacers between the biologically important sequences.

Egg laying in the gastropod mollusc Aplysia is controlled in part by egg-laying hormone (ELH). ELH is a 36-amino-acid peptide which is amidated at its carboxyl terminus. Sequencing of the gene encoding ELH has revealed some interesting features (Scheller et al. 1983). The carboxyl terminal amidation of ELH is apparently brought about by the mechanism outlined earlier (i. e. with a C-terminally glycine-extended ELH acting as a biosynthetic intermediate). ELH shares a common precursor with four small neuroactive peptides known as α-, β-, γ-, and δ-bag cell factors. The ELH/bag cell factor precursor is synthesized by neurons clustered near the abdominal ganglion. Once activated, these so-called bag cells release ELH and the various bag cell factors in a co-ordinate manner. These factors diffuse into the general circulation and give rise to the behaviour that culminates in egg laying. Interestingly, the bag cells themselves are thought to be activated by 34-amino-acid peptides A and B secreted by the atrial gland, an organ found within the reproductive tract. Like ELH, A and B peptide are C-terminally amidated. The genes encoding A and B peptide have also been sequenced and both have been found to be related to the ELH precursor. Within the three genes an intriguing series of insertions and deletions ensures that the A and B peptide precursors do not contain ELH or the bag cell factors. Similarly, the ELH/bag cell factor precursor does not give rise to A and B peptide.

Why is the multihormone precursor such a common feature in peptide hormone biosynthesis? In the case of the ELH/bag cell factor precursor the function appears to be the precise control of behaviour leading to egg laying. Since all the bioactive peptides are released together, the resulting behaviour is carefully coordinated. The release of ACTH and β-endorphin from the anterior pituitary is also under coordinate control, possibly for similar reasons. Coordinate release of multiple copies of yeast α-mating factor may be a means of maximizing the biological signal while minimizing the energy expenditure involved in precursor biosynthesis.

Conclusion

As explained earlier, the study of POMC continues to be one of the most important models for peptide hormone biosynthesis. The study of this multihormone precursor will enable us to propose general rules for the processing not only of other peptide hormone precursors but also putative prohormones predicted by cDNA techniques. Several mechanisms have already emerged which have enabled us to predict probable sites of amidation, glycosylation and phosphorylation (Table 1). The understanding of the mechanism of C-terminal amidation has been a particularly exciting development. It is the most commonly observed post-translational modification and a phylogenetically ancient mechanism. C-terminal amidation is an absolute requirement for the expression of biological activity for several neuropeptides. It is highly probable that many of the as yet undiscovered peptide neurotransmitters will be found to be C-terminally amidated. In the absence of specific assays it seems likely that the discovery of many neuropeptides will come through a study of this common post-translational modification. Indeed, a vasoactive peptide named neuropeptide Y has been isolated from extracts of gut using a chemical assay for C-terminally amidated amino acids (Tatemoto et al. 1982).

Acknowledgements. The original research reported in this article was supported by grants from the Medical Research Council of Canada (MT-1658 and MT-6733), the United States Public Health Service (HDO4365) and a Scholarship from the Fond de la recherche en santé du Québec. I would like to thank Dr. S. Solomon for his continuing support and encouragement. Thanks are also due to Susan James and Isabel Lehmann for technical assistance and Dianne Prowse and Sherville Walrond for preparation of the manuscript.

References

Amara SG, Jones V, Rosenfeld MG, Ong ES, Evans RM (1982) Alternative RNA processing in calcitonin gene expression generates mRNAs encoding different polypeptide products. Nature 298: 240–244

Bause E, Legler G (1981) The role of the hydroxy amino acid in the tryplet sequence. Asn-Xaa-Thr (Ser) for the N-glycosylation step during glycoprotein biosynthesis. Biochem J 195: 639–644

Bennett HPJ, Lowry PJ, McMartin C, Scott AP (1974) Structural studies of α-melanocyte-stimulating hormone and a novel β-melanocyte-stimulating hormone for the neurointermediate lobe of the pituitary of the dogfish *Squalus acanthias*. Biochem J 141: 439–444

Bennett HPJ, Browne CA, Solomon S (1981) Biosynthesis of phosphorylated forms of corticotropin related peptides. Proc Natl Acad Sci USA 78: 4713–4717

Bennett HPJ, Brubaker PL, Seger MA, Solomon S (1983) Human phosphoserine 31 corticotropin$_{1-39}$: isolation and characterization. J Biol Chem 258: 8108–8112

Brownstein MJ, Russell JT, Gainer H (1980) Synthesis, transport and release of posterior pituitary hormones. Science 207: 373–378

Chang T-L, Loh YP (1984) *In vitro* processing of pro-opiocortin by membrane-associated and soluble converting enzyme activities from rat intermediate secretory granules. Endocrinology 114: 2092–2099

Chen CLC, Dionne FT, Roberts JL (1983) Regulation of the pro-opiomelanocortin mRNA levels in rat pituitary by dopaminergic comppounds. Proc Natl Acad Sci USA 80: 2211–2215

Chou PY, Fasman GD (1978) Empirical predictions of protein conformation. Ann Rev Biochem 47: 251–276

Chrétien M, Seidah NG (1981) Chemistry and biosynthesis of pro-opiomelanocortin. Mol Cell Biochem 34: 101–127

Docherty K, Carroll RJ, Steiner DF (1982) Conversion of proinsulin to insulin: involvement of a 31 500 molecular weight thiol protease. Proc Natl Acad Sci USA 79: 4613–4617

Dores RM (1982) Evidence for a common precursor for α-MSH and β-endorphin in the intermediate lobe of the pituitary of the reptile *Anolis carolinersis*. Peptides 3: 925–935

Eberwine JH, Roberts JL (1984) Glucocorticoid regulation of pro-opiomelanocortin gene transcription in the rat pituitary. J Biol Chem 259: 2166–2170

Eipper BA, Mains RE (1980) Structure and biosynthesis of pro-adrenocorticotropin/endorphin and related peptides. Endocrin Rev 1: 1–27

Eipper BA, Mains RE (1981) Further analysis of post-translational processing of β-endorphin in rat intermediate pituitary. J Biol Chem 256: 5689–5695

Eipper BA, Mains RE, Glembotski CC (1983) Identification in pituitary tissue of a peptide α-amidation activity that acts on glycine-extended peptides and requires molecular oxygen, copper and ascorbic acid. Proc Natl Acad Sci USA 80: 5144–5148

Eskridge EM, Shields D (1983) Cell-free processing and segregation of insulin precursors. J Biol Chem 258: 11487–11491

Farquhar MG, Palade GE (1981) The Golgi apparatus (complex) – (1954–1981) – from artifact to center stage. J Cell Biol 91: 77 s–103 s

Glembotski CC (1981) Subcellar fractionation studies on the post-translational processing of pro-adrenocorticotropic hormone/endorphin in rat intermediate pituitary. J Biol Chem 256: 7433–7439

Glembotski CC (1982) Characterization of the peptide acetyltransferase activity of bovine and rat intermediate pituitaries responsible for the acetylation of β-endorphin and α-melnotropin. J Biol Chem 257: 10501–10509

Horikawa S, Takai T, Toyosato M, Takahashi H, Noda M, Kakidani H, Kubo T, Hirose T, Inayama S, Hayashida H, Miyata T, Numa S (1983) Isolation and structural organization of the human preproenkephalin B gene. Nature 306: 611–614

Hoshina H, Hortin G, Boime I (1982) Rat pro-opiomelanocortin contains sulfate. Science 217: 63–64

Julius D, Blair L, Brake A, Sprague H, Thorner J (1983) Yeast α-factor is processed from a larger precursor polypeptide: The essential role of a membrane-bound dipeptidyl aminopeptidase. Cell 32: 839–852

Julius D, Schekman R, Thorner J (1984) Glycosylation and processing of prepro-α-factor through the yeast secretory pathway. Cell 36: 309–318

Kawauchi H (1983) Chemistry of pro-opiocortin-related peptides in the salmon pituitary. Arch Biochem Biophys 227: 343–350

Kilpatrick DL, Gibson KD, Jones BN (1983) Is adrenal proenkephalin glycosylated? Arch Biochem Biophys 224: 402–404

Kurjan J, Herskowitz I (1982) Structure of a yeast pheremone gene (MF α): A putative α-factor precursor contains four tandem copies of mature α-factor. Cell 30: 933–943

Lazure C, Seidah NG, Pélprat D, Chrétien M (1983) Proteases and post-translational processing of pro-hormones: A review. Can J Biochem 61: 501–505

Lennarz WJ (ed) (1980) The biochemistry of glycoproteins and proteoglycans. Plenum, New York

Le Roith D, Shiloach J, Roth J (1982) Is there an earlier phylogenetic precursor that is common to both the nervous and endocrine systems? Peptides 3: 211–215

Lowry PJ, Scott AP (1975) The evolution of vertebrate corticotropin and melanocyte stimulating hormone. Gen Comp Endocrinol 26: 16–23

Lowry PJ, Bennett HPJ, McMartin C, Scott AP (1974) The isolation and amino acid sequence of an adrenocorticotropin from the pars distalis and a corticotropin-like intermediate lobe peptide from the neurointermediate lobe of pituitary of the dogfish *Squalus acanthias*. Biochem J 141: 427–437

Mains RE, Eipper BA, Glembotski CC, Dores RM (1983) Strategies for the biosynthesis of bioactive peptides. Trends Neurosci 6: 229–235

McLean C, Lowry PJ (1981) Natural occurance but lack of melanotrophic activity of γ-MSH in fish. Nature 290: 341–343

Meyer DL, Krause E, Dobberstein R (1982) Secretory protein transportation across membranes – the role of the "docking protein". Nature 297: 647–653

Mizuno K, Matsuo H (1984) A novel protease from yeast with specificity towards paired basic residues. Nature 309: 558–560

Moore DD, Walker MD, Diamond DJ, Conkling MA, Goodman HM (1982) Structure, expression and evolution of growth hormone genes. Recent Prog Horm Res 38: 197–225

Moore H-P., Gumbiner B, Kelly RB (1983) A subclass of proteins and sulphated macromolecules secreted by AtT-20 (mouse pituitary tumor) cells is sorted with adrenocorticotropin. J Cell Biol 97: 810–817

Noda M, Terenishi Y, Rakahashi H, Toysota M, Notake M, Nakanishi S, Numa S (1982) Nature 297: 431–434

Nozaki M, Gorbman A (1984) Distribution of immunoreactive sites for several components of proopiocortin in the pituitary and brain of adult lamprey *Petromyzon marinus* and *Entospherus tridertatus*. Gen Comp Endocrinol 53: 335–352

Scheller RH, Jackson JF, McAllister LB, Rothman BS, Mayeri E, Axel R (1983) A single gene encodes multiple neuropeptides mediating a stereotyped behaviour. Cell 32: 7–22

Seger MA, Bennett HPJ (1983) Specific labelling of pro-opiomelanocortin (POMC)-related peptides with ^{32}P using casein kinase. In: Hruby VH, Rich DH (eds) Peptides: structure and function. Proceedings of the Eighth American Peptide Symposium. Pierce Chemical, Rockford, pp 253–256

Seger MA, Bennett HPJ (1985) Processing of the N-terminal fragment of pro-opiomelanocortin in the intermediate lobe of the pituitary. (to be published)

Sly WS, Fischer HD (1982) The phosphomannosyl recognition system for intracellular and intercellular transport of lysosomal enzymes. J Cell Biochem 18: 67–85

Tatemoto K, Carlquist M, Mutt V (1982) Neuropeptide Y – a novel brain peptide with structural similarities to peptide YY and pancreatic polypeptide. Nature 296: 659–660

Uhler M, Herbert E, D'Eushachio P, Ruddle FD (1983) The mouse genome contains two nonallelic pro-opiomelanocortin genes. J Biol Chem 258: 9444–9453

Walter P, Blobel G (1982) Signal recognition particle contains a 7S RNA essential for protein translocation across the endoplasmic reticulum. Nature 299: 691–698

ACTH and Related Peptides in Lung Cancer*

J. G. Ratcliffe

University of Manchester, Department of Chemical Pathology, Hope Hospital,
Salford M6 8HD, Great Britain

Introduction

The study of ACTH and related peptides in lung cancer has had an important influence in
formulating current concepts of ACTH biosynthesis and the wider aspects of ectopic hormone production. The fact that tumours arising in tissues traditionally regarded as "nonendocrine", such as lung, can synthesise and secrete hormones has only been clearly recognised by clinicians in the last 25 years. For some years after the characterisation of the
ectopic ACTH syndrome the phenomenon was thought to be very rare. Thus, a critical
analysis of nonendocrine tumours associated with Cushing's syndrome in 1968 showed
there to be only about 133 acceptable cases of the syndrome in the literature (Azzopardi
and Williams 1968). The majority were due to small cell and carcinoid tumours of the
lung, with fewer cases due to endocrine tumours of foregut origin, phaeochromocytoma
and related tumours and certain ovarian tumours. Increased clinical awareness and the
application of newer methods of hormone assay have fundamentally changed the per-
spective on the prevalence of ACTH production by extrapituitary tissues, both tumorous
and nontumorous. The overt ectopic ACTH syndrome with gross hypercortisolaemia and
hypokalaemic alkalosis is still unusual in patients with lung cancer, but biochemical evidence suggests that ACTH production by lung tumours is common.

This chapter describes the structure, prevalence and control of ACTH-related peptides
in lung cancer. For more detailed consideration of the wider subject of ectopic hormone
production, the reader is referred to reviews by Rees and Ratcliffe (1974), Sherwood and
Gould (1979) and Ratcliffe (1982). The mechanism of ectopic hormone production has
been well reviewed by Baylin and Mendelsohn (1980).

Structure of ACTH and Related Peptides

The nature of the association between adrenocortical hyperfunction and an extrapituitary
tumour was clarified by the demonstration in these patients of elevated plasma ACTH activity together with biologically active ACTH in the tumours and decreased pituitary
ACTH. Subsequently, it was shown that the condition can occur in hypophysectomised

* Some of the work described in this paper was supported by the Scottish Hospitals Endowment
  Research Trust and the Cancer Research Campaign

patients, and arteriovenous ACTH gradients across the tumour were demonstrated. These data provided strong evidence that extrapituitary tumours, including lung tumours, can synthesise and release ACTH-like peptides. More recently synthesis and secretion of such peptides by tumour cell cultures in vitro has been demonstrated.

However, proof of the exact chemical structure of these peptides is hampered by the limited availability of surgically resected human tumours associated with the ectopic ACTH syndrome. Small cell lung cancer is not usually treated surgically, and the interpretation of structural data on peptides derived from autopsy material is questionable. Furthermore, tumour concentrations of ACTH-like peptides are much lower than in the pituitary, so that it is rare for sufficient material to be available for detailed structural analysis. Much of the present information therefore depends on a combination of chromatographic and radioimmunoassay analysis, supplemented on occasion by bioassays, isoelectric focusing and, rarely, amino acid composition or sequencing.

There is now good evidence that tumours associated with the ectopic ACTH syndrome produce a wide range of peptides related to ACTH. In general, the primary amino acid sequences of these peptides closely resemble their normal counterparts, but the relative abundance of the several molecular forms in tumours differs from the normal pituitary and varies from tumour to tumour. Ectopic ACTH-related peptides are more heterogeneous than eutopic ACTH, with greater proportions of high-molecular-weight forms, and fragments.

The following peptides have been characterised in varying detail in tumours associated with the ectopic ACTH syndrome:

1. ACTH and its C- and N-terminal fragments
2. Lipotrophin (LPH) and its fragments
3. High-molecular-weight precursor forms of ACTH

Adrenocorticotropic Hormone

By 1969 it was clear that lung and other tumours associated with the ectopic ACTH syndrome contained a material very similar to 1–39 pituitary ACTH in chemical, immunological and biological test systems (Liddle et al. 1969). Subsequently, detailed chemical analysis of ACTH from a malignant thymic carcinoid tumour indicated that the peptide resembled the 2–38 sequence of ACTH (Lowry et al. 1976). It was considered likely that this was due to cleavage of 2–38 ACTH from its precursor by endopeptidases not present or inactive in the normal pituitary gland. The alternative explanation, that the tumour synthesised 1–39 ACTH with subsequent exopeptidase cleavage at C- and N-termini, was less plausible because exopeptidase action would result in incomplete loss of terminal residues followed by partial loss of penultimate residues. Hence amino acid analysis would give fractional values and several N- and C-terminals. These predictions were not borne out by the experimental data. Others have reported amino acid residues in ectopic ACTH not normally found in human pituitary ACTH though the overall composition was very similar. Thus, although minor differences from authentic 1–39 ACTH may occur, no major amino acid differences have been identified in ectopic ACTH.

However, analysis of partially purified extracts showed that tumours, especially bronchial carcinoids, often had a gross excess of C- to N-terminal ACTH immunoactivity (up to 50-fold) and a less pronounced excess of N-terminal to bioactive ACTH 2- to 3-fold). These apparent differences between eutopic and ectopic ACTH were resolved by chromatography of tumour extracts, which demonstrated that in addition to material coeluting with 1–39 ACTH, most tumours contained significant amounts of peptides with only C-terminal or N-terminal ACTH immunoactivity.

The C-terminal peptide fragment was isolated and partially characterised from a bronchial carcinoid as 18–39 ACTH (corticotrophin-like intermediate lobe peptide, CLIP) by amino acid composition and N- and C-terminal analysis (Ratcliffe et al. 1973). The N-terminal peptide appears to be the N-terminal 15 amino acid sequence of ACTH (Orth et al. 1973). In some cases it does not cross-react in the αMSH assay, suggesting lack of C-terminal amidation or N-terminal acetylation, whereas in others it cross-reacts in assays specific for αMSH.

Since αMSH and CLIP are peptides normally confined to species with a distinct intermediate lobe of the pituitary, it seems likely that peptidases are involved with specificities resembling those of the intermediate rather than the anterior lobe. In man a distinct intermediate lobe is normally only present during fetal life and pregnancy.

Lipotrophin

Following the recognition that βLPH formed the C-terminal portion of the ACTH precursor proopiocortin (POC) in the anterior pituitary and contained both the βMSH and β-endorphin sequences, these peptides were sought in ectopic ACTH-producing tumours. Both βLPH and γLPH (1–58 βLPH) and β-endorphin (61–91 βLPH) are consistently found but the relative proportions vary, as with ACTH and its related fragments (Tanaka et al. 1978). γLPH is often the major tumour component, and in contrast to the normal pituitary, LPH fragments occur, e.g. α-endorphin (61–76 βLPH), βhMSH (41–58 βLPH) or βpMSH (37–58 βLPH). βMSH, like αMSH and CLIP, is a typical intermediate lobe peptide, so its presence in tumours again emphasises the resemblance of tumour metabolism to the intermediate rather than to the anterior lobe of the pituitary.

High-Molecular-Weight ACTH

In the early 1970s it became evident that tumours and plasma from patients with ectopic ACTH syndrome contained ACTH-immunoactive peptide(s) larger than 1–39 ACTH (Yalow and Berson 1973). This was originally termed "big" ACTH. It is more acidic than 1–39 ACTH and has little steroidogenic activity. HMW ACTH was often the predominant molecular form of ectopic ACTH, whereas it was quantitatively a minor component in normal pituitaries. It could be converted to authentic ACTH ("little" ACTH) by trypsinisation with evolution of biological activity. These data suggested that HMW ACTH was a precursor in the biosynthesis of tumour ACTH.

The existence of intermediate forms between big and little ACTH in ectopic ACTH-producing tumours had been suggested from gel chromatography profiles, but they were not well characterised at the time when Mains and Eipper (1976) reported an extensive series of experiments on an ACTH-producing mouse pituitary cell line (AtT-20). By pulse chase experiments combined with SDS-polyacrylamide gel electrophoresis they showed that labelled amino acids were first incorporated into a 31 000-dalton molecule, which was then converted to 23 000-, 13 000- and 4500-dalton molecules. The 31K, 23K and 13K molecules are glycoproteins and the last is a glycosylated form of 1–39 ACTH but not its precursor (summarised in Eipper and Mains 1980).

Similar studies with ectopic ACTH-producing tumours showed a similar pattern with three HMW ACTH peaks in addition to 1–39 ACTH, the largest and intermediate forms being glycosylated. These data indicate that HMW forms of pituitary and tumour ACTH are not qualitatively different. Cell-free synthesis of ACTH-like peptides directed by mRNA from mouse pituitary tumour cells or ectopic ACTH-producing tumours has confirmed the synthesis of ACTH in HMW form. The elegant application of recombinant DNA technology has allowed the determination of the nucleotide sequence of complementary DNA for a bovine ACTH-LPH precursor, and hence the full peptide sequences (Nakanishi et al. 1979).

Although the general pattern of biosynthesis of the ACTH precursor in ectopic ACTH-producing tumours closely resembles that demonstrated in the mouse pituitary tumour and bovine pituitary, there are subtle differences in processing, so that the relative proportions of the HMW species differ in the ectopic situation. In ectopic ACTH-producing tumour extracts, the largest species with ACTH i.r. corresponds to the 31 K pituitary component and thus probably represents the full precursor molecule (POC). In some tumours this contains almost all the ACTH immunoreactivity (i.r.) and there is only a small amount of 1–39 ACTH. In others, the profile resembles that found in normal pituitaries in that the 31 K species is only a minor component and most of the ACTH i.r. corresponds to 1–39 ACTH (Pullan et al. 1980). Using assays for the N-terminal portion of POC (γ_3MSH), extracts of ectopic ACTH tumours show two peaks of activity, which may correspond to glycosylated and nonglycosylated forms of 1–76 N-POC. Alternatively, the second peak of smaller molecular size may correspond to glycosylated γ_3MSH (51–76 N-POC) (Bertagna et al. 1983). In contrast, human pituitary exctracts show a single major peak of i.r. γ_3MSH corresponding to the 16 K fragment (i.e. glycosylated 1–76 N-POC), although Ekman et al. (1982) report some heterogeneity with peptides of smaller molecular size and uncertain structure (? 1–62, 1–48, 1–30 N-POC).

The profiles of ACTH-related peptides in plasma also show subtle differences in the eutopic and ectopic situations. In the ectopic ACTH syndrome, plasma i.r. ACTH occurs in two peaks: one corresponds to 1–39 ACTH and the other is a broad heterogeneous peak corresponding to the 22 K fragment (i.e., N-POC/ACTH) (Ratter et al. 1983). The latter material is absent in plasma in which the ACTH is from the pituitary. The profiles of plasma i.r. N-POC (pro-γMSH) also reveal differences. In the ectopic ACTH syndrome, plasma N-POC is heterogeneous with peaks corresponding to 31 K (precursor POC), 22 K (N-POC-ACTH), 6–10 K (? nonglycosylated 1–76 N-POC) and γ_3MSH (51–76 N-POC) (Hale et al. 1984). In contrast, when plasma ACTH is derived from the pituitary, there is a single peak corresponding to 1–76 N-POC but with an apparently larger molecular

Table 1. Processing of HMW ACTH in ectopic ACTH syndrome[a]

	Tumour	Plasma
PRO-opiocortin (31 K precursor)	+	+
N-POC/ACTH (22 K)		+
N'POC (poorly glycosylated) (?6–10 K)	+	+
γ_3-MSH		+

[a] In the anterior pituitary the 31 K and 22 K components are minor, and N-POC is present as glycosylated 1–76 N-POC (16 K). In plasma i.r. ACTH derived from the pituitary there is no 22 K component, but a major component corresponding to glycosylated 1–76 N-POC. + indicates major presence of the peptide

size than the equivalent peak in "ectopic" plasma. It may represent glycosylated 1–76 N-POC.

A summary of the processing of HMW ACTH components in the ectopic ACTH syndrome is shown in Table 1.

Prevalence of ACTH Production by Lung Tumours

Estimates of the prevalence depend critically on the criteria accepted for hormone production. Clinically overt ectopic ACTH syndrome with evidence of gross hypercortisolism and hypokalaemic alkalosis occurs only in 2% on 3% of patients with small cell lung cancer and extremely rarely with the other histological tumour types. If estimates are based on biochemical rather than clinical criteria the prevalence increases substantially. Thus, impaired suppression of corticosteroids is reported in about 50% of patients with small cell lung carcinoma in the absence of clinically apparent syndromes. If lung tumours are extracted and hormone concentrations measured by sensitive immuno- or bioassays in the extract, then positive ACTH and LPH levels are detected in the majority (90%) of small cell and carcinoid tumours at levels within the range found in proven cases of ectopic ACTH syndrome (>10 ng/g wet weight) (Bloomfield et al. 1977). Tumour ACTH and LPH concentrations correlate well, suggesting that both hormones are expressed together as in the normal pituitary. The ACTH content also is well correlated with the presence of secretory granulation in these tumours.

In contrast to small cell carcinomas and bronchial carcinoids, the production of ACTH by other types of lung cancer is less well established. The clinical syndrome is very rare in non-small cell carcinomas. ACTH and LPH levels in extracts of epidermoid, adeno- and large cell carcinomas are low and often near the limit of detection. In part such levels may represent retained circulating hormone and/or nonspecific assay effects. Occasionally, lung tumours with relatively high ACTH levels which are initially classified on conventional histology as non-small cell carcinomas are later reclassified as combination tumours with carcinoid elements. Data from tumour cell culture studies confirm that ACTH secretion occurs in small cell carcinoma lines irrespective of whether there is in vivo evidence of ectopic ACTH secretion, but perhaps surprisingly, also in 20% on 30% of other histological tumour types (Sorenson et al. 1981; Luster et al. 1983). The explanation of these discrepancies is not yet clear. Of course tumour hormone levels do not necessarily correlate with the rate of hormone synthesis or secretion, and in vitro findings may not accurately reflect the in vivo situation.

Recent evidence also suggests that the ACTH precursor is synthesised in normal porcine lung and thyroid tissue (Clements et al. 1982). Thus, acid extracts of these tissues contained significant amounts of immunoactive ACTH, βLPH and β-endorphin. However, the ACTH-like bioactivity was only 4% of the ACTH-like immunoactivity and the latter was very heterogeneous. The content of βLPH was much greater than that of immunoactive ACTH or β-endorphin, suggesting that the common ACTH precursor is processed predominantly to peptides other than ACTH and β-endorphin. This pattern is distinct from that described for anterior pituitary and intermediate lobe. These data suggest that the ACTH precursor is normally expressed in at least some cell types in the lung and thyroid but that the pattern of processing is such as to produce very low levels of ACTH and β-endorphin. In the neoplastic cell there may be a loss of tissue-specific processing so that relatively more ACTH and β-endorphin is produced. The pattern in ectopic ACTH-producing lung tumours may thus differ only quantitatively from both the normal lung and normal pituitary. The role of ACTH-related peptides in the normal lung is unknown but the pattern of peptides described above raises the possibility of physiological function for peptides in addition to ACTH and β-endorphin. The lung cells producing these peptides may have chemoreceptor functions and participate in proliferative and metaplastic responses to injury (e.g. inflammation, smoking) by releasing locally active peptides and amines.

Overall, the evidence points to the conclusion that ACTH-like peptides are synthesised commonly in lung tumours but that the pattern of processing, storage and secretion differs between tumour types, with small cell carcinomas most commonly storing and secreting higher amounts of peptides closely resembling authentic ACTH. Non-small cell tumours may also frequently express the ACTH precursor gene, but they less commonly process, store and secrete authentic ACTH. A high prevalence of production of ACTH-like peptides in lung tumours is perhaps more understandable if the ACTH precursor is produced by some normal non-neoplastic lung cells.

Circulating Levels of ACTH-Like Peptides in Lung Cancer

In patients with the classical manifestations of the ectopic ACTH syndrome, plasma ACTH levels are uniformly elevated (>100 ng/l), and approximately two-thirds of patients have immunoactive ACTH levels greater than are found in pituitary-dependent Cushing's syndrome. A plasma ACTH level of greater than 250 ng/litre in a patient with Cushing's syndrome is highly suggestive of ectopic production and should lead to an intensive search for an extrapituitary tumour (Ratcliffe et al. 1972). It is important to recognise that about half the patients with ACTH-producing bronchial carcinoids or thymomas present with Cushing's syndrome *before* the tumour is diagnosed.

As predicted from recent knowledge of ACTH biosynthesis, plasma LPH and N-POC levels are also elevated in the syndrome (Hope et al. 1981). The ratio of βLPH (or βMSH-like peptides) to ACTH also tends to be increased, possibly due to the slower metabolic clearance of LPH. The mean βMSH/ACTH concentration ratio in the ectopic ACTH syndrome usually exceeds 2, whereas the ratio is approximately unity with pituitary ACTH excess. The ratio of $\beta:\gamma$LPH also tends to be increased with an ectopic source.

There is wide variation in estimates of the prevalence of elevated circulating ACTH immunoactivity in patients with lung cancer who do not have the overt ectopic ACTH syndrome. Gewirtz and Yalow (1974) and Ayvazian et al. (1975) reported that 21 of 24 patients with untreated carcinoma of lung had afternoon plasma ACTH concentrations above 150 ng/litre, whereas only 6% of normal controls had such levels. Two of the three patients with low plasma ACTH were alive and well 3 and 5 years after tumour detection, even without antitumour therapy. In a subsequent study, plasma ACTH was elevated in 66 of 136 patients and was more often elevated (59/107) in those with extensive than limited disease (11/29). In those with extensive disease, patients with adenocarcinoma exhibited elevated values less frequently than patients with other tumour types. In about two-thirds of patients suitable for surgical resection (non-small cell) plasma ACTH was elevated above 150 ng/litre (Yalow et al. 1979).

However, a significant proportion (20% on 30%) of patients with nonmalignant lung disease also showed elevated ACTH levels. These data suggest that plasma ACTH levels have only limited value in screening for lung cancer, but that elevated levels generally indicate a poor prognosis.

Similar results were reported by Wolfson and Odell (1977), who found that 72% of patients with untreated lung cancer without ectopic ACTH syndrome (histology unspecified) have elevated ACTH levels and 61% had elevated βLPH. Again, about 30% of patients with nonmalignant lung disease had elevated levels.

More recent studies have suggested that plasma or serum immunoactive levels are elevated only in a minority. Even in small cell carcinoma, elevated levels were found only in 24%–30% and were poorly related to stage of disease and clinical course (Gropp et al. 1980; Hansen et al. 1980). The stronger association with small cell carcinoma (though not with its histological subtypes) is more compatible with ACTH measurements in tumours, and with pathological evidence of positive immunostaining for ACTH in 19% of small cell tumours but not other histological types.

Our own studies confirm that the prevalence of high levels of plasma ACTH immunoactivity in patients with lung cancer is relatively low, is mainly associated with extensive small cell cancer and is related to the specificity of the ACTH assay (Ratcliffe et al. 1982). Thus, using an unextracted plasma or serum assay for total ACTH i. r. we found high levels in 24% of patients with small cell cancer and 3% of patients with non-small cell cancer. In patients with small cell cancer, levels were high in 12% with limited disease and 32% with extensive disease. In contrast, using an ACTH assay after plasma extraction, with porous glass which measures mainly 1–39 ACTH, no lung cancer patients had elevated levels. These last findings were confirmed using assay for βMSH and LPH after prior extraction.

We have attempted to analyse some of the factors which may account for these discrepancies, but they do not appear to be due to differences in type or timing of samples, reference ranges or nature and extend of tumour. Differences in assay specificity may be important with some assays detecting high-molecular-weight species and fragments in addition to 1–39 ACTH. Our finding of a reduced prevalence with the extracted ACTH assay is compatible with the idea that high-molecular-weight species (e. g. 31 K, 22 K) are important circulating components of ACTH immunoactivity in patients with lung tumours. The low prevalence reported with a radioreceptor assay which measures predominantly 1–39 ACTH also supports this contention. Further definition of the nature of circulating ACTH in lung cancer with application of assays specific for the several components (high-molecular-weight forms, N-POC, biologically active ACTH and ACTH fragments) is clearly required.

The source of circulating ACTH-like immunoactivity in lung cancer is also unresolved. It could be derived from the pituitary or the non-neoplastic lung tissue in addition to the tumour itself. However, the pituitary seems an unlikely source, since the predominant hormonal form following pituitary stimulation corresponds to biologically active ACTH and should therefore be associated with signs and symptoms of cortisol excess. Small amounts of ACTH are present in normal lung tissue and in non-neoplastic diseased lung. Lung tissue from a "smoking" dog without invasive carcinoma but with atypical histological changes contained ACTH mainly in the high-molecular-weight form, whereas lung tissue from smoking dogs without histological changes did not. Thus, the lung injured by smoking may make an important contribution to circulating immunoreactive ACTH. Furthermore, Auerbach et al. (1961) showed that heavy smokers without obvious lung cancer frequently showed histological changes of carcinoma – in situ. Torstensson et al. (1980) reported that acute administration of dexamethasone suppresses ACTH levels by a similar proportion of initial values in patients with lung cancer and nonmalignant lung disease, suggesting that the tumour is not the only source of plasma ACTH.

An additional factor which must be taken into account in interpreting circulating hormone levels is the presence of antibodies. Sera from patients with small cell cancer may contain immune complexes, comprising macromolecular IgG, C1q-binding activity, complement and glycoprotein high-molecular-weight ACTH components but not authentic ACTH (Havemann et al. 1979). Thus suggests that high-molecular-weight forms of ACTH released into the circulation are autoantigenic, though it is not clear whether autoantibody formation is restricted to small cell carcinoma. The ability to detect elevated ACTH immunoactivity in lung cancer may thus be related to the presence of ACTH autoantibodies as well as assay specificity.

Control of Secretion of ACTH-Like Peptides

The biosynthesis and release of ACTH in tumours associated with the ectopic ACTH syndrome is generally considered to be autonomous, but there is increasing evidence that this is not always so. Indeed, if ACTH-like peptides were produced by normal lung some control mechanisms would be expected, and it would be surprising if these were lost in all lung tumours. Wide variations in plasma ACTH levels occur over weeks or months in some cases of the ectopic ACTH syndrome. Occasionally in bronchial carcinoid tumours, metyrapone or dexamethasone induces corticosteroid responses, suggesting either that tumour ACTH secretion is sensitive to corticosteroids or that the hypothalamic-pituitary-adrenal axis is not always suppressed. Few systematic observations are available on this point. A dramatic acute increase of plasma ACTH levels has been observed after cortisone or hydrocortisone but not dexamethasone in a patient with ectopic ACTH syndrome due to a hepatic tumour (Himsworth et al. 1977). Several agents have been found to modify ACTH release from ectopic ACTH-producing tumours in vitro. Norepinephrine, serotonin, TRH, LRH, and prostaglandin E1 may increase ACTH release and/or cyclic AMP concentrations. Median eminence extracts give a dose-response relationship with ACTH released or cAMP accumulated in dispersed tumour cells. cAMP may not be the sole intracellular messenger, since a high calcium concentration also enhances ACTH release. Such data suggest that membranes of ectopic ACTH-producing tumours have multiple receptors for hormones and biogenic amines (Hirata et al. 1979).

The role of peptides with corticotropin-releasing activity is of particular interest. Tumour peptides with CRF activity have been demonstrated in several tumours which also produce ACTH. Tumour CRF activity may have an important function in maintaining and controlling tumour ACTH secretion, and could influence neoplastic differentiation and growth. Although the nature of tumour CRF-like substances is not well characterised, a dose-response relationship similar to that of hypothalamic CRF has been demonstrated by Yamamoto et al. (1976). Other workers have reported no release of N-POC (pro-γMSH), ACTH, and LPH from ectopic ACTH-producing tumours in vitro in response to graded doses of rat stalk median eminence extracts or synthetic AVP, although the peptides were secreted in a pulsatile fashion. (Ratter et al. 1983). The apparently autonomous secretion of tumour ACTH may be attributable in part to the local effect of tumour CRF-like substances.

Occasionally, tumours may produce peptides with CRF activity without inducing concomitant tumour ACTH secretion. Thus, Cushing's syndrome has been reported in a patient with a medullary thyroid carcinoma in which tumour peptides with CRF activity were identified as similar to porcine gastrin-releasing peptides 1-27 and 1-14 (Howlett et al. 1985). Gastrin-releasing peptide (GRP) is the mammalian equivalent of amphibian bombesin. GRP has weak intrinsic CRF-like activity but more powerful CRF-41 potentiating potency, and the tumour extract stimulated ACTH release from dispersed rat anterior pituitary cells in vitro.

Clearly no conclusion can be drawn on how the secretion of tumour ACTH-like peptides is controlled, but there are sufficient data to suggest that in some cases at least it is not autonomous. The relationship with tumour CRF-like substances is intriguing. If the substances affecting hormone secretion, cell differentiation and growth can be identified, they could form the basis of new therapeutic approaches to the control of lung cancer.

Acknowledgements. Mrs. Pauline Bullock provieded skilled secretarial assistance in preparing this manuscript.

References

Auerbach O, Stout AP, Hammond EC, Garfunkel L (1961) Changes in bronchial epithelium in relation to cigarette smoking and in relation to lung cancer. N Eng J Med 265: 253–260

Ayvazian LF, Schneider B, Gewirtz G, Yalow RS (1975) Ectopic production of big ACTH in carcinoma of the lung. Am Rev Respir Dis 111: 279–287

Azzopardi JG, Williams ED (1968) Pathology of nonendocrine tumours associated with Cushing's syndrome. Cancer 22: 274–286

Baylin SB, Mendelsohn G (1980) Ectopic (inappropriate) hormone production by tumours: mechanisms involved and the biological and clinical implications. Endocr Rev 1 (1): 45–77

Bertagna X, Seurin D, Pique L, Luton JP, Bricaire H, Girard F (1983) Peptides related to the NH_2-terminal end of proopiocortin in man. J Clin Endocrinol Metab 56: 489–495

Bloomfield GA, Holdaway IM, Corrin B, Ratcliffe JG, Rees GM, Ellison M, Rees LH (1977) Lung tumours and ACTH production. Clin Endocrinol 6: 95–104

Clements JA, Funder JW, Tracy K, Morgan FJ, Campbell DJ, Lewis A, Hearn MTW (1982) Adrenocorticotrophin, β-Endorphin, and β-lipotropin in normal thyroid and lung: possible implications for ectopic hormone secretion. Endocrinology 111: 2097–2102

Eipper BA, Mains RE (1980) Structure and biosynthesis of pro-adrenocorticotropin/endorphin and related peptides. Endocr Rev 1 (1): 1–27

Ekman R, Hakanson R, Larsson I, Sundler F, Thorell JI (1982) Radioimmunoassay of pro-γ-melanotropin, the amino terminal fragment of proopiolipomelanocortin. Endocrinology 111: 578–583

Gewirtz G, Yalow RS (1974) Ectopic ACTH production in Carcinoma of the lung. J Clin Invest 53: 1022–1032

Gropp C, Havemann K, Scheuer A (1980) Ectopic hormones in lung cancer patients at diagnosis and during therapy. Cancer 46: 347–354

Hale AC, Ratter SJ, Tomlin SJ, Lytras N, Besser GM, Rees LH (1984) Measurement of immunoreactive γ-MSH in human plasma. Clin Endocrinol 21: 139–148

Hansen M, Hansen HH, Hirsch FR, Arends J, Christenson J, Christenson JM, Hummer L, Kuhl C (1980) Hormonal polypeptides and amine metabolites in small cell carcinoma of the lung with special reference to stage and subtypes. Cancer 45: 1432–1437

Havemann K, Gropp C, Scheuer A, Scherfe T, Gramse M (1979) ACTH-like activity in immune complexes of patients with oat-cell carcinoma of the lung. Br J Cancer 39: 43–50

Himsworth RL, Bloomfield GA, Coombes RC, Ellison ML, Gilkes JJH, Lowry PJ, Setchell KDR, Slavin G, Rees LH (1977) Big ACTH and calcitonin in an ectopic hormone-secreting tumour of the liver. Clin Endocrinol 7: 45–62

Hirata Y, Yoshimi H, Matsukura S, Imura H (1979) Effects of hypothalamic extract and other factors on release of ACTH and Adenosine 3'5' monophosphate levels in dispersed non-pituitary tumour cells. J Clin Endocrinol Metab 49: 317–321

Hope J, Ratter SJ, Estivariz FE, McLoughlin L, Lowry PJ (1981) Development of a radioimmunoassay for an amino terminal peptide of proopiocortin containing the γMSH region: measurement and characterization in human plasma. Clin Endocrinol 15: 221–227

Howlett TA, Price J, Hale AC, Doniach I, Rees LH, Wass JAH, Besser GM (1985) Pituitary ACTH-dependent Cushing's syndrome due to ectopic production of a bombesin-like peptide by a medullary carcinoma of thyroid. Clin Endocrinol 22: 91–101

Liddle GW, Nicholson WE, Island DP, Orth DN, Abe K, Lowder SC (1969) Clinical and laboratory studies of ectopic humoral syndromes. Recent Prog Horm Res 25: 283–314

Lowry PJ, Rees LH, Tomlin S, Gillies G, Landon J (1976) Chemical characterisation of ectopic ACTH purified from a malignant thymic carcinoid tumour. J Clin Endocrinol Metab 43: 831–835

Luster W, Gropp C, Havemann K (1983) Peptide hormone synthesising lung tumour cell lines: establishment and first characterisation of biosynthetic products. Acta Endocrinol [Suppl] (Copenh) 253: 24–25

Mains RE, Eipper BA (1976) Biosynthesis of adrenocorticotropic hormone in mouse pituitary tumour cells. J Biol Chem 251: 4115–4120

Nakanishi S, Inoue A, Kita T, Nakamura M, Chang ACY, Cohen S, Numa A (1979) Nucleotide sequence of cloned cDNA for bovine corticotropin-β-lipotropin precursor. Nature 278: 423–427

Orth DN, Nicholson WE, Mitchell WM, Island DP, Liddle GW (1973) Biologic and immunologic characterisation and physical separation of ACTH and ACTH fragments in the ctopic ACTH syndrome. J Clin Invest 52: 1756–1769

Pullan PT, Clement-Jones V, Corder R, Lowry PJ, Besser GM, Rees LH (1980) ACTH, LPH and related peptides in the ectopic ACTH syndrome. Clin Endocrinol 13: 437–445

Ratcliffe JG (1982) Ectopic production of hormones in malignant disease. In: O'Riordan JLH (ed) Recent advances in endocrinology and metabolism. Churchill Livingstone, Edinburgh, pp 187–209

Ratcliffe JG, Knight RA, Besser GM, Landon J, Stansfeld AG (1972) Tumour and plasma ACTH concentrations in patients with and without the ectopic ACTH syndrome. Clin Endocrinol 1: 27–44

Ratcliffe JG, Scott AP, Bennett HPJ, Lowry PJ, McMartin C, Strong JA, Walbaum PR (1973) Production of a corticotrophin-like intermediate lobe peptide and of corticotrophin by a bronchial carcinoid tumour. Clin Endocrinol 2: 51–55

Ratcliffe JG, Podmore J, Stack BHR, Spilg WGS, Gropp C (1982) Circulating ACTH and related peptides in lung cancer. Br J Cancer 45: 230–236

Ratter SJ, Gillies G, Hope J, Hale AC, Grossman A, Gaillard R, Cook D, Edwards CRW, Rees LH (1983) Proopiocortin related peptides in human pituitary and ectopic ACTH secreting tumours. Clin Endocrinol 18: 211–218

Rees LH, Ratcliffe JG (1974) Ectopic hormone production by non-endocrine tumours. Clin Endocrinol 3: 263–299

Sherwood LM, Gould VE (1979) Ectopic hormone syndromes and multiple endocrine neoplasia. In: DeGroot LJ (ed) Endocrinology, vol 3. Grune and Stratton, New York, pp 1733–1766

Sorenson GD, Pettengill OS, Brinck-Johnson T, Cate CC, Maurer LH (1981) Hormone production by cultures of small cell carcinoma of the lung. Cancer 47: 1289–1296

Tanaka K, Nicholson WE, Orth DN (1978) The nature of the immunoreactive lipotropins in human plasma and tissue extracts. J Clin Invest 62: 94–104

Torstensson S, Thoren M, Hall K (1980) Plasma ACTH in patients with bronchogenic carcinoma. Acta Med Scand 207: 353–357

Wolfson AR, Odell WD (1979) ProACTH: use for early detection of lung cancer. Am J Med 66: 765–772

Yalow RS, Berson SA (1973) Characteristics of 'big' ACTH in human plasma and pituitary extracts. J Clin Endocrinol Metab 36: 415–423

Yalow RS, Eastridge CE, Higgins G, Wolf J (1979) Plasma and tumour ACTH in carcinoma of the lung. Cancer 44: 1789–1792

Yamamoto H, Hirata Y, Matsukura S, Imura H, Nakamura M, Tanaka A (1976) Studies on ectopic ACTH producing tumours. IV CRF-like activity in tumour tissues. Acta Endocrinol (Copenh) 82: 183–187

Physalaemin-Like Immunoreactivity from Human Lung Small Cell Carcinoma: Isocratic Reversed-Phase HPLC Analysis of the Chemically Modified Peptide

L. H. Lazarus and O. Hernandez

Laboratory of Behavioral and Neurological Toxicology, Bio-organic Chemistry Laboratory of Molecular Biophysics, National Institute of Environmental Health Sciences, Research Triangle Park, NC 27709, USA

Introduction

The metastatic human lung tumor, small cell carcinoma (SCC), exhibits ectopic (inappropriate) synthesis of numerous peptide hormones, such as ACTH, [Arg[8]]-vasopressin, hCGβ, calcitonin, oxytocin, prolactin, and growth hormone (Rees et al. 1974; Rees 1976; Baylin and Mendelsohn 1980). Furthermore, tumors resected from patients (Wood et al. 1981), or passaged in nude (athymic) mice (Erisman et al. 1982), or cultured into numerous cell lines (Moody et al. 1981, 1983 a, b; Sorenson et al. 1982) were found to contain immunoreactivity to the amphibian peptide bombesin and more recently to physalaemin (Erisman et al. 1982; Lazarus et al. 1983), which is the prototype of the tachykinin family of peptides discovered by Erspamer et al. (1964). The presence of immunoreactivity to bombesin was detected in human fetal and neonatal lungs (Wharton et al. 1978) and in the lungs of children (Cutz et al. 1981; Track and Cutz 1982) and appears to be associated with a mammalian form of bombesin similar to porcine gastrin-releasing peptide, a 27 amino acid residue peptide (Iwanago 1983; Price et al. 1983; Tsutsumi et al. 1983). On the other hand, the physalaemin-like immunoreactive material (PLIM), initially discovered in the gastrointestinal tract (Lazarus and DiAugustine 1980; Lazarus et al. 1980), was also present in neuronal (Lazarus et al. 1980) and pulmonary tissue from diverse mammlian species (Lazarus et al. 1982). Further studies showed however, that PLIM differed from the amphibian peptide (Lazarus et al. 1982) and the purified immunoreactive material was in fact an octapeptide with homology to the N-terminal region of physalaemin (Wilson et al. 1984).

The initial evidence for physalaemin-like immunoreactivity (PSLI) in SCC initially relied on radioimmunoassays and immunohistochemical localization in fixed tissue sections (Erisman et al. 1982). Reversed-phase HPLC using linear elution gradients afforded two peaks of immunoreactivity in the tumor extracts, which coincide with the retention time of reduced and oxidized synthetic physalaemin (Lazarus et al. 1983). Those data also indicated that tumor PSLI contained an oxidizable residue and a trypsin-sensitive bond, and cross-reacted with three antisera specific to the N-terminal region of physalaemin. Moreover, the HPLC-isolated peptide contracted guinea pig ilial smooth muscle in vitro in a dose-dependent manner comparable to physalaemin at the same concentrations (Lazarus et al. 1983). However, recent studies on the behaviour of amphibian peptides on reversed-phase HPLC demonstrated that isocratic elution conditions amplify minimal structural differences between peptides, thereby significantly affecting their interaction with the hydrophobic bonded phase of the column packing material (Hernandez et al. 1984). It seems, therefore, that the use of gradient elution systems to predict amino acid composition of small peptides (Meek 1980) has limited application to resolving natural peptides which may differ by a single amino acid, as in the case of the phyllolitorins (Yasuhara et al. 1983) and angiotensin analogues (Klickstein and Wintraub 1982).

Thus, we sought to reassess the structural relationship between the tumor immunoreactive peptide and physalaemin. Furthermore, due to the limited quantities of tumor peptide, which would preclude complete structural analyses, we selectively modified specific amino acid residues in order to establish their presence in the unknown peptide and isolated the modified peptides using an isocratic solvent system. The data in this communication shows that the tumor peptide structurally resembles, but is not identical to, the amphibian peptide physalaemin, as suggested by Erspamer (1983).

Materials and Methods

Tumor

A human lung SCC, propagated in nude mice and donated by G. Sato, was extracted as previously described (Erisman et al. 1982). The freeze-dried extracts (stored at -20 °C) were resolubilized in 1.0 N formic acid at 90 °C (Lazarus et al. 1983) containing 1 mM EDTA and 10 mM 2-mercaptoethanol and clarified by centrifugation.

Radioimmunoassay

Physalaemin antiserum PS-XII-170-1 (PS-1) 1:40000 final dilution) was used in a RIA as described previously (Lazarus and DiAugustine 1980). The antibody recognizes a sequence in the amino terminal region of physalaemin, most probably the tripeptide -Asp-Pro-Asn- (Lazarus and DiAugustine 1980; Lazarus et al. 1980, 1982), although the involvement of pGlu and Ala has not been ruled out. The HPLC fractions were either sampled directly or dried first in a Speed-Vac (Savant) and redissolved in 0.01 N formic acid before immunoassay.

Reversed-Phase HPLC

Vydac ODS (5 μm, 300 Å pore) columns fitted with a precolumn were used on either Waters Associates or Laboratory Control Data HPLC equipment, the latter with a Rheodyne injector. The isocratic solvents used were 0.01 N trifluoacetic acid, pH 2.07 (Bennett et al. 1981), and 0.02 M triethylamine acetate, pH 5.03 or pH 6.91, containing 1 mM EDTA and 22% acetonitrile (v/v) as the organic modifier. The solvents were pumped through the column at a flow rate of 1.0 ml/min with a back pressure of 1900 psi. The eluent was monitored at 214 nm and fractions collected at 0.5-min intervals. Controls against cross-contamination of the HPLC system were followed as previously published (Erisman et al. 1982), since picogram quantities of the synthetic peptide can yield spurious and erroneous results (Lazarus et al. 1980), as described by Lazarus et al. 1982). Further steps were also taken: purification of microgram quantities of chemically modified peptides and isolation of the tumor peptide were carried out in different laboratories; only nanogram amounts (approx. 10 ng) of synthetic peptide were used after the analysis of the tumor material; and the injector was flushed with 60% acetonitrile after each completed run.

Chemical Modifications

1. Alkylation of Lysine Residues. The method of Means and Feeney (1968) was used. Physalaemin (1 µg) was dissolved in 50 µl 0.1 *M* sodium borate, pH 8.8, containing 10% (v/v) acetone. Sodim borohydride (ca. 1 mg) was added and the mixture was sonicated for

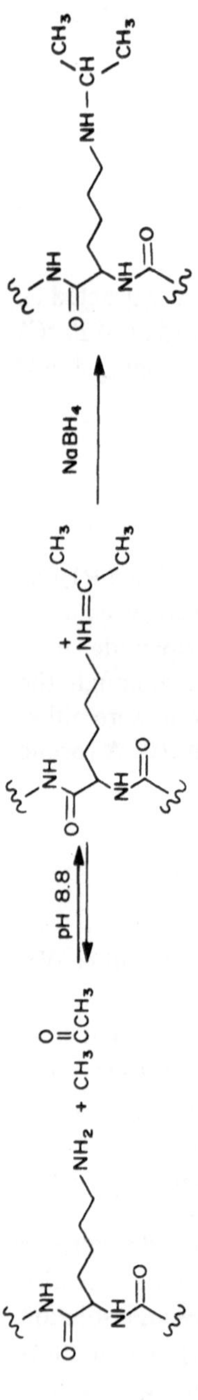

Fig. 1. *N*-Alkylation of lysine residues

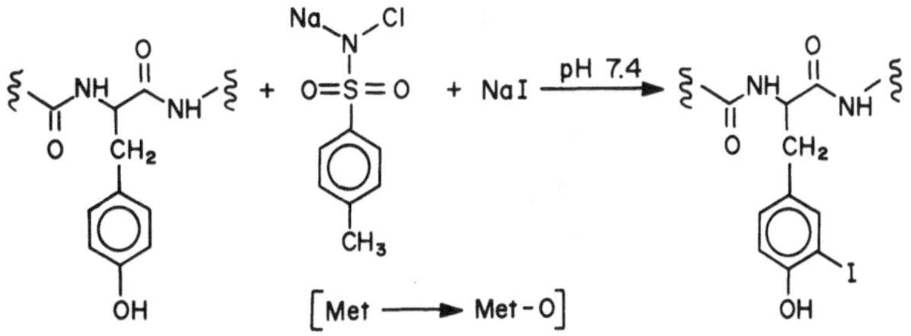

Fig. 2. Iodination of tyrosine residues

1 min. After 1 h at room temperature, the reaction was quenched with acetic acid (5 μl) and analyzed directly by HPLC. The reaction sequence is illustrated in Fig. 1.

2. Iodination of Tyrosine Residues. The method of Lazarus et al. (1977) was used. Physalaemin (1 μg) was dissolved in 0.4 M phosphate, pH 7.4, followed by sodium iodide (12.8 μg, 2 μl stock solution in buffer) and chloramine-T (76 μg, 2 μl stock solution) and the mixture was sonicated for 1 min. The reaction was quenched with solid sodium metabisulfite (ca. 1 mg) followed by acetic acid (5 μl), and analyzed for immunoreactivity by HPLC. The iodination of tyrosine in a peptide and the oxidation of methionine to methionine sulfoxide are shown in Fig. 2.

3. Tumor Samples. The reactions described above were repeated using approximately 2.5 ng physalaemin-equivalents of the tumor peptide previously isolated by HPLC and taken to dryness in a Speed-Vac.

For structural verification, samples of *N*-isopropyl- and iodinated physalaemin were purified by HPLC and analyzed by FAB-mass spectrometry. Both drivatives gave molecular ion(s) consistent with the proposed structure. The use of sodium deuteride (NaBD₄) afforded labeled peptide which showed a molecular ion one mass unit higher, consistent with the incorporation of only one deuterium.

Results

Our results on the analysis of PSLI from SCC are based on the modification of the HPLC behavior of peptides introduced by changing the pH of the eluting solvents and chemical modification of amino acid side chains and the subsequent separation by an isocratic solvent system with identification by use of a sensitive radioimmunoassay (Lazarus and Di-Augustine 1980).

Tumor Peptide Isolation

The extraction of SCC PSLI in the presence of a reducing agent enabled the isolation of a fully reduced peptide with a retention time of 7.5 min at pH 5.03; this represents a capacity factor (k′) of 2.56 as given in Table 1. These values are smaller than those obtained for physalaemin, which had a retention time of about 11 min and a k′ of 3.92 (Fig. 3).

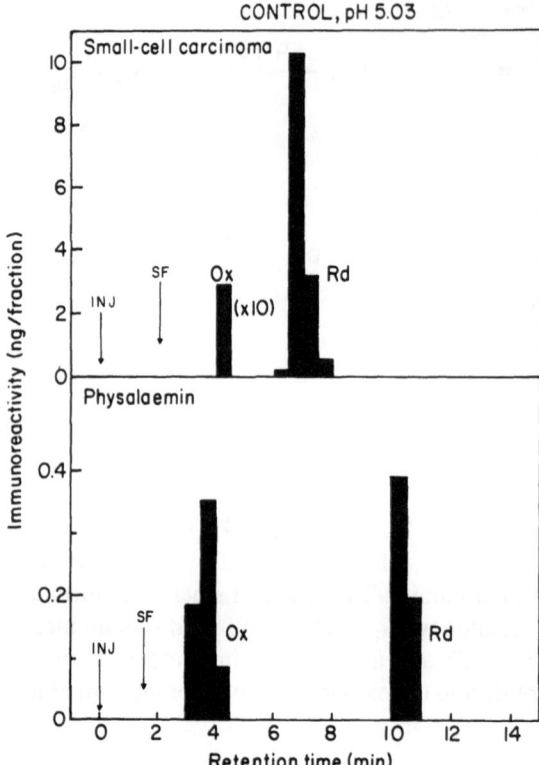

Fig. 3. Isocratic elution of PSLI and physalaemin on reversed-phase HPLC in 0.02 M triethamine acetate, pH 5.03, 1 mM EDTA and containing 22% (v/v) acetonitrile. The 0.5-ml fractions were collected at a flow rate of 1.0 ml per min. Aliquots of the fractions were either immunoassayed directly or dried and redissolved before assaying in the RIA. Refer to the *Methods* section for complete details. Abbreviations: *INJ*, sample injected; *SF*, solvent front; *Ox*, oxidized (methionine sulfoxide) peptide; *Rd*, reduced peptide

Table 1. Capacity factors (k')[a] for the physalaemin-like immunoreactivity from small-cell carcinoma of the human lung and amphibian physalaemin

Sample	pH		
	2.07	5.03	6.91
	Physalaemin-like immunoreactivity		
Control			
Oxidized	0.87	1.11	1.04
Reduced	1.85	2.56	2.96
Alkylated			
Oxidized	–	–	3.37
Reduced	4.31	–	4.74
Iodinated (oxidized)	7.17	–	–
	Amphibian physalaemin		
Control			
Oxidized	1.01	1.26	0.91
Reduced	3.23	3.92	3.00
Alkylated			
Oxidized	4.57	–	–
Reduced	5.15	–	4.47
Iodinated (oxidized)	7.08	–	7.63

[a] The capacity factor (k') is defined as $(V_t-V_o)/V_o$, where V_t is the retention time of the peptide and V_o is the solvent front

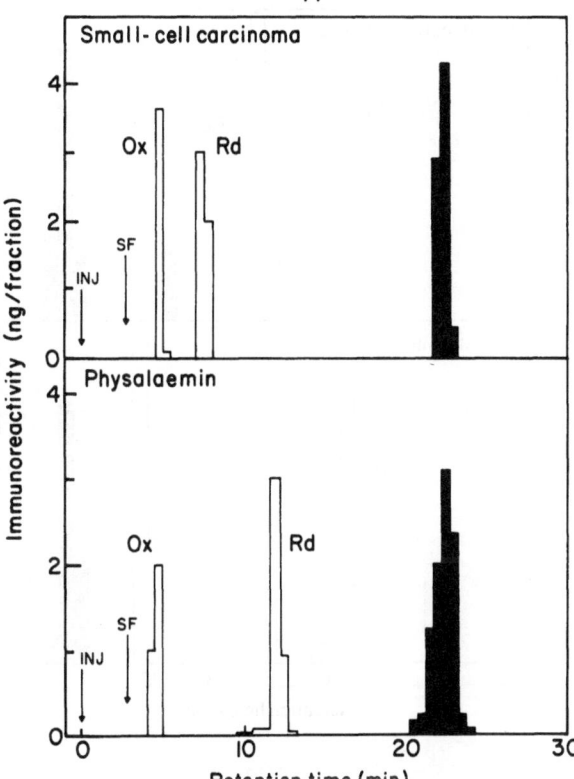

Fig. 4. Separation of iodinated PSLI and physalaemin on reversed-phase HPLC by isocratic elution in 0.01 N TFA, pH 2.07, containing 22% (v/v) acetonitrile. Refer to the legend in Fig. 1 for details. *Solid bar graphs* represent the experimental samples and *open bars,* the untreated controls

Rechromatography of the isolated PSLI yielded another peak of immunoreactivity with a k' of 1.1, which is the elution position of an oxidized peptide (Table 1).

Change in pH of the Elution Solvent

Modification of the ionic nature of the constituent side chains of a peptide are influenced by the pH of the elution solvent (Molnár and Horváth 1977; Hernandez et al. 1984). As seen in Table 1, the capacity factors for PSLI and physalaemin at pH 2.07 and 6.91 undergo shifts relative to their values at pH 5.03. In fact, the αpH of physalaemin (the ratio of k' values at pH 5.03 and pH 6.91) is 1.31, which is very close to the published value of 1.29 (Hernandez et al. 1984).

Chemical Modification of Peptide Side Chains

The iodination of the tyrosine residue in physalaemin substantially increased its retention time (Fig. 4), which was the result of the introduction of a single atom of iodine per peptide molecule. A similar increase in retention time was also seen with PSLI. In comparison with their oxidized forms, both peptides exhibited a similar change in k', the Δk' (Hernandez et al. 1984) values being 6.07 and 6.30 for physalaemin and PSLI, respectively. Mass

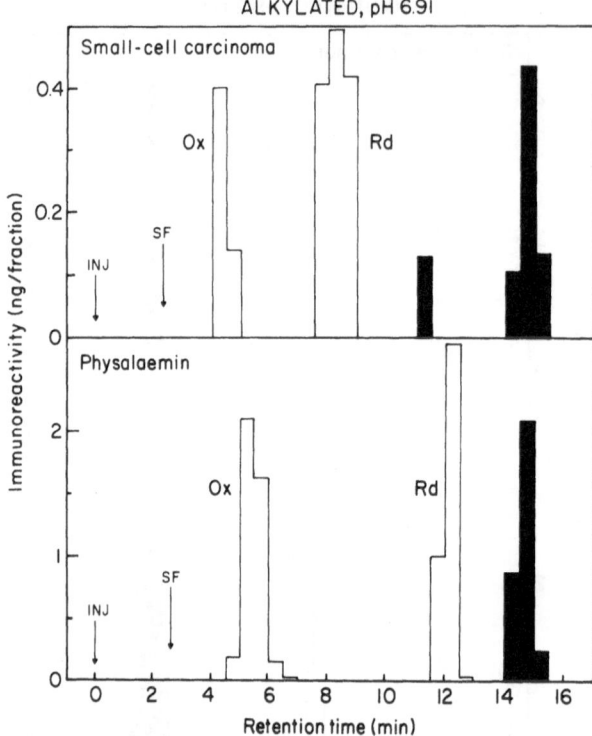

Fig. 5. Reversed-phase HPLC separation of alkylated PSLI and physalaemin in an isocratic solvent containing 0.02 M triethylamine acetate, pH 6.91, mM EDTA and 22% (v/v) acetonitrile. Details of the HPLC run are given in the legends to Figs. 3 and 4

spectrometry verified that physalaemin was a sulfoxide derivative and monoiodinated. HPLC analysis indicated that the reaction went to 95% completion.

The reductive alkylation with acetone reagent specifically introduces a single molecule of the isopropyl group on lysine residues (Means and Feeney 1968). This modification produces a more hydrophobic peptide, as shown in Fig. 5. Although the k' values and retention times for physalaemin and PSLI are similar, the Δk' reveals a significant difference: at pH 2.07, the Δk' values for physalaemin and PSLI are 1.92 and 2.46, respectively.

Discussion

The resolving power of isocratic elution conditions for peptide hormones on reversed-phase bonded supports was illustrated by Molnár and Horváth (1977), and proved effective in separating a series of angiotensin-related peptides (Klickstein and Wintroub 1982). The retention of peptides on the solid support, however, depends upon several interacting factors, such as the pH of the eluent which modifies the ionization of constituent side chains of the peptide (Molnár and Horváth 1977; Hernandez et al. 1984), the composition of the peptide in terms of its "polarity" and chain length (Molnár and Horváth 1977), and the nature of the organic modifier in the eluting solvent (Hernandez et al. 1984). Applying these criteria to the identification of PSLI, our results allowed discrimination between PSLI and the amphibian peptide physalaemin. The complete resolution of tumor PSLI and amphibian physalaemin on reversed-phase HPLC using isocratic conditions is contrasted to their coelution observed in a gradient elution system (Lazarus et al. 1983). The isocratic separation of peptides enhances the extent and magnitude of the interactions between peptides and the HPLC hydrophobic bonded phase (Molnár and Horváth 1977) which leads to increased selectivity and greater resolution. The gradient elution of peptides, on the other hand, appears to minimize differences between structurally related peptides and would be of limited use for predicting peptide composition as proposed by Meek (1980). In fact, gradient elution was unable to resolve the two naturally occurring nonapeptide phyllolitorins, which differ in one hydrophobic residue (Leu or Phe) (Yasuhara et al. 1983) and a series of angiotensin analogues (Molnár and Horváth 1977).

The absence of a substantial shift in the k' value of PSLI between acid and neutral pH indicates that the peptide does not contain a His residue. At a neutral pH the imidazole ring of histidine is largely un-ionized, which leads to increased retention, relative to acid pH, in the reversed-phase column (Hernandez et al. 1984). Under these conditions Arg and Lys residues remain completely ionized and their role in affecting selectivity is minimal.

The introduction of hydrophobic groups on side chains of amino acids increased the retention of physalaemin and PSLI on the HPLC column matrix to a similar extent even though the peptides are structurally different. It appears that the strong hydrophobicity imparted to the peptides by these reagents cancels out inherent differences that exist between them. The covalent coupling of the nonpolar isopropyl moiety on the ε-amino group of lysine increases its hydrophobic character; that is, the introduction of the hydrocarbon side chain increases the area on the peptide surface that can interact with the hydrophobic side arms of the stationary support (Molnár and Horváth 1977; Hernandez et al. 1984). The proteolytic susceptibility of PSLI also suggested the presence of a positively charged amino acid residue (Lazarus et al. 1982). Thus, we conclude from the known specificity of the reductive alkylation for lysine (Means and Feeney 1968) and the specificity of the iodination reaction which was confirmed by FAB mass spectrometry on physalaemin that PSLI probably contains Lys and Tyr residues.

The number of these amino acid residues and their position in the sequence of the physalaemin-like peptide in small cell carcinoma of the lung cannot be precisely determined at this time. However, it appears to contain certain structural homologies with the amphibian peptide: (a) cross-reactivity with antibodies specific for the physalaemin amino acid sequence -Asp-Pro-Asn- (Lazarus and DiAugustine 1980; Lazarus et al. 1980, 1982); (b) the presence of Lys and Tyr residues according to selective chemical reactions with an isopropyl group and iodine, respectively; (c) oxidation of a labile sulfur-containing residue, presumably methionine; and (d) the contraction of isolated guinea pig ilial smooth muscle in a dose-dependent manner and at concentrations equivalent to the physalaemin standard (Lazarus et al. 1982). These bioactivity data indicate that PSLI probably contains the generalized tachykinin sequence -Phe-X-Gly-Leu-Met, since the phenylalanyl residue and the C-terminal tripeptide are required for bioactivity (Erspamer and Melchiorri 1973; Bertaccini 1976, 1980). Although the C-terminal amide is not essential for biological activity (Bertaccini 1976), the presence of -Met-NH$_2$ as the C-terminal residue awaits further study. Although physalaemin has a Lys residue preceding the highly conserved phenylalanine, which is followed by tyrosine (Table 2), it is entirely possible, but not proven, that this sequence similarly exists in PSLI. A close examination of the sequences of the known tachykinins given in Table 2 shows that the fourth position from the C-terminus is always a strongly hydrophobic amino acid with relative lipophilicities ranging from 1.46 to 2.24 (Molnár and Horváth 1977).

In summary, our data detailed in this communication and the recent discoveries of preprotachykinins containing the sequences of both substance P and the newly identified substance K (Nawa et al. 1983, 1984), in addition to neurokinin β (neuromedin K) (Kanga-

Table 2. Tachykinin family of peptides

Peptide	Sequence[a]			Source
Physalaemin	pGlu-Ala-Asp-Pro-Asn-Lys	Phe-Tyr-Gly-Leu-Met-NH₂		Amphibian
Uperolein	Pro Ala			Amphibian
Phyllomedusin	pGlu-Asn Arg	Ile		Amphibian
Kassinin	Asp-Val-Pro-Lys-Ser-Asp-Glu	Val		Amphibian
[Lys⁵,Thr⁶]-Physalaemin	Lys-Thr			Amphibian
[Glu²,Pro⁵]-Kassinin	Asp-Glu-Pro-Lys-Pro-Asp-Glu	Val		Amphibian
Hylambatin	Asp-Pro-Pro Asp-Arg		Met	Amphibian
Eledoisin	Pro-Ser-Lys-Asp-Ala	Ile		Molluscan
Substance P	Arg-Pro-Lys Gln-Gln	Phe		Mammalian
Neuromedin K (neurokinin β)	Asp-Met-His-Asp-Phe	Val		Mammalian
Neuromedin L (neurokinin α)	His-Lys-Thr-Asp-Ser	Val		Mammalian
Substance K	Arg-His-Lys-Thr-Asp-Ser	Val		Mammalian

[a] The *boxed area* indicates sequences common to and required for tachykinin bioactivity; the *straight line* denotes amino acids identical to physalaemin, which is the prototype peptide of the tachykinins

wa et al. 1983) and neurokinin α (neuromedin L) (Kimura et al. 1983), illustrate the proliferation of bioactive peptides that belong to the tachykinin family in mammalian tissue (Table 2). This confirms the view of Erspamer (1983) "that mammalian tissues contain not only generic counterparts of amphibian peptides, but also peptides identical, or nearly so, to the skin [i. e., amphibian] peptides." However, the poignant admonition of Mutt (1983), that only the actual purification and structural analysis of an unknown peptide provides sufficiently convincing proof of peptide homology, must not be understated, particularly in this study on an immunoreactive peptide from a neoplasm and in any area of peptide biochemistry.

Acknowledgements. We greatly appreciated the technical assistance of Mark E. Allen, Melanie A. Harper, Don J. Harvan, Beverly J. Irons, and Michael P. Walker, the discussions with William E. Wilson, and the typing skills of Peggy Ellis.

References

Baylin SB, Mendelsohn G (1980) Ectopic (inappropriate) hormone production by tumors: mechanisms involved and the biological and clinical implications. Endocrin Rev 1: 45–77

Bennett HPJ, Browne CA, Solomon S (1981) The use of perfluorinated carboxylic acids in the reversed-phase HPLC of peptides. J Liquid Chromatogr 3: 1353–1365

Bertaccini G (1976) Active polypeptides of nonmammalian origin. Pharmacol Rev 28: 127–177

Bertaccini G (1980) Peptides of the amphibian skin active on the gut. I. Tachykinins (substance P-like peptides) and ceruleins. Isolation, structure, and basic functions. In: Glass GBJ (ed) Gastrointestinal hormones. Raven, New York, pp 317–340

Cutz E, Chan W, Track NM (1981) Bombesin, calcitonin and leu-enkephalin immunoreactivity in endocrine cells of human lung. Experientia 37: 765–767

Erisman MD, Linnoila RI, Hernandez O, DiAugustine RP, Lazarus LH (1982) Human lung small-cell carcinoma contains bombesin. Proc Natl Acad Sci USA 79: 2379–2383

Erspamer V (1983) Amphibian skin peptides in mammals - looking ahead. Trends Neurosci 6: 200–201

Erspamer V, Melchiorri P (1973) Active polypeptides of the amphibian skin and their synthetic analogues. Pure Appl Chem 35: 463–494

Erspamer V, Anastasi A, Bertaccini G, Cei JM (1964) Structure and pharmacological actions of physalaemin, the main active polypeptide of the skin of *Physalaemus fuscumaculatus*. Experientia 20: 489–490

Hernandez H, Dermott K, Lazarus LH (1984) High-performance liquid chromatography of amphibian peptides. Selectivity changes induced by pH. J Liquid Chromatogr 7: 893–905

Iwanaga T (1983) Gastrin-releasing peptide (GRP)/bombesin-like immunoreactivity in the neurons and paraneurons of the gut and lung. Biomed Res 4: 93–104

Kangawa K, Minamoto N, Fukuda A, Matsuo H (1983) Neuromedin K: a novel mammalian tachykinin identified in porcine spinal cord. Biochem Biophys Res Commun 114: 533–540

Kimura S, Okada M, Sugita Y, Kanazawa I, Munekata E (1983) Novel neuropeptides, neurokinin α and β, isolated from porcine spinal cord. Proc Jpn Acad [B] 59: 101–104

Klickstein LB, Wintroub BU (1982) Separation of angiotensins and assay of angiotensin-generating enzymes by high-performance liquid chromatography. Anal Biochem 120: 146–150

Lazarus LH, DiAugustine RP (1980) Radioimmunoassay for the tachykinin peptide physalaemin: Detection of a physalaemin-like substance in rabbit stomach. Anal Biochem 197: 350–357

Lazarus LH, Perrin MH, Brown MR (1977) Mast cell binding of neurotensin: I. Iodination of neurotensin and characterization of the interaction of neurotensin with mast cell receptor sites. J Biol Chem 252: 7174–7179

Lazarus LH, Linnoila RI, Hernandez O, DiAugustine RP (1980) A neuropeptide in mammalian tissues with physalaemin-like immunoreactivity. Nature 287: 555–558

Lazarus LH, DiAugustine RP, Soldato CM (1982) A substance with immunoreactivity to the peptide physalaemin in mammalian respiratory tissue. Exp Lung Res 3: 329–341

Lazarus LH, DiAugustine RP, Jahnke GD, Hernandez O (1983) Physalaemin: An amphibian tachykinin in human lung small-cell carcinoma. Science 219: 79–81

Means GE, Feeney RE (1968) Reductive alkylations of amino groups in proteins. Biochem 7: 2192–2201

Meek JL (1980) Prediction of peptide retention times in high-pressure liquid chromatography on the basis of amino acid composition. Proc Natl Acad Sci USA 77: 1632–1636

Molnár I, Harváth C (1977) Separation of amino acids and peptides on nonpolar stationary phases by high-performance liquid chromatography. J Chromatogr 142: 623–640

Moody TW, Pert CB, Gazdar AF, Carney DN, Minna JD (1981) High levels of intracellular bombesin characterize human small-cell lung carcinoma. Science 214: 1246–1248

Moody TW, Bertness V, Carney DN (1983 a) Bombesin-like peptides and receptors in human tumor cells lines. Peptides 4: 683–686

Moody TW, Russell EK, O'Donohue TL, Linden CD, Gazdar AF (1983 b) Bombesin-like peptides in small cell lung cancer: Biochemical characterization and secretion from a cell line. Life Sci 32: 487–493

Mutt V (1983) New approaches to the identification and isolation of hormonal polypeptides. Trends Neurosci 6: 357–360

Nawa H, Hirose T, Takashima H, Inayama S, Nakanishi S (1983) Nucleotide sequences of cloned cDNAs for two types of bovine brain substance P precursor. Nature 306: 32–36

Nawa H, Doteuchi M, Igano K, Inouye K, Nakanishi S (1984) Substance K: A novel mammalian tachykinin that differs from substance P in its pharmacological profile. Life Sci 34: 1153–1160

Price J, Penman E, Bourne GL, Rees LH (1983) Characterization of bombesin-like immunoreactivity in human fetal lung. Regul Pept 7: 315–322

Rees LH (1976) Concepts in ectopic hormone production. Clin Endocrinol [Suppl] 5: 363s–372s

Rees LH, Bloomfield GA, Rees GM, Corrin B, Franks LM, Ratcliffe JG (1974) Multiple hormones in a bronchial tumor. J Clin Endocrinol Metab 3: 1090–1097

Sorenson GD, Bloom SR, Ghatei MA, Del Prete SA, Cate CA, Pettengill OS (1982) Bombesin production by human small cell carcinoma of the lung. Regul Pept 4: 59–66

Track NS, Cutz E (1982) Bombesin-like immunoreactivity in developing human lung. Life Sci 30: 1553–1556

Tsutsumi Y, Osamura RY, Watanabe K, Yanaihara N (1983) Simultaneous immunohistochemical localization of gastrin-releasing peptide (GRP) and calcitonin (CT) in human bronchial endocrine-type cells. Virchows Arch [A] 400: 163–171

Wharton J, Polak JM, Bloom SR, Ghatei MA, Solcia E, Brown MR (1978) Bombesin-like immunoreactivity in the lung. Nature 273: 769–771

Wilson WE, Harvan DJ, Hamm C, Lazarus LH, Klapper DG, Yajima H (1985) Physalaemin-like peptides: purification and elucidation of structure using atom bombardment-tandem mass spectrometry. In: Proceedings of the Ninth American Peptide Symposium. Pierce Chemical, Rockford

Wood SM, Wood JR, Ghatei MA, Lee YC, O'Shaughnessy D, Bloom SR (1981) Bombesin, somatostatin and neurotensin-like immunoreactivity in bronchial carcinoma. J Clin Endocrinol Metab 53: 1310–1312

Yasuhara T, Nakajima T, Nokihara K, Yanaihara C, Yanaihara N, Erspamer V, Erspamer GF (1983) Two new frog skin peptides, phyllolitorins, of the bombesin-ranatensin family from *Phyllomedusa sauvagei*. Biomed Res 4: 407–412

Calcitonin in Human Malignancies

G. Milhaud

Service de Médecine Nucléaire, Hôpital Saint-Antoine, Paris, France

Introduction

The production of calcitonin by small cell carcinoma is an important aspect of the ectopic production of peptides by tumoral tissue. Interestingly, small cell carcinoma of the lung can also produce other hormones, such as ACTH or ADH, and enzymes, such as histaminase; and this list is substantially extended elsewhere in this volume. I would like therefore to discuss some aspects of hormone production by this tumor; I will principally refer to calcitonin, as I have been engaged in the study of the pathologic secretion of this hor-

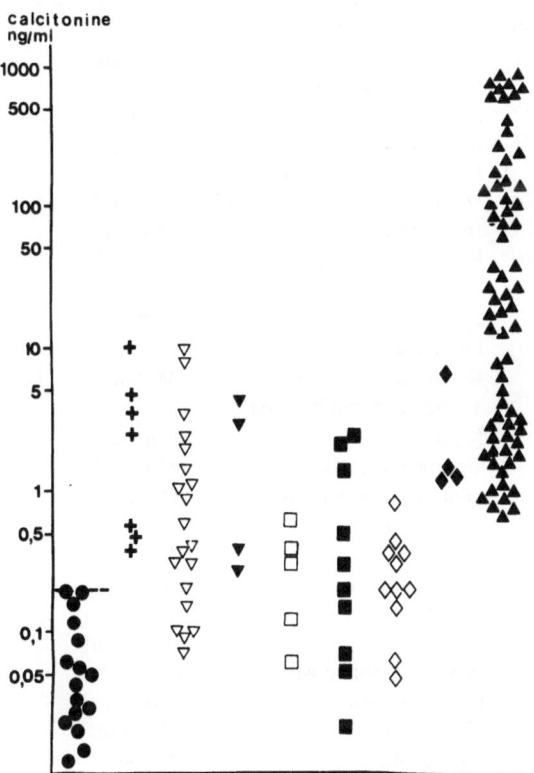

Fig. 1. Circulating calcitonin levels in tumorous conditions: ▲, medullary cancer; ◆, trabecular cancer; ◇, other thyroid cancers; ■, carcinoid; □, phe-ochromocytoma; ▼, melanoma; ▽, bronchial cancer; +, hepatoma; ●, normal

mone since the discovery in my laboratory in 1967 of the secretion of calcitonin by a then obscure thyroid tumor: medullary carcinoma of the thyroid (Milhaud et al. 1968).

Our work and that of Meyer and Abdel Bari (1968), which appeared some weeks later, were the first demonstrations of the pathologic secretion of calcitonin. Since then this hormone has become accepted as the specific marker for medullary carcinoma of the thyroid and for another distinct anatomopathologic entity, poorly differentiated follicular carcinoma of the thyroid.

Calcitonin ist also secreted by a large number of tumors arising in different tissues. We initially presented the first case of ectopic production of calcitonin by a bronchial carcinoid Milhaud et al. (1970), and this was followed by the documentation of the same phenomenon in intestinal carcinoid Milhaud et al. (1972). Work from our laboratory and from many others has demonstrated the diversity of the tissues in which ectopic production of this hormone can occur. Figure 1 is an illustration of the different tumors we have found to be associated with high levels of circulating calcitonin. Table 1 gives a summary of the research performed on the ectopic production of the hormone in the last few years. Certainly the ectopic production of calcitonin is a subject of continuing interest.

At this juncture, several questions should be addressed. One important problem is that of definition. Since calcitonin has been found in normal lung extracts and has been localized within the pulmonary endocrine cells of the human lung, should we still speak of "ectopic" production of the hormone?

Table 1. Disorders with ectopic production of calcitonin

Date	Subject of research	Group
1970	Bronchial carcinoid	Milhaud
1971	Intestinal carcinoid	Milhaud
	Phaeochromocytoma	Milhaud
1972	Struma ovarii	Robboy
1973	Mucosal neuroma	Voelkel
	Bronchial carcinoma	Silva
1974	Melanoma	Milhaud
	Insulinoma	Milhaud
1976	Mammary cancer	Hillyard
1977	Ganglioneuroma	Milhaud
	Sympathoblastoma	Milhaud
	Laryngeal apudoma	Milhaud
	Renal clear cell cancer	Milhaud
	Testicular dysembryoma	Milhaud
	Myeloma	Milhaud
	Oesophageal cancer	Milhaud
	Pleural epithelial cancer	Dubrisay
	Thymic carcinoid	Abe
	Pancreatic cancer	Abe
	Gastric cancer	Abe
	Hepatic carcinoid	Himsworth
1978	Vipoma	Rambaud
	Pancreatic somatostatinoma	Galmiche
1979	Paraganglioma	White
1982	Hepatoma	Maguire
	Lymphoma	Pfluger

Furthermore, are small cell carcinomas of the lung of two types, one arising from normal calcitonin-producing cells and the other, from cells which subsequently produce the hormone ectopically?

Specific receptors for calcitonin exist in normal lung and in cell lines derived from pulmonary tumors. What are the functions of these receptors and to what extent are they affected by the local production of high levels of the hormone by tumors?

Conversely, it is tempting to suggest an induction of the tumors by a receptor defect initiating a hypersecretion which, as is well known in the field of endocrinology, could lead to hypertrophy and malignancy. Furthermore, we know now from the work on insulin that polypeptide hormones can be internalized, and thus the presence of very low amounts of calcitonin could be due to the presence of the hormone bound to the plasma membrane. In some tissues, on occasion, could the localization of the hormone in cells de-

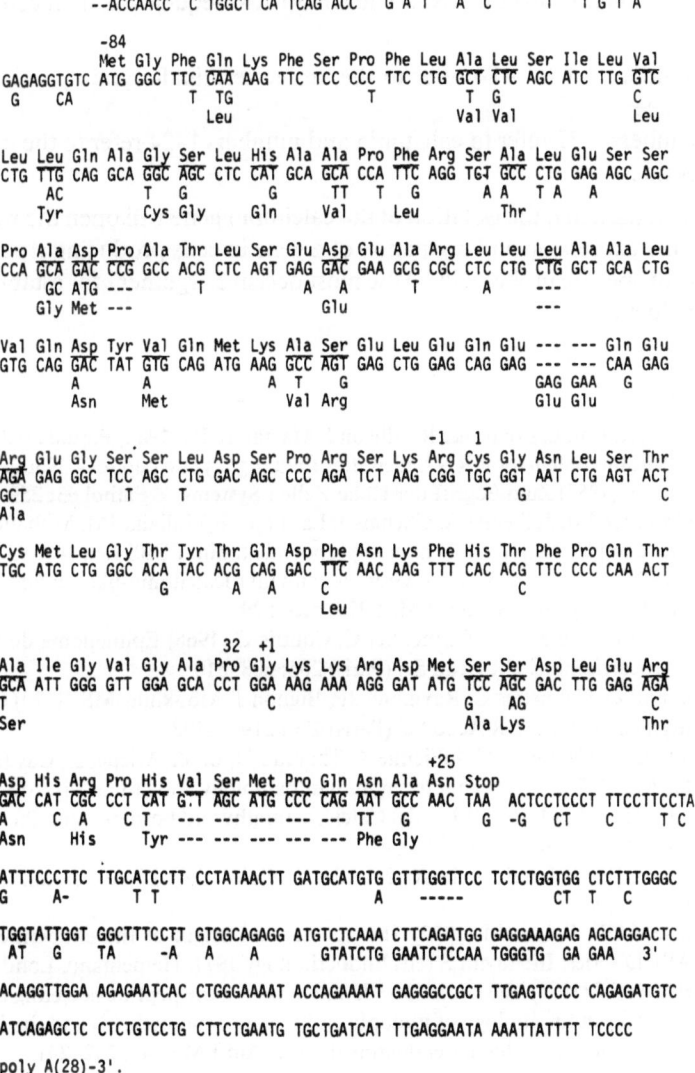

Fig. 2. Complete sequence of human pre-pro-calcitonin

termined by immunocytochemical methods be due to detection of internalized molecules or of their fragments?

Another question deals with the various theories which have been proposed to explain the secretion of calcitonin by a variety of cells. These concepts rely on common morphological criteria, such as the clear cell theory of Pages (1955) and Feyrter (1952), on common histochemical properties (the apudomas) Pearse and Polak (1972), or on a common embryonic origin, the neural crest (Weichert 1970). The problem inherent in all these concepts remains the large numbers of exceptions encountered. Furthermore, they do not shed any light on the molecular events leading to the derepression of a gene. They do, however indicate (with reference in this particular case to the neural crest theory) a certain propensity of some tissues to derepression of a particular gene or familly of genes, leading to the expression of peptide hormones.

Finally, the elucidation of the partial structure of the human calcitonin mRNA in Craiggs laboratory (Craig et al. 1982) and of the entire structure in our own (Le Moullec et al. 1984) represents a major advance. Figure 2 illustrates the complete sequence of human calcitonin precursor mRNA. The amino acid sequence is deduced from the nucleotide sequence.

Numbers -1 to -84 refer to the amino-terminal-cryptic peptide and to the leader sequence;
Numbers 1-32 refer to calcitonin and numbers 1-24 refer to the carboxy terminal-cryptic peptide.

It is certain that the isolation of the calcitonin gene will open the way to pertinent research on the expression and regulation of a gene in cells producing a peptide. Furthermore, the use of specific probes will be the most decisive argument for establishing so-called ectopic production.

References

Craig RK, Hall L, Edbrooke R, Allison J, Macintyre IK (1982) Partial nucleotide sequence of human calcitonin precursor mRNA identifies flanking cryptic peptides. Nature 295: 345-347
Feyrter F (1952) Zum Begriff der Helle-Zellen-Systeme. Z Pathol 63: 259
Le Moullec JM, Jullienne A, Chenais J, Lasmoles F, Guliana JM, Milhaud G, Moukhtar MS (1984) The complete sequence of human preporcalcitonin. FEBS Lett 167: 93-97
Meyer JS, Abdel Bari W (1968) Granules and thyrocalcitonin-like activity in medullary carcinoma of the thyroid gland. N Engl J Med 278: 523-529
Milhaud G, Tubiana M, Parmentier C, Coutris G (1968) Epithélioma de la thyroïde sécrétant de la thyrocalcitonine. CR Acad Sci (Paris) 266: 608-610
Milhaud G, Calmettes C, Raymond JP, Bignon J, Moukhtar MS (1970) Carcinoïde sécrétant de la thyrocalcitonine. CR Acad Sci (Paris) 270: 2195-2198
Milhaud G, Calmettes C, Jullienne A, Tharaud D, Bloch-Michel H, Cavaillon JP, Colin R, Moukhtar MS (1972) A new chapter in human pathology: calcitonin disorders and therapeutic use. In: Talmage RV, Munson PL (eds) Calcium, farathyroid hormone and the calcitonins, Excerpta Medica, Amsterdam, pp 56-70
Pages A (1955) Essai sur le système des "cellules claires" de Feyrter. Thèse de doctorat Dehan, Montpellier
Pearse AGE, Polak JM (1972) The neural crest origin of the endocrine polypeptide cells of the APUD series. In: Taylor S (ed) Endocrinology 1971. Heinemann, London, pp 145-152
Weichert RF (1970) The neural ectodermal origin of the peptide-secreting endocrine glands. A unifying concept for etiology of multiple endocrine adenomathosis and the inappropriate secretion of peptide hormones by nonendocrine tumors. Am J Med 49: 232-241

Differential Expression of the Human Calcitonin – CGRP Gene in Medullary Thyroid Carcinoma and Lung Carcinoma Cell Lines

R. K. Craig, M. R. Edbrooke, J. H. Riley, J. H. McVey, and D. Parker

Cancer Research Campaign Endocrine Tumour Molecular Biology Group,
The Courtauld Institute of Biochemistry, The Middlesex Hospital Medical School,
Mortimer Street, London W1P 7PN, Great Britain

Introduction

The application of molecular techniques has provided new insight into the structure and expression of peptide hormone genes (Craig and Hall 1983). Application of this technology to the study of calcitonin gene expression has proved particularly intriguing. Studies on the structure and expression of the rat calcitonin gene have demonstrated the generation by RNA processing of alternative mRNA species in an apparently tissue-specific manner (Amara et al. 1982). Each mRNA encodes a polyprotein subsequently cleaved and modified by enzymes within the secretory pathway to yield calcitonin or the calcitonin gene-related peptide (CGRP). The application of immunocytochemical techniques using antiserum raised against a synthetic fragment of rat CGRP demonstrates a wide distribution of CGRP-producing cells and points to a possible role as a neurotransmitter or neuromodulator molecule (Rosenfeld et al. 1983). In man, calcitonin is synthesized normally at low levels by C cells of the thyroid (Austin and Heath 1981), though at elevated levels ectopically by lung carcinoma (Coombes et al. 1974) and at grossly elevated levels eutopically in medullary thyroid carcinoma (Milhaud et al. 1974). In common with rat, differential expression of the human calcitonin gene also results in the expression of two different mRNA species encoding polyproteins, proteolytically cleaved within the secretory pathway to produce biologically active peptides, calcitonin, or CGRP (Craig et al. 1982; Edbrooke et al. 1985).

Here, using cDNA hybridization probes specific to calcitonin and CGRP mRNA, we describe the differential expression of the human calcitonin gene in three human lung carcinoma cell lines, DMS 53, DMS 153 and BEN, all previously shown on the basis of immunological evidence to synthesize varying amounts of calcitonin (Pettengill et al. 1980; Sorenson et al. 1981; Ham et al. 1980). We also compare the size and distribution of calcitonin- and CGRP poly(A)-containing RNA transcripts with those present in medullary thyroid carcinoma tissue, and demonstrate the presence of CGRP circulating in the plasma of medullary thyroid carcinoma patients, in extracts of medullary carcinoma, and secreted into medium by one lung carcinoma cell line (DMS 153), as judged by a human CGRP radioimmunoassay.

Recent Results in Cancer Research. Vol 99
© Springer-Verlag Berlin · Heidelberg 1985

Methodology

All constructions of cDNA recombinants, and nucleotide sequence analysis of cloned cDNA sequences, have been described in detail elsewhere (Allison et al. 1981; Craig et al. 1982; Edbrooke et al. 1985). Total RNA was isolated from snap-frozen medullary thyroid carcinoma tissue as described by Hall et al. (1979), whilst total RNA was isolated in a similar manner from semi-confluent cell monolayers after direct lysis of the cells at 37 °C into 20 mM Tris-HCl ph 7.6 containing 1 mM EDTA, 2% (w/v) SDS and 250 g/ml proteinase K. Poly (A)-containing RNA was then isolated by oligo (dT)-cellulose chromatography as described by Craig et al. (1976).

The differential distribution of calcitonin and CGRP RNA transcripts was determined using 1.1% (w/v) agarose gel electrophoresis and RNA blotting onto Biodyne membranes as described elsewhere (Taylor et al. 1984; Edbrooke et al. 1985). Membranes were probed with recloned or gel-purified calcitonin or CGRP cDNA-specific restriction fragments, radiolabelled using a nick-translation kit and (α-^{32}P)dCTP (400 Ci/mmol) from Amersham International, UK. to a specific activity of 1–3.2×10^8cpm/µg (see Rigby et al. 1977).

Human CGRP was custom synthesized by Penisula Laboratories. USA whilst Tyr-CGRP$_{25-37}$-amide and Tyr-CGRP$_{1-8}$ were synthesized by Dr. Richard Titmus, Celltech Ltd., UK. Rat CGRP and Katacalcin were obtained from Peninsula Laboratories, and human calcitonin and salmon calcitonin were obtained from Ciba-Geigy, Basle, Switzerland and Sigma, respectively. Where necessary, human CGRP was purified by ion exchange and reverse-phase chromatography by Dr. Mike Eaton of Celltech Ltd., and the structure and purity were confirmed by mass spectrometry (M – Scan Ltd.). Radioimmunoassay with antisera raised in rabbits against hCGRP or Tyr-CGRP$_{25-37}$-amide conjugated to ovalbumin (Reichlin 1980) was performed essentially as described by Girgis et al. (1980) except that bound ^{125}I-labelled tracer was quantitatively recovered by precipitation using 10 µl goat anti-rabbit after the addition of 1 µl pre-immune rabbit serum carrier to each 400 µl assay (see Craig et al. 1976). The amount of tracer recovered was determined by gamma counting.

Results

Differential Expression of Calcitonin and CGRP RNA Transcripts

We have recently described (Edbrooke et al. 1985) the nucleotide sequence of a 1615-bp cDNA fragment inserted into a recombinant plasmid phT-B 58 isolated from a cDNA library constructed from poly(A)-containing RNA isolated from human medullary thyroid carcinoma tissue (Allison et al. 1981). The sequence inserted into phT-B 58 represents a partially processed polyadenylated calcitonin gene transcript, and, as can be seen from Fig. 1, contains sequence in common with cDNA present in two recombinants (phT-B 3, phT-B 6) which we have shown previously to contain calcitonin mRNA sequence (Craig et al. 1982). The additional cDNA sequence cloned into phT-B 58 represents a transcribed intron, the CGRP polyprotein coding sequence and the 3' untranslated region of the CGRP mRNA. The open reading frame within the human CGRP exon (see Fig. 2) encodes 53 amino acids followed by a termination codon. Examination of this sequence shows that human CGRP is a peptide of 37 amino acids (1 to 37) flanked at the amino terminal end by paired basic amino acids (-1, -2), and at the carboxyl terminal end by a glycine residue ($+1$), four basic amino acids ($+2$ to $+5$) and a tetrapeptide ($+6$ to $+9$).

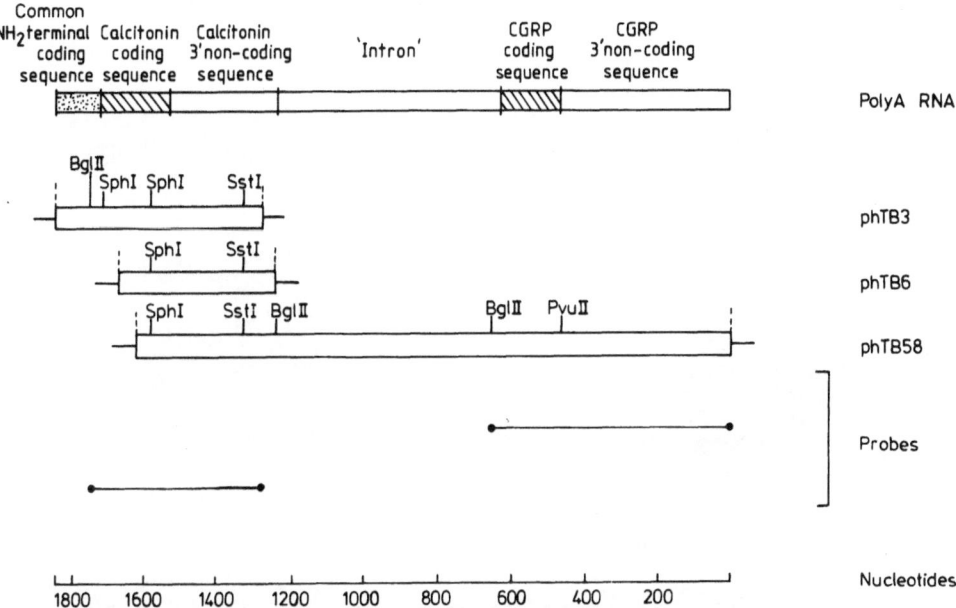

Fig. 1. Schematic representation of overlapping cDNA sequences cloned within the recombinant plasmids phTB 3, phTB 6 and phTB 58 (see Craig et al. 1982; Edbrooke et al. 1984). The cDNA was prepared and cloned from total cellular poly(A)-containing RNA isolated from human medullary thyroid carcinoma (MCT) tissue as described previously (Allison et al. 1981). Indicated are restriction sites separating regions of sequence used as sequence-specific hybridization probes, and the relative positions of calcitonin, CGRP, and the common amino terminal peptide within known sequence. *Vertical dashed lines* denote Pst I sites separating cDNA and plasmid sequence

```
                                                 -6  -5  -4  -3 |-2  -1 | 1   2   3   4   5   6   7   8
                                                 Ile Ile Ala Gln|Lys Arg| Ala Cys Asp Thr Ala Thr Cys Val
5'---CAG ATC TTC TCT TCT TTC TCC ATC CTG CAA ATC AGA ATC ATT GCC CAG|AAG AGA| GCC TGT GAC ACT GCC ACC TGT GTG
                                                 T G      C                       T   C A               C
                                                 Val Thr                          Ser     Asn


     9  10  11  12  13  14  15  16  17  18  19  20  21  22  23  24  25  26  27  28  29  30  31  32  33  34
     Thr His Arg Leu Ala Gly Leu Leu Ser Arg Ser Gly Gly Val Val Lys Asn Asn Phe Val Pro Thr Asn Val Gly Ser
     ACT CAT CGG CTG GCA GGC TTG CTG AGC AGA TCA GGG GGT GTG GTG AAG AAC AAC TTT GTG CCC ACC AAT GTG GGT TCC
     C                               G   G   A                       G                               C   T
                                                                     Asp


     35  36  37 |+1 |+2  +3  +4  +5 |+6  +7  +8  +9
     Lys Ala Phe|Gly|Arg Arg Arg Arg|Asp Leu Gln Ala STOP
     AAA GCC TTT|GGC|AGG CGC CGC AGG|GAC CTT CAA GCC TGA GCA GCT GAA CGA CTC AAG AAG GTC ACA ATA AAG CTG AAC
     G           |  C|C C                       G   T
     Glu
```

```
         STOP
     TCC TTT TAA TGT---3'
```

Fig. 2. The amino acid sequence of human calcitonin gene-related peptide (1–37) and flanking peptides as predicted by nucleotide sequence analysis of cDNA inserted into recombinant plasmid phTB 58

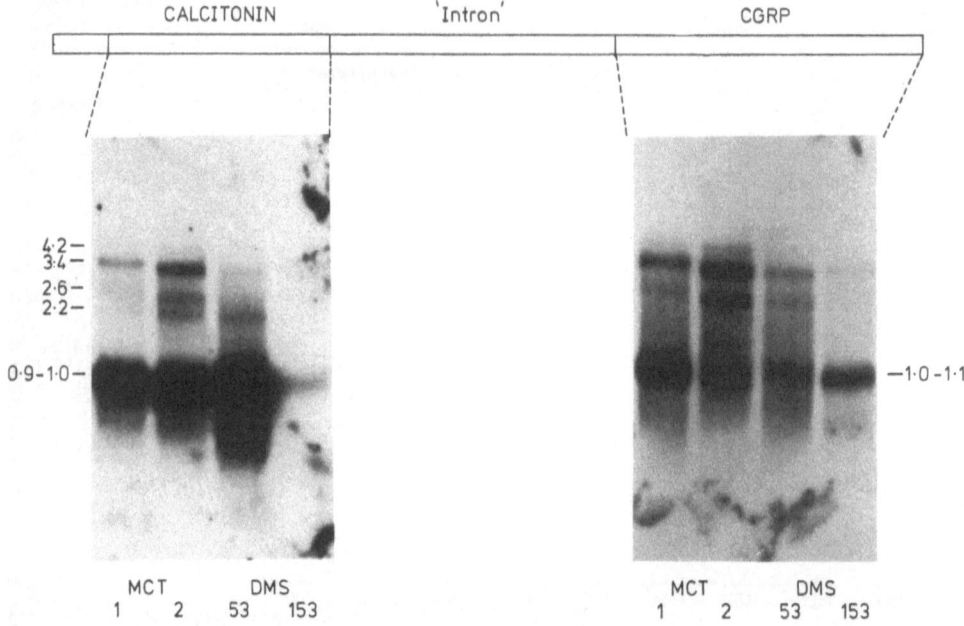

Fig. 3. Relative distribution of human calcitonin- and CGRP-containing RNA sequences in total poly(A)-containing RNA isolated from two medullary thyroid carcinomas (MCT$_1$ and MCT$_2$) and two small cell carinoma cell lines (DMS 53 and DMS 153), as determined by RNA blotting using calcitonin- and CGRP-specific cDNA hybridization probes. The relative size of the RNA species is indicated (kb), as determined by the comparative electrophoretic mobility of DNA fragments of known size

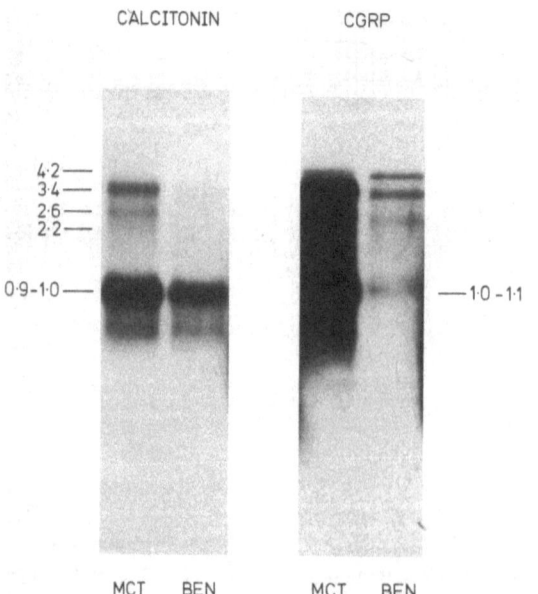

Fig. 4. Relative distribution of human calcitonin- and CGRP-containing RNA sequences in total poly(A)-containing RNA isolated from the BEN cell line as determined by RNA blotting (see Fig. 3)

At the amino terminal end a further five amino acids (-3 to -7) are encoded, the first of which (arginine -7), has been assigned on the basis of the position of the splice junction, and by analogy with the rat calcitonin gene (Amara et al. 1982; Rosenfeld et al. 1983). The presence of the glycine residue ($+1$) reflects the requirement in vivo for an amidated carboxyl terminal phenylalanine (Bradbury et al. 1982), whilst the basic amino acids identify sites of proteolytic cleavage. Amidated human CGRP has a calculated Mr 3786, and differs from rat CGRP by four amino acids (alanine 1; aspartic acid 3; asparagine 25; lysine 25).

Using sequence-specific cDNA hybridization probes (see Fig. 1), we have investigated by RNA blotting the relative distribution of calcitonin and CGRP poly(A)-containing RNA in three cell lines and two human medullary thyroid carcinomas. Analysis of the relative distribution of calcitonin poly(A)-containing RNA transcripts by agarose gel electrophoresis and RNA blotting shows (Fig. 3 and 4) that both medullary thyroid carcinomas (MCT_1 and MCT_2) expressed an abundance of calcitonin mRNA (0.9–1.0 kb), whilst high levels of calcitonin mRNA were also present in poly(A)-containing RNA isolated from the DMS 53 (Fig. 3) and BEN cell lines (Fig. 4), but traces only in the DMS 153 cell line. Four less abundant larger RNA species containing calcitonin transcripts (4.2 kb, 3.4 kb, 2.6 kb and 2.2 kb) were also present in the medullary thyroid carcinoma and DMS 53 RNA preparations, whilst similar species were present at lesser abundance in RNA isolated from the BEN cell line. Renalysis of the same filters using a CGRP-specific cDNA probe revealed a different distribution of RNA transcripts. In MCT_1 a predominant band of 1.0–1.1 kb, the CGRP mRNA, was observed, and three of the four larger RNA species. The 2.2 kb transcript did not hybridize CGRP sequences. In MCT_2 the distribution of CGRP sequences was similar except that the processed CGRP mRNA was present at low abundance in a diffuse band (0.9–1.3 kb). In the cell lines, CGRP transcripts were present at relatively high levels in the DMS 153 cells (a single band of 1.0–1.1 kb in size), but at low levels only in the BEN and DMS 53 cells, again as a diffuse 0.9- to 1.3-kb band indicative of considerable heterogeneity. The larger CGRP RNA transcripts were also apparent in the BEN and DMS 53 cells, and at low levels only, in the DMS 153 cells. The results described above demonstrate the differential expression of the human calcitonin/CGRP gene in medullary thyroid carcinoma and lung carcinoma in a similar manner to that described previously for the rat calcitonin gene.

Identification of CGRP in Plasma and Tissue Extracts

We have also determined by radioimmunoassay the presence of CGRP in plasma from patients with medullary thyroid carcinoma, in an extract from medullary thyroid carcinoma tissue, and in lung cell line secretions. Antiserum has been raised in rabbits against synthetic amidated human CGRP and against a tyrosinated amidated C-terminal CGRP fragment (Tyr-$CGRP_{25-37}$-amide) conjugated to chick ovalbumin. A summary of these results is shown in Fig. 5 using antiserum raised against the amidated tyrosinated C-terminal CGRP fragment. The assay is sensitive in the 1-ng range as judged by the displacement of ^{125}I-Tyr-$CGRP_{25-37}$-amide tracer using highly purified synthetic CGRP. The assay is specific to CGRP, since the iodinated tracer is not displaced by human clacitonin, salmon calcitonin, katacalcin, Tyr-$CGRD_{1-8}$, or PDA-4 (C-terminal-flanking peptide of the CGRP polyprotein; see Fig. 2). The antibody also shares antigenic determinants with synthetic rat CGRP. Using this assay we have identified high levels of CGRP in an extract of medullary thyroid carcinoma tissue (MCT_1), a tissue known to contain high levels of CGRP

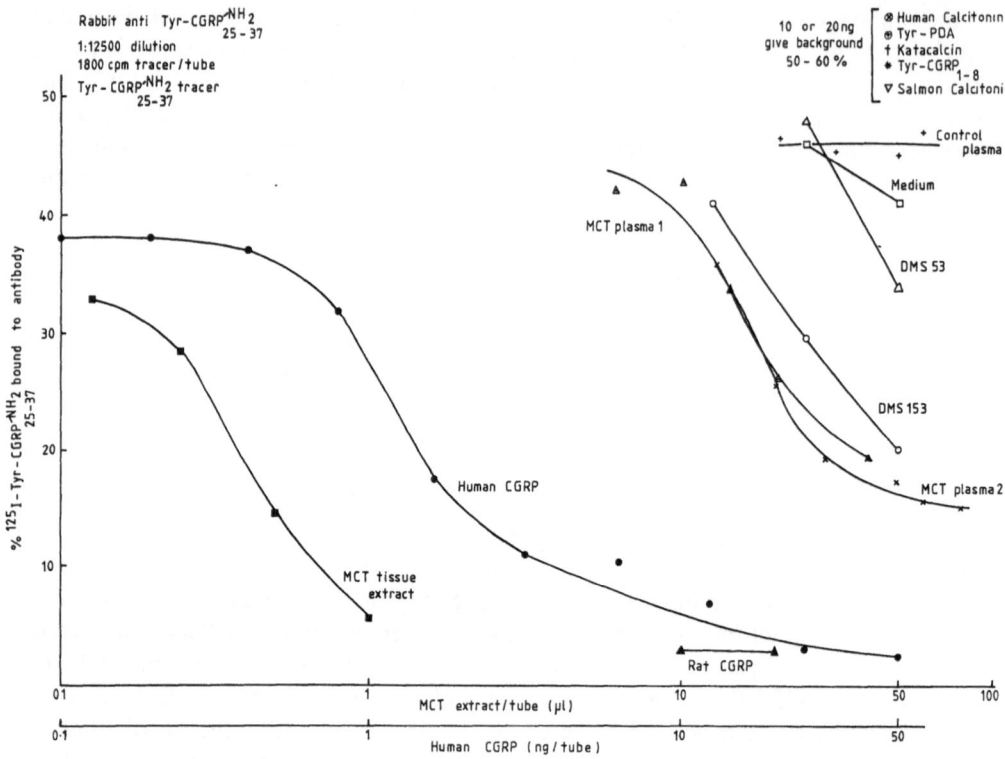

Fig. 5. Identification of human CGRP in plasma and a thyroid tissue extract from medullary thyroid carcinoma patients and in secretions of small cell carcinoma cell lines by radioimmunoassay. MCT tissue extract (■——■); hCGRP standard (●——●); MCT plasma samples (△——△, ×——×); DMS 153 (○——○) and DMS 53 (△——△) cell line secretions; normal human plasma (+——+); cell growth medium (□——□); and rat CGRP (▲——▲)

mRNA (see Fig. 3). Measurable but varying amounts (3.8–35.9 nmol/litre) of CGRP were identified in eight of nine samples taken from the circulating plasma of medullary thyroid carcinoma patients known to have elevated calcitonin levels. Measurable amounts of CGRP-immunoreactive material was also present in DMS 153 cell line secretions (10.0 nmol/litre), but only trace amounts in secretions from the BEN (data not shown) and DMS 53 cell lines. No CGRP (within the sensitivity range of the assay) was detected in plasma from two healthy adults. In all samples analysed, CGRP when detectable gave rise to displacement curves essentially parallel to those obtained using the synthetic human CGRP standard.

Discussion

Our recent studies on the structure and expression of the human calcitonin-CGRP gene in medullary thyroid carcinoma and lung carcinoma (Edbrooke et al. 1985) have defined the amino acid sequence of human CGRP and demonstrated that this peptide, like rat CGRP (Fisher et al. 1983), has a potent action on the cardiovascular system and implicates at least two separate post-transcriptional mechanisms in the processing of a common RNA transcript: (a) selective utilization of splice sites resulting in CGRP mRNA synthesis; and (b) specific cleavage followed by polyadenylation resulting in calcitonin mRNA synthesis. Here we have extended our observations at the molecular level to show that in addition to medullary thyroid carcinoma, and the DMS 53 and 153 small cell carcinoma cell lines, the human calcitonin gene is also expressed in an additional human bronchial carcinoma cell line, the BEN cell line, confirming the observation of otherworkers based on radioimmunoassay alone (Ham et al. 1980). In addition, using a radioimmunoassay based on human CGRP, we have demonstrated that CGRP immunoreactive material is secreted by the DMS 153 cell line; that CGRP can be present at high levels in medullary thyroid carcinoma tissue; and that CGRP was present in elevated, i. e., measurable, levels in the plasma of all but one of the medullary thyroid carcinoma cases examined. Recently we have also localized, by immunocytochemical techniques at the light level, CGRP-producing cells in sections of paraffin-embedded medullary thyroid carcinoma tissue (D. Williams, R. K. Craig unpublished work). Our observations have a number of fundamental implications, and are of potential clinical value.

At the molecular level, the availability of a number of human cell lines producing high levels of either calcitonin or CGRP mRNA from a common RNA transcript forms the basis of complementation assays and sources of "regulatory" factors, which, in combination with appropriate manipulation of the human calcitonin-CGRP gene should now allow definitive analysis of the post-transcriptional switching mechanism involved in the tissue-specific expression of the calcitonin gene. Analysis of the transcriptional mechanisms involved in calcitonin gene expression in lung cell lines may also ultimately provide insight into the molecular basis of abnormal or ectopic hormone gene expression in lung carcinoma. The availability of a radioimmunoassay for human CGRP, and our identification using RIA of VGRP in lung cell secretions and elevated plasma CGRP in patients with medullary thyroid carcinoma, indicate that the measurement of plasma CGRP levels may be of value in the management of medullary thyroid carcinoma (Hill et al. 1973) and lung carcinoma. Recently other workers (Morris et al. 1984) have also reported the amino acid sequence of the human CGRP peptide purified using a rat CGRP antibody from an extract of medullary thyroid carcinoma tissue. The sequence reported is in agreement with our own. In addition the authors also describe elevated circulating plasma levels in a number of cases of medullary thyroid carcinoma using a radioimmunoassay based on antiserum raised against a rat tyrosinated CGRP fragment.

Acknowledgements. We are grateful to: Drs. Richard Titmus and Mike Eaton of Celltech Ltd., for peptide synthesis and purification; Dr. Bruce Ponder, CRC Medullary Thyroid Carcinoma Study Group, for some plasma samples; Dr. Carmel Hillyard for advice on antibody production; Profs. Olive Pettengill and George Sorenson for the DMS 153 and 53 cell-lines; and Dr. Morag Ellison of the Ludwig Institute of Cancer Research, Sutton, UK for the BEN cell line. We are also grateful to Prof. P. N. Campbell for his continued support and encouragement. This work was supported by the Cancer Research Campaign, and in part by the Wellcome Trust.

References

Allison J, Hall L, MacIntyre I, Craig RK (1981) The construction and partial characterization of plasmids containing complementary DNA sequences to human calcitonin precursor polyprotein. Biochem J 199: 725–731

Amara SG, Jonas V, Rosenfeld MG, Ong ES, Evans RM (1982) Alternative RNA processing in calcitonin gene expression generates mRNAs encoding different polypeptide products. Nature 298: 240–244

Austin LA, Heath H (1981) Calcitonin. N Engl J Med 304: 269–278

Bradbury AF, Finnie MDA, Smyth DG (1982) Mechanism of C-terminal amide formation by pituitary enzymes. Nature 298: 686–688

Coombes RC, Greenberg PB, Hillyard C, MacIntyre I (1974) Plasmaimmunoreactive-calcitonin in patients with non-thyroid tumours. Lancet I: 1080–1083

Craig RK, Hall L (1983) Cloning and expression of polypeptide hormones. Genet Eng 4: 57–125

Craig RK, Brown PA, Harrison OS, McIlreavy D, Campbell PN (1976) Guinea-pig milk-protein synthesis. Biochem J 160: 57–74

Craig RK, Hall L, Edbrooke MR, Allison J, MacIntyre I (1982) Partial nucleotide sequence of human calcitonin precursor mRNA identifies flanking cryptic peptides. Nature 295: 345–347

Edbrooke MR, Parker D, McVey JH, Riley JH, Sorenson GD, Pettengill OS, Craig RK (1985) Expression of the human calcitonin/CGRP gene in lung and thyroid carcinoma. EMBO J 4: 715–724

Fisher LA, Kikkawa DO, Rivier JE, Amara SG, Evans RM, Rosenfeld MG, Vale WW, Brown MR (1983) Stimulation of noradrenergic sympathetic outflow by calcitonin gene-related peptide. Nature 305: 534–536

Girgis SI, Galan Galan F, Arnett TR, Rogers RM, Bone Q, Ravazzola M, MacIntyre I (1980) Immunoreactive human calcitonin-like molecule in the nervous systems of protochordates and a cyclostome, myxine. J Endocrinol 87: 375–382

Hall L, Craig RK, Campbell PN (1979) mRNA species directing synthesis of milk proteins in normal and tumour tissue from human mammary gland. Nature 277: 54–56

Ham J, Ellison ML, Lumsden J (1980) Tumour calcitonin. Biochem J 190: 545–550

Hill CS, Ibanez ML, Samaan NA, Ahearn MJ, Clark RL (1973) Medullary (solid) carcinoma of the thyroid gland: An analysis of the MD Anderson Hospital experience with patients with the tumour, its special features and its histogenesis. Medicine 52: 141–171

Milhaud G, Calmette C, Taboulet J, Julienne A, Moukhtar MS (1974) Hypersecretion of calcitonin in neoplastic conditions. Lancet I: 462–463

Morris HR, Panico M, Etienne T, Tippins J, Girgis SI, MacIntyre I (1984) Isolation and characterization of human calcitonin gene-related peptide. Nature 308: 746–747

Pettengill OS, Sorenson GD, Wurster-Hill DH, Curphey TJ, Noll WW, Cate CC, Maurer LH (1980) Isolation and growth characteristics of continuous cell lines from small-cell carcinoma of the lung. Cancer 45: 906–918

Reichlin M (1980) Use of glutaraldehyde as a coupling agent for proteins and peptides. Methods Enzymol 70: 159–165

Rigby PWJ, Dieckmann M, Rhodes C, Berg P (1977) Labelling deoxyribonucleic acid to high sepcific activity in vitro by nick translation with DNA polymerase I. J Mol Biol 113: 237–251

Rosenfeld MG, Mermod J-J, Amara SG, Swanson LW, Sawchenko PE, Rivier J, Vale WW, Evans RM (1983) Production of a novel neuropeptide encoded by the calcitonin gene via tissue-specific RNA processing. Nature 304: 129–135

Sorenson GD, Pettengill OS, Brinck-Johnsen T, Cate CC, Maurer LH (1981) Hormone production by cultures of small-cell carcinoma of the lung. Cancer 47: 1289–1296

Taylor JB, Craig RK, Beale D, Ketterer B (1984) Construction and characterization of plasmid containing complementary DNA to mRNA encoding the N-terminal amino acid sequence of the rat glutathione transferase Ya subunit. Biochem J 219: 223–231

Calcitonin in Lung Cancer*

C. Gropp, W. Luster, and K. Havemann

Klinik Bergisch-Land, Im Saalscheid 5, 5600 Wuppertal-Ronsdorf 21, FRG

Introduction

Calcitonin (CT), a calcium-regulating peptide, is eutopically produced in patients with medullary carcinoma of the thyroid and ectopically by a number of nonthyroid tumors. Ectopic production of human calcitonin (hCT) has been found mainly in carcinoma of the lung (Silva et al. 1974; Gropp et al. 1980), breast cancer (Coombes et al. 1974), pheochromocytoma (Milhaud et al. 1974), and islet cell tumors of the pancreas (Milhaud et al. 1974; Deftos et al. 1975). Apart from C cell carcinoma it is only in lung cancer that immunoreactive hCT has been found to be elevated in high frequency and in these tumors hCT has been used as a tumor marker to monitor therapy (Gropp et al. 1980). Besides the hCT another calcitonin, the salmon calcitonin (sCT), has been detected in such animals as eel, salamander, chicken, pigeon, and rat (Perez Cano et al. 1981, 1982). In these animals human and salmon calcitonin can be detected simultaneously. Recently Fischer et al. (1983) have described sCT also in human thyroid glands and in human brain. In this paper a more detailed analysis of hCT present in sera, tissue extracts, and supernatants of permanent lung cancer cell lines is presented. Furthermore, initial data on the determination of sCT in lung cancer are described.

Methods

Patients

Serum samples from 101 patients with small cell lung cancer (SCLC), from 16 patients with non-small cell lung cancer (NSCLC), from 18 patients with C cell carcinoma of the thyroid, and from age- and sex-matched controls were assayed for hCT and sCT.

Assays for Calcitonin

hCT was determined in the specimens by means of a commercial radioimmunoassay using an antiserum against synthetic hCT (17–32), as described previously (Luster et al. 1982).

---

* Supported by the Deutsche Forschungsgemeinschaft

sCT was determined by a radioimmunoassay using an antiserum against sCT. The double-antibody radioimmunoassay was obtained from Diagnostic System Laboratories (Webster, Texas). The sensitivity of the assays, expressed as the lowest detectable dose measured with acceptabel precision, is 25 pg/ml for hCT and 0.65 ng/ml for sCT. The normal range extends from undetectable to 120 pg/ml for hCT and from undetectable to 0.9 ng/ml for sCT.

Extraction of Calcitonin from Tissue Material

Tumor tissue pieces were homogenized in 3 vols. of distilled water or in distilled water containing the following proteinase inhibitors: 0.03 mg/ml phenylmethylsulfonyl fluoride, 0.03 mg/ml p-mercuric benzoate, and 0.01 M EDTA, and centrifuged for 20 min at 20000 g.

Gel Filtration

Three-milliliter samples of serum tissue extracts and culture medium were subjected to gel filtration on a 1.5 × 90 cm column of AcA 54 (LKB Stockholm, Sweden), equilibrated with 80 mM KH$_2$ PO$_4$, 0.4 M EDTA, and 0.1% NaN$_3$, ph 7.4, or with the same buffer containing proteinase inhibitors. Protein was eluted in the presence of the above buffers with a flow rate of 8 ml/h. Four-milliliter fractions were collected, lyophilized and resuspended in 400 μl distilled water to estimate calcitonin. The column was calibrated with blue dextran (Pharmacia, Uppsala, Sweden), aldolase (Boehringer, Mannheim, Federal Republic of Germany), bovine serum albumin, ovalbumin, chymotrypsinogen A (Pharmacia, Uppsala, Sweden), myoglobin whale, cytochrome c, cyanocobalamin, bromophenol blue (Serva, Heidelberg, Federal Republic of Germany), and ^{125}I-calcitonin (Immunonuclear, Stillwater, Minn).

Affinity Chromatography

Whole serum, lyophilized fractions of immunoreactive calcitonin obtained by gel filtration, or synthetic human calcitonin were applied at 22 °C to 1-ml columns of concanavalin A-Sepharose (Pharmacia, Uppsala, Sweden), which were equilibrated with 50 mM sodium phosphate buffer, ph 7.0, containing 0.2 M NaCl and 0.02% NaN$_3$. The columns were washed with a ten-fold volume of the same buffer. For elution of specifically bound protein 1 ml buffer containing 0.2 M methyl D-mannoside (Goldstein et al. 1965) was applied to the column. Exactly the same procedure was follwed with protein A-Sepharose CL-4 B. Specifically bound protein was eluted with 0.1 M glycine-HCl, pH 3.0 (Hjelm et al. 1972).

Polyacrylamide Gel Electrophoresis

Polyacrylamide gel electrophoresis in the presence of sodium dodecyl sulfate was performed according to Weber and Osborn (1969). Aliquots 50 μl of lyophylized fractions obtained by gel filtration were run on 10% polyacrylamide gels. Electrophoresis was also carried out in parallel, in the absence and in the presence of 10 mM dithiothreitol. Gels were cut into 1-mm slices and each slice was shaken for 16 h at 4 °C in 0.2 ml of radioimmunoassay buffer prior to the calcitonin assay (Goltzman et al. 1978).

Cell Cultures

Cell lines from SCLC and NSCLC were established as described by Luster et al. (this volume). In addition, six permanent C cell carcinoma cell lines established in our laboroatory (Luster et al. 1984) were assayed for hCT and sCT.

Results and Discussion

Serum Levels of hCT and sCT

hCT and sCT were determined in the sera of 101 patients with small cell lung cancer, 16 with non-small cell lung cancer, and 18 with C cell carcinoma (Fig. 1). hCT was elevated in 20% and sCT in 54% of cases of small cell lung cancer, with much higher levels for sCT. With few exceptions, patients with non-small cell lung cancer did not show raised levels of either hCT and sCT. In C cell carcinoma 72% of patients were positive for hCT and only 39% for sCT, with much higher levels for hCT (Fig. 2). As is shown in Fig. 3, there was no correlation between the two calcitonins in patients, sera. Elevated sCT immunoreavitivity was seen in patients with low as well as with high hCT immunoreactivity.

hCT and sCT in Culture Medium

High amounts of sCT-like material were detectable in the incubation medium of several lung cancer cell lines (Fig. 4). In the medium of the lung cancer cell lines, the sCT immu-

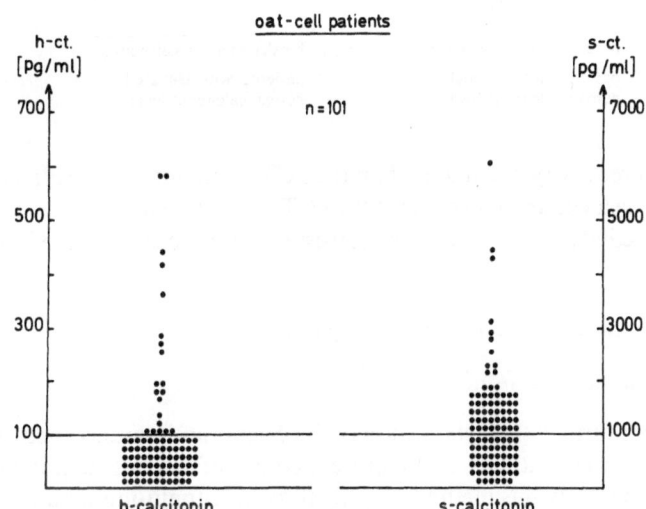

Fig. 1. Human calcitonin (hCT) and salmon calcitonin (sCT) values in sera of 101 patients with small cell lung cancer.

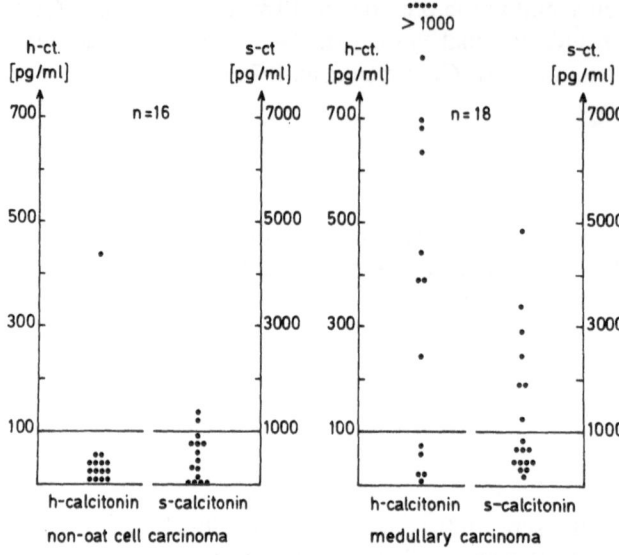

Fig. 2. Human calcitonin *(hCT)* and salmon calcitonin *(sCT)* values in sera of patients with non-oat cell carcinoma and medullary carcinoma of the thyroid

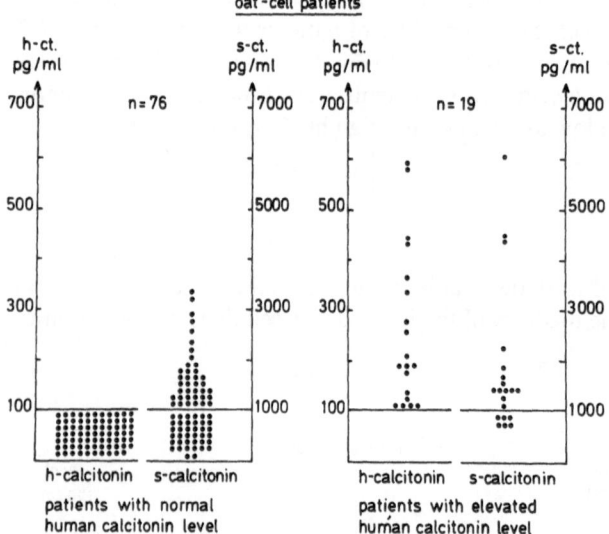

Fig. 3. Salmon calcitonin serum levels *(sCT)* in oat cell lung cancer patients with normal and elevated human calcitonin *(hCT)* serum values

noreactivity was higher than the hCT immunoreactivity. In contrast, in the medium of the C cell carcinoma cell lines the hCT immunoreactivity was higher than the sCT immunoreactivity. These results are consistent with those recorded in patients' sera.

Biochemical Characterization of Calcitonin

Human Calcitonin

For further characterization of the calcitonin-immunoreactive proteins we investigated sera, tissue extracts, and culture medium of various lung tumors using gel filtration techniques. By this method, usually three calcitonin-immunoreactive proteins with molecular

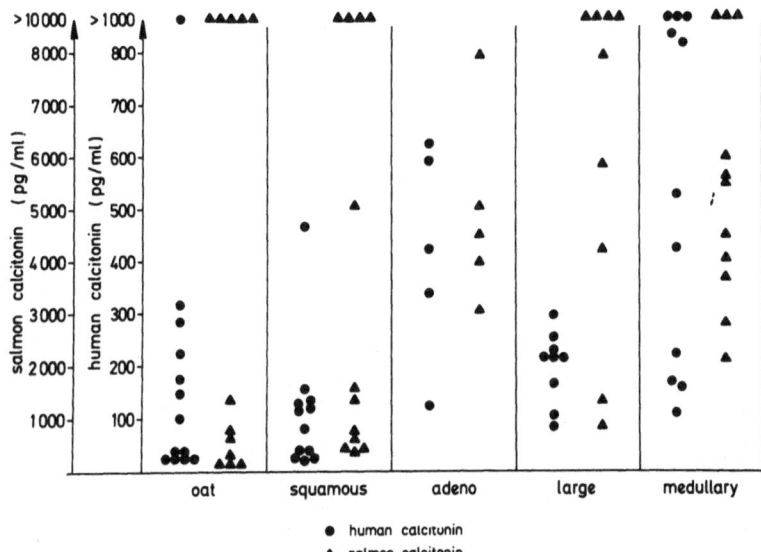

Fig. 4. Human and salmon calcitonin in culture medium of lung cancer and medullary cancer cell lines

Table 1. Characterization of human calcitonin immunoreactive material

Immuno-reactivity	Gel filtration for determination of the molecular-weight	Protein A affinity chromatography	Lectin chromatography	Ion exchange chromatography	SDS stability	Stability against proteolytic activities	Iso-electric point
Calcitonin	100000 48000 20000	No interaction with immunoglobulins	No glycoprotein component	Specific enrichment of the 20000-dalton fraction	Degradation to 17000-dalton core protein	17000-dalton protein most stable	5.5–6.0

weights of about 100000, 48000, and 20000 could be separated. Figure 5 shows these three calcitonin-containing fractions in the serum of a patient with small cell lung cancer. These three fractions were detectable before therapy, during therapy when a partial remission was achieved, and at the time of relapse. The high-molecular-weight calcitonins could also be demonstrated by this method in the culture medium from small cell and from non-small cell lung cancer (Fig. 6). Usually there was more higher-molecular-weight calcitonin in the medium than in the patients' serum. There were also differences between different tumors, in that in some tumors only one or two of the higher molecular weight calcitonin fractions were present. No significant amounts of low-molecular-weight physiological calcitonin were detectable. In contrast, in the medium of medullary carcinoma cell lines large amounts of low-molecular-weight calcitonins were also demonstratable as well as high-molecular-weight forms.

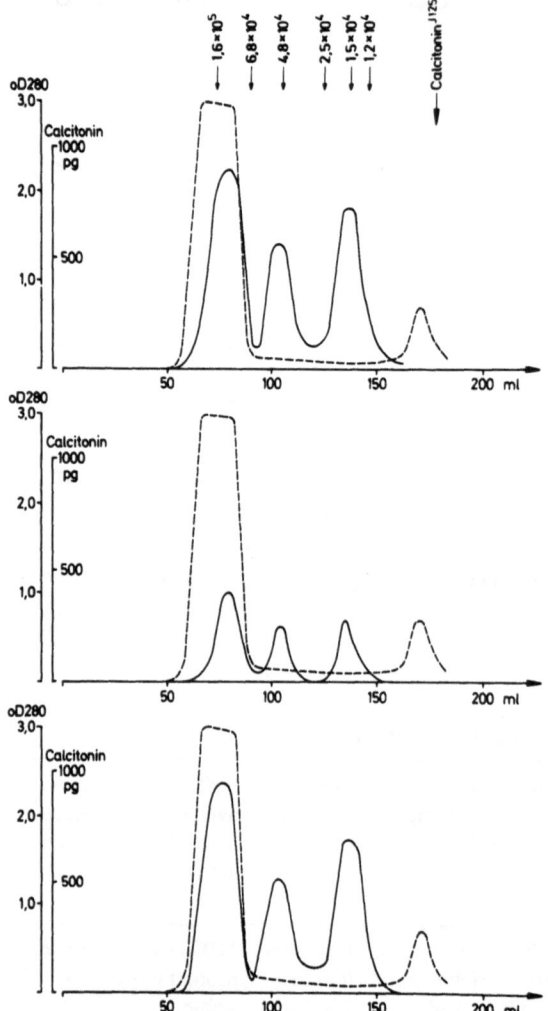

Fig. 5. AcA 54 column chromatography of one patient's serum before therapy *(above)*, during therapy *(middle)*, and at the time of tumor progression *(below)*. Fractions were assayed for calcitonin. This result is representative for six patients' sera

In further experiments the different calcitonin fractions were incubated in the presence of sodium dedecyl sulfate and then separated by polyacrylamide gel electrophoresis. Denaturation of the various fractions resulted in a 17000-dalton immunoreactive band. An additional 3500-dalton calcitonin appeared in the gels, which has a similar molecular weight to physiologic calcitonin (Fig. 7).

Incubation experiments with proteolytic enzymes such as trypsin showed that by this procedure the high-molecular-weight calcitonin fractions of 100000 and 48000 daltons were degraded to the relatively stable 17000-dalton calcitonin-immunoreactive protein. This 17000-dalton protein seems to be a relatively stable core protein, which may represent the calcitonin prohormone synthesized by the lung tumor cells (Jacobs et al. 1981; Desplan et al. 1980). Further characterizations are summarized in Table 1.

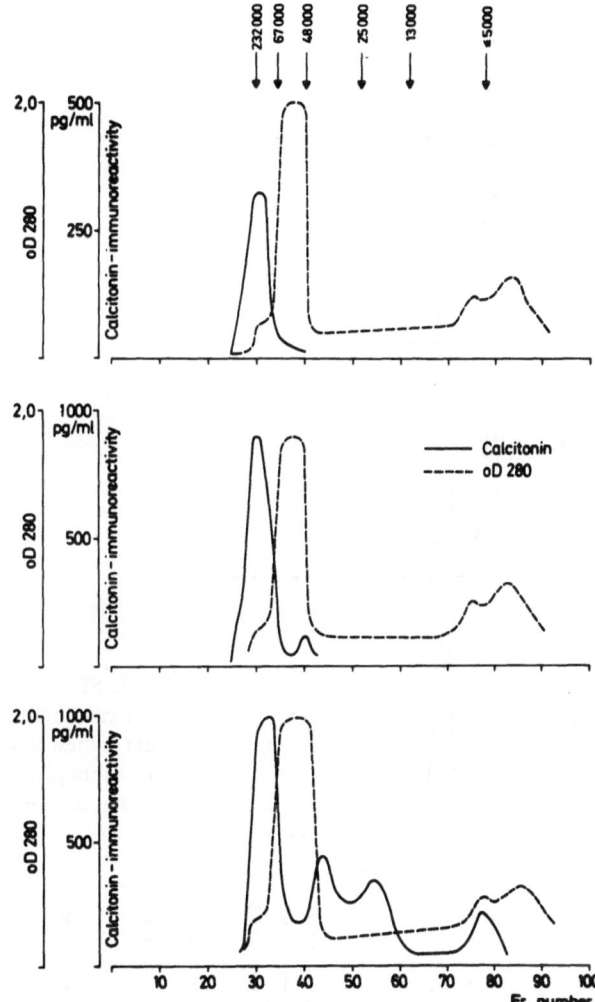

Fig. 6. AcA 54 column chromatography of the culture medium of a squamous cell *(above)*, a large cell *(middle)*, and a small cell lung cancer cell line *(below)*. Fractions were assayed for human calcitonin

Salmon Calcitonin

In parallel to hCT, sCT immunoreactivity was determined in the culture medium of lung cancer cell lines after fractionation of the medium by gel chromatography. As shown in Fig. 8, this method for sCT allowed detection of immunoreactive fractions with molecular weights of 100000, 25000, 13000, and 5000.

In summary, our results show for the first time the coexistence of elevated levels of hCT- and sCT-like material in the serum of patients with SCLC. Both calcitonin immunoreactivities have been shown to be secreted into the incubation medium by cultured lung cancer cells. Here, in contrast to studies with medullary carcinoma cells, the values for sCT-like material are higher than for the hCT-like material. The same results were found in vivo, where in lung cancer patients the sCT values were higher than the hCT values and in medullary cancer patients the reverse was true. Our studies also confirm reports from Fischer et al. (1983), who demonstrated both hCT and sCT in human thyroid and brain. From our studies the nature of the sCT-like material is not clear. Recent investiga-

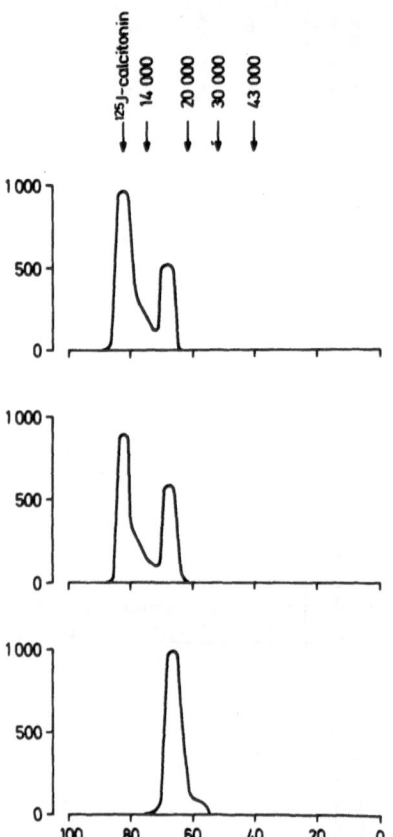

Fig. 7. SDS polyacrylamide gel electrophoresis of the three calcitonin fractions obtained by gel filtration, from highest to lowest molecular weight. Gels were cut into 1-mm pieces and calcitonin was estimated. These findings are representative of six sera and two tumor extracts

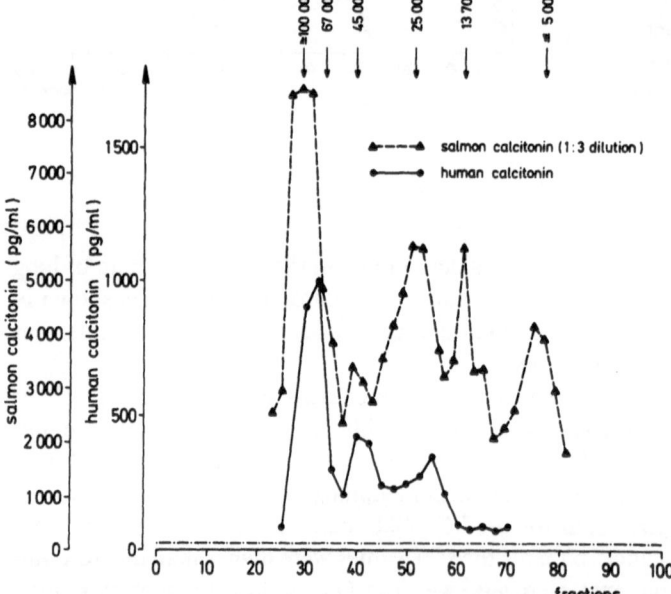

Fig. 8. sCT immunoreactivity determined in culture medium of lung cancer cell lines after fractionation by gel chromatography

tions of the hCT in lung and medullary cancer patients and from the chromatography studies suggest that we detected by radioimmunoassay an sCT precursor molecule from which the sCT may derive. In any case, the sCT-like material may be of further interest as a tumor marker in lung cancer patients. Further studies for the biochemical characterization of the higher-molecular sCT fractions are planned.

References

Coombes RC, Hillyard C, Grennberg PB, MacIntyre I (1974) Plasma immunoreactive calcitonin in patients with non-thyroid tumors. Lancet I: 1080–1083

Deftos LJ, Ross BA, Bronzent D, Parthemore JG (1975) Immunochemical heterogeneity of calcitonin in plasma. J Clin Endocrinol Metab 40: 409–412

Desplan C, Benicourt C, Jullienne A, et al. (1980) Cellfree translation of mRNS coding for human and murine calcitonin. FEBS Lett 117: 89–92

Fischer JA, Tobler PH, Henke H, Ischopp FA (1983) Salmon and human calcitonin-like peptides coexist in the human thyroid and brain. J Clin Endocrinol Metab 57: 1314–1316

Goldstein IJ, Hollerman CE, Smith EE (1965) Proteincarbohydrate interaction. II. Inhibition studies on the interaction of concavalin A with polysaccharides. Biochemistry 4: 876–883

Goltzman D, Tischler AS (1978) Characterization of calcitonin released by medullary thyroid carcinoma in tissue culture. J Clin Invest 61: 449–458

Gropp C, Havemann K, Scheuer A (1980) Ectopic hormones in lung cancer patients at diagnosis and during therapy. Cancer 46: 347–354

Hjelm H, Hjelm K, Sjöquist J (1972) Protein A from Staphylococcus aureus, its isolation by affinity chromatography and its use as immunosorbent for isolation of immunoglobulins. FEBS Lett 28: 73–76

Jacobs JW, Lund PK, Potts JT, Jr, Bell NH, Habener JF (1981) Procalcitonin is a glycoprotein. J Biol Chem 256: 2803–2807

Luster W, Gropp C, Sostmann H, Kalbfleisch H, Havemann K (1982) Demonstration of immunoreactive calcitonin in sera and tissues of lung cancer patients. Eur J Cancer Clin Oncol 18: 1275–1283

Luster W, Gropp C, Kern HF, Havemann K (1984) Biosynthesis of hormone immunoreactive proteins by non-small cell lung tumor cells in vitro. Acta Endocrin 105: [Suppl] 264: 55–56

Milhaud G, Calmette C, Taboulet J, Julienne A, Moukhtar MS (1974) Hypersecretion of calcitonin in neoplastic conditions. Lancet I: 462–463

Perez Cano R, Galan Galan F, Girgis SI, Arnett TR, MacIntyre I (1981) A human calcitonin-like molecule in the ultramobranchial body of the amphibia (Rana pipiens). Expeientia 37: 1116–1117

Perez Cano R, Girgis SI, Galan Galan F, MacIntyre I (1982) Identification of both human and salmon calcitonin-like molecules in birds suggesting the existence of two calcitonin genes. J Endocrinol 92: 351–355

Silva OL, Becker KL, Primack A, Doppmann I, Snider RH (1974) Ectopic section of calcitonin by oat-cell carcinoma. N Engl J Med 290: 1122–1124

Weber K, Osborn M (1969) The reliability of molecular weight determination by dodecyl sulfate-polyacrylamide gel electrophoresis. J Biol Chem 244: 4406–4412

Parathyroid Hormone and PTHmRNA in a Human Small Cell Lung Cancer

H.-J. Schmelzer, R.-D. Hesch, and H. Mayer

Gesellschaft für Biotechnologische Forschung mbH, Mascheroder Weg, 3300 Braunschweig, FRG

Introduction

Hypercalcemia is often associated with malignant diseases. The frequency of ectopic production of PTH as a cause of humoral hypercalcemia in association with malignant diseases is controversial. Some studies have suggested that patients with tumors and hypercalcemia have elevated iPTH in the serum or in tumor extracts (Sherwood et al. 1967; Benson et al. 1974), whereas others have been unable to detect iPTH in either tumor extracts or serum (Powell et al. 1973). Further studies have shown that humoral hypercalcemia is due to an altered form of PTH, which is unrecognized or only partially recognized by the radioimmunoassay (RIA) for PTH but which is still able to bind to the PTH receptor and is measurable in an in vitro bioassay (Goltzman et al. 1981). Up to now it has not been possible to distinguish between de novo synthesis of PTH by the ectopic tumor cells and the accumulation of exogenous PTH in the tumor tissue.

To determine the capacity of tumor tissue for de novo synthesis of PTH we selected a hypercalcemia-associated tumor from a patient with elevated serum iPTH and examined it for the presence of mRNA coding for PTH in a wheat germ cell free translation system. The specificity of this assay depends on the selective precipitation of an in vitro-synthesized [^{35}S]methionine-labelled protein by an anti-PTH-polyclonal antibody and the molecular weight estimation of this precipitated protein by SDS-PAGE. The sensitivity is sufficient to allow as little as 30 pg translatable PTHmRNA to be detected.

Materials and Methods

Parathyroid glands, ectopic tumor tissue, and control tissues were collected and frozen at $-70\,^{\circ}\mathrm{C}$. In a male patient hypercalcemia (serum Ca^{2+} 5.0/mmol/liter; normal 2.15–2.75 mmol/liter) and elevated iPTH in serum (1100 pg/ml; normal 100–300 pg/ml) as measured by a midregion-specific (against amino acids 44–68) antiserum were associated with a central bronchial tumor with lymphogenic and hematogenic metastases. The tumor was histologically a small cell lung cancer. The size of the parathyroid glands was normal after autopsy. RNA was prepared using the guanidinium hydrochloride procedure according to Hu et al. (1979). PTH was extracted from the upper phase of the gradient according to Aurbach (1959). The wheat germ system was prepared according to Roberts et al. (1973). Immunoprecipitation in conjunction with *Staphylococcus* protein A

Table 1. Amounts of iPTH and RNA in various tissues

Tissue	Wet weight	iPTH/g tissue to 44–68; 28–28 AA	Total RNA mg
Human ectopic tumor	0.3 g	4 ng	0.07
Human parathyroid adenoma	0.5 g	3 ng	0.5
Human blood	10 g	–	0.1
Porcine parathyroid glands	6 g	0.95 ng	5.5
Rat parathyroid and thyroid glands	0.8 g	–	0.75
Porcine lung	3 g	0	–
Porcine bronchial tissue	3 g	0	–

was carried out according to MacSween and Eastwood (1981) with a polyclonal antibody recognizing human, bovine, and rat PTH in RIA (gift from Professor Ziegler, University of Heidelberg). The RIA for PTH, using antibodies recognizing two different regions of PTH, was carried out according to Hehrmann et al. (1977). [^{14}C]Protein markers were obtained from Amersham.

Measurements of PTH in the Tumor by RIA

Because of the limited amounts of tissue protein available, DNA and RNA were separated in a single step by centrifugation of the homogenate in a CsCl gradient. PTH was quantified after acetone, acetic acid, and TCA extraction by antibodies recognizing two different regions of the PTH molecule. The concentration of PTH in the tumor was similar to that in the human adenoma. In porcine bronchial and lung homogenate no iPTH could be detected at 10 times higher protein concentration in the assay (Table 1). Proteolysis of [^{125}I]PTH tracer could mimic a displacement can be excluded by the following criteria. The homogenate of negative controls showed no elevated activity. The extraction procedure at low pH inactivates cellular proteases. Without treatment of the blood samples the RIA is used in the clinic routinely.

Immunoprecipitation of Pre-pro PTH from In Vitro Translated mRNA of the Tumor

mRNA from porcine parathyroid glands was used as standard in the wheat germ cell free translation system. A number of translation products exhibited apparent molecular weights between 5 kd and 100 kd. After immunoprecipitation of the translation products by an anti-PTH antiserum plus protein A a single protein of 13 kd became visible. PTH 3 ng extracted from porcine glands competed in the immunoprecipitation with the [^{35}S]methionine-incorporated 13-kd protein, which then remains in the supernatants of the immunoprecipitate (Fig. 1). Proteins are not normally correctly processed in the in vitro translation system. As shown by Habener et al. (1980), the earliest in vitro translation product of bovine PTHmRNA is pre-pro PTH, a protein containing a leader sequence consisting of 25 additional amino acids preceding the 90 aminoacids of pro PTH. Thus the 13-kd Protein corresponds well to the size expected for pre-pro PTH. As a positive homologous control, in vitro translation of mRNA from a human adenoma also showed a single

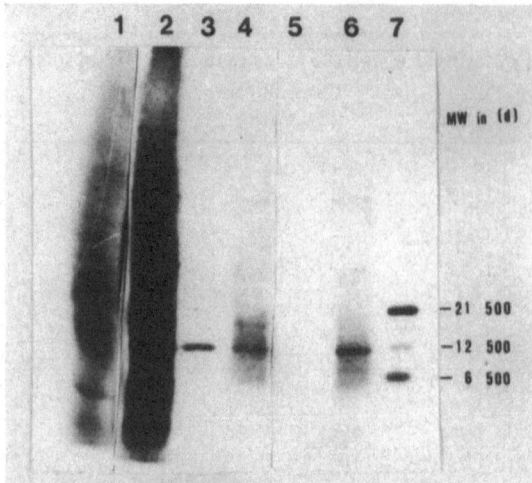

Fig. 1. Autoradiogram of PAGE 14–20% of [³⁵S]methionine-labeled translation products in wheat germ system of RNA from porcine parathyroid glands. *1* internal control; *2* translation products of 10 µg total porcine RNA; *3* immunoprecipitate of position 2; *4* supernatant of the immunoprecipitate of position 3; *5* immunoprecipitate of position 3 with 1 ng PTH; *6* supernatant of the immunoprecipitate of position 5. Exposure time: 1 Week

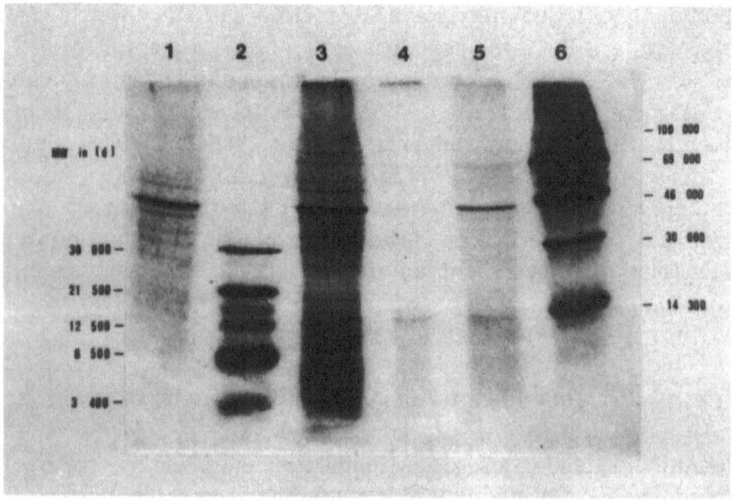

Fig. 2. Autoradiogram of PAGE 14–20% of [³⁵S]methionine-labeled translation products in wheat germ system of RNA from human parathyroid adenoma. *1* internal control; *2* MWmarker; *3* translation products of 10 µg total RNA of human parathyroid adenoma; *4* immunoprecipitate of position 3; *5* supernatant of the immunoprecipitate of position 4; *6* MWmarker. Exposure time: 1 week

13-kd protein on immunoprecipitation (Fig. 2). Two further controls were carried out to confirm the specificity of the immunoprecipitation. When 300 µg BSA was added to the reaction mixture it did not compete with the 13-kd protein in the antibody precipitation. In vitro translation products directed by mRNA prepared from human blood were not

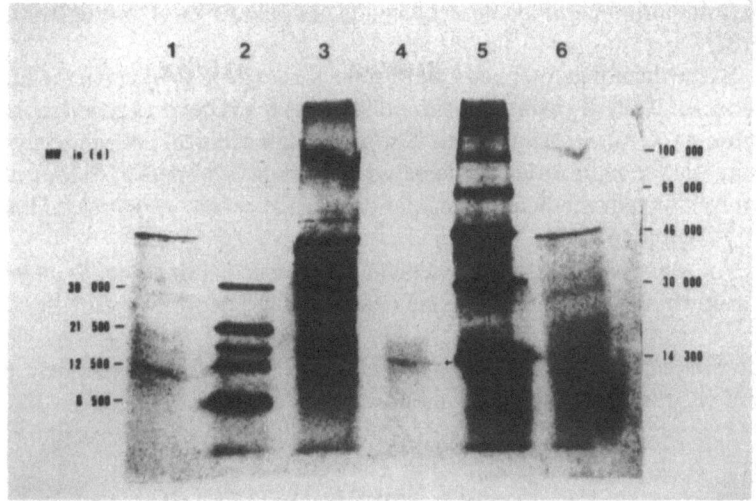

Fig. 3. Autoradiogram of PAGE 14–20% of [^{35}S]methionine-labeled translation products in wheat germ system of RNA from ectopic tumor. *1* internal control; *2* MWmarker; *3* translation products of 10 µg total tumor RNA; *4* immunoprecipitate of position 3; *5* MWmarkers; *6* supernatant of the immunoprecipitate of position 4. Exposure time: 2 weeks

precipitated by the antibody (results not shown). The tumor RNA was translated to yield a number of products in the apparent molecular weight range of 5–70 kd. A specific protein of 13 kd was immunoprecipitated (Fig. 3). Because of the limited amount of mRNA from this tumor we were not able to show that cold PTH could compete with the [^{35}S]methionine-incorporated 13-kd protein in the immunoprecipitation.

Discussion

A number of peptide hormones are produced by the tumor cells of small cell and non-small cell lung carcinomas. In hypercalcemia-associated tumors several mediators of bone resorption have been described, including prostaglandins (Tashjian et al. 1972), osteoclast-activating factor (Mundy et al. 1974), and tumor-derived transforming growth factors (Ibbotson et al. 1983). Prostaglandins and osteoclast-activating factors are mainly local bone-resorbing factors, whereas transforming growth factors produced by many tumors are powerful bone resorbers; parathyroid hormone and parathyroid hormone-like activity are also responsible for bone resorption.

In a series of tumors associated with hypercalcemia Simpson et al. (1983) could not identify a PTHmRNA using a sequence-specific cDNA probe. No correlation of iPTH and PTHmRNA was domonstrated by the authors.

In this contribution we have been able to demonstrate in a small cell lung carcinoma, the presence of PTH by RIA, and using an in vitro translation system, the presence of PTH-specific mRNA in a small cell lung carcinoma. The identification of specific translation products can be used as a criterion for ectopic hormone production, in addition to immunological detection of the hormone in question. The final criterion for establishing the identity or nonidentity of tumor-derived parathyroid hormone with glandular parathyroid hormone will be the sequencing of cloned PTHcDNA derived from ectopic tumor mRNA.

It is evident that the parathyroid gene is a single-copy gene in the human genome (Vasicek et al. 1983). By using somatic cell hybrids it has been assigned to the short arm of chromosome 11 (11 p) (Mayer et al. 1983). Through classical linkage analysis it has been localized on 11 p in the order; centromere-PTH-β-globin-insulin (Antonavakis et al. 1983). It is not yet known whether a disregulation of PTH occurs in or around the structural gene for PTH hormone itself.

The expression of the PTH gene in the ectopic tumor provides an interesting system for a study of the correlation of gene regulation and neoplastic growth.

Acknowledgements. We wish to thank Mrs. Enters for her excellent technical assistance and J. Collins for reading the manuscript.

References

Antonavakis SE, Phillips JA III, Mallonee RL, Karzazian HH Jr, Fearon ER, Waber PG, Kronenberg HM, Ullrich A, Meyers DA (1983) β-Globin locis is linked to the parathyroid hormone (PTH) locus and lies between the insulin and PTH loci in man. Proc Natl Acad Sci USA 80: 6615–6619

Aurbach GP (1959) Isolation of parathyroid hormone after extraction with phenol. J Biol Chem 234: 3179–3181

Benson RC Jr, Riggs BL, Pickard BM, Arnaud CD (1974) Radioimmunoassay of parathyroid hormone in hypercalcemia patients with malignant disease. Ann J Med 56: 821–826

Goltzman D, Stewart AF, Broadus AE (1981) Malignancy-Associated hypercalcemia: evaluation with a cytochemical bioassay for parathyroid hormone. J Clin Endocrinol Metab 53: 899–904

Habener JF, Rosenblatt M, Kemper B, Kronenberg HM, Rich A, Potts JT Jr (1980) Pre-pro parathyroid hormone: amino acid sequence, chemical synthesis, and some biological studies of the precursor region. Proc Natl Acad Sci USA 75: 2616–2620

Hehrmann R, Nordmeyer H, Wilke R, Hesch RD (1977) Radioimmunoassay of hPTH; characterisation of two new antisera from sheep and clinical results. Acta Endocrinol (Copenh) [Suppl] 208: 117–118

Hu CP, Slate DL, Gravel R, Ruddle FH (1979) Biological detection of specific mRNA molecules by microinjection. Proc Natl Acad Sci USA 76: 4503–4506

Ibbotson KJ, D'Sonza SM, Ng KW, Osborne CK, Niall M, Martin TJ, Mundy GR (1983) Tumor-derived growth factor increases bone resorption in a tumor associated with humoral hypercalcemia of malignancy. Science 221: 1292–1294

MacSween JM, Eastwood SL (1981) Recovery of antigen from staphylococcal protein A – antibody adsorbens. Methods Enzymol 73: 459–471

Mayer H, Breyel E, Bostock C, Schmidtke J (1983) Assignment of the human parathyroid hormone gene to chromosome 11. Hum Genet 64: 283–285

Mundy GR, Raisz LG, Cooper RA, Schechter GP, Salmon SE (1974) Evidence for secretion of an osteoclast stimulating factor in myeloma. N Engl J Med 291: 1041–6

Powell D, Singer FR, Murray TM, Minkin C, Potts JT Jr (1973) Nonparathyroid humoral hypercalcemia in patients with neoplastic diseases. N Engl J Med 289: 176–181

Roberts BE, Paterson BM (1973) Efficient translation of Tobacco mosaic virus RNA and rabbit globin 9 S RNA in a cellfree system from commercial wheat germ. Proc Natl Acad Sci USA 70: 2330–2334

Sherwood LM, O'Riordan JLH, Aurbach GD, Potts JT Jr (1967) Production of parathyroid hormone by nonparathyroid tumors. J Clin Endocrinol Metab 27: 140–146

Simpson EL, Mundy GR, D'Sonza SM, Ibbotson KJ, Bockmann R, Jacobs JW (1983) Absence of parathyroid hormone messenger RNA in nonparathyroid tumors associated with hypercalcemia. N Engl J Med 309: 325–330

Tashjian AH Jr, Voelkel EF, Levine L, Goldhaber P (1972) Evidence that the bone resorption-stimulating factor produced by mouse fibrosarcoma cells is prostaglandin E_2 a new model for the hypercalcemia of cancer. J Exp Med 136: 1329–1343

Vasicek TJ, McDevitt BE, Freeman MW, Fennick BJ, Hendy GN, Potts JT Jr, Rich A, Kronenberg HM (1983) Nucleotide sequence of the human parthyroid hormone gene. Proc Natl Acad Sci USA 80: 2127–2131

The Endocrine Lung Tumor Cell: In Vitro Studies

The Pathogenesis of Hormone-Producing Tumors of the Lung*

E. M. McDowell, K.-P. Keenan, and B. F. Trump**

Department of Pathology, University of Maryland, School of Medicine, 10 S. Pine Street, Baltimore, MD 21202, USA

Introduction

Pulmonary neoplasms, and in particular bronchogenic carcinomas, are one of the most critical problems in oncology. A phenotypic variable associated with these tumors is their ability to synthesize and secrete biochemical markers of diverse kinds. These markers fall into several different categories, including peptide hormones, nonpeptide hormones and hormone-like substances, regulatory peptides, enzymes, structural proteins, and other less well understood antigenic substances. Hormones and regulatory peptides have been associated with all the histological types of bronchogenic carcinoma.

Our purpose in this brief review is to summarize some of the information regarding the pathogenesis of the various types of bronchogenic neoplasms. We will attempt to relate knowledge concerning normal development and maintenance of the tracheobronchial epithelium (TBE) and its regeneration following injury, to the pathogenesis of preneoplastic and neoplastic lesions. A more detailed discussion of this subject is given elsewhere (McDowell and Trump 1983).

The Adult Epithelium

Morphology

The surface of the normal TBE is a pseudostratified columnar epithelium. This means that cells of all types rest on the basal lamina but not all cell types reach the lumen. Cells which reach to the lumen are tall and columnar, whereas those that do not are short cells. This simple fact is often overlooked because there are relatively few normal areas in the bronchi of adult humans, even in putatively healthy individuals.

Ciliated cells have an electron-lucent cytoplasm and reach the lumen. Numerous cilia project from their apices into the lumen. *Secretory (mucous) cells* also reach the lumen but

---

  * This is contribution # 1749 from the Cellular Pathobiology, Laboratory, Department of Pathology, University of Maryland, Baltimore, MD 21201, USA Parts of this work were supported by Grant HL-24722 from the National Heart, Lung and Blood Institute (NIH) and by Contract N01-CP-15738 from the National Cancer Institute
 ** Dr. Trump is and American Cancer Society Professor of Clinical Oncology

have a more electron-dense cytoplasm, well-developed rough endoplasmic reticulum (RER) and Golgi apparatus, and variable numbers of mucous granules. *Dense-core granulated* (DCG) *cells* (endocrine cells) sometimes reach the lumen, but the majority of these cells are short. Their cytoplasm is filled with variable numbers of dense-cored granules. *Basal cells* are short cells which do not reach the lumen. They usually contain more keratin than the other cell types and are attached to the basal lamina by hemidesmosomes. In addition to these mature cell types, a few immature *preciliated cells* may be seen. Preciliated cells are very numerous during normal development and became numerous during regeneration (Section: "Epithelial Development and Regeneration Following Injury"). They often contain a few mucous granules. The submucosal bronchial glands have ciliated ducts, the linings of which resemble the surface TBE. The ciliated ducts pass into collecting ducts which are lined by mitochondria-rich cells. The collecting ducts receive glandular elements, the mucous tubules (lined with mucous cells), which terminate in serous tubules (Meyrick et al. 1969). The glandular epithelium also contains myoepithelial cells (Meyrick and Reid 1970) and a few DCG cells (Bensch et al. 1965), but there are no basal cells in the submucosal glands.

Cell Kinetics

Cell kinetic studies of human (Harris et al. 1975) and animal TBE (Boren and Paradise 1978) show that the adult epithelium has very low mitotic and labeling indices. However, certain cells have the ability to divide and do so rapidly following injury. This property resides in columnar secretory and short basal cells, but because of their numerical superiority the secretory cells appear to be much more important than basal cells in terms of replication, at least in the hamster trachea (Keenan et al. 1982a–c). It is a commonly held thesis that basal cells of TBE are analogous to those in the skin and represent an undifferentiated or "stem" cell population, the division of which gives rise to other cell types during regeneration following injury and to neoplastic cells during tumorigenesis. In the skin, which is a truly stratified rather than pseudostratified epithelium, the basal cells are poorly differentiated. They divide to give rise to the more highly differentiated suprabasal layers, including the stratum corneum. Basal cells of human skin are reactive for low-molecular-weight 45K and 46K keratin proteins. The higher molecular weight keratins, 55K and 63K, are associated only with the suprabasal and cornified layers (Banks-Schlegel et al. 1981; Said et al. 1983; Thomas et al. 1984). In contrast, basal cells of the normal human bronchus contain both low- *and* high-molecular-weight keratins 45K, 46K, 55K, and 63K (Said et al. 1983), suggesting that bronchial basal cells are *not* analogous to epidermal basal cells. Nevertheless, the view continues to be held that basal cells of TBE are undifferentiated stem cells, although they show differentiation in the form of high-molecular-weight keratins (in contrast to basal cells of the skin) and although several investigators have described considerable mitotic potential associated with secretory cells of TBE (Boren and Paradise 1978; Keenan et al. 1982a–c; Jeffery et al. 1982).

The kinetic status of DCG cells in the adult TBE is presently obscure. There are no definitive data on the ability of DCG cells of the normal adult TBE to divide – although mitoses have been observed in these cells in the respiratory tract of developing fetuses (McDougall 1978; Sorokin et al. 1982).

Cell kinetics of the bronchial glands have not been analyzed in detail. It is clear, however, that in this locus basal cells are absent and both mucous cells and serous cells can divide (Meyrick 1977). Therefore, basal cells cannot be considered candidates for the cell of origin of neoplasms which truly arise in the bronchial submucosal glands.

Epithelial Development and Regeneration Following Injury

In 1962, R. A. Willis published "The Borderland of Embryology and Pathology." In that text he stressed that regenerating epithelia in adults display structures and properties essentially similar to those of the corresponding immature tissues of the fetus. The TBE is no exception, and regeneration of the mucociliary epithelium following diverse forms of injury shares many similarities with normal development.

Fetal and Neonatal Development

The primordial TBE is composed of tubes, derived from the endoderm. At first the tubes are lined by a single layer of unspecialized columnar cells (Leeson and Leeson 1964; Conen and Balis 1969; Cutz and Conen 1972). Later cell specialization occurs and the epithelium becomes pseudostratified. At that time DCG cells, ciliated cells, secretory cells, and basal cells can be recognized.

We are presently studying the development of the tracheal epithelium in fetal and neonatal Syrian golden hamsters. The gestation period is 16 days. Up to day 11 in utero, the primordial trachea observed was lined by a single layer of primitive columnar cells. Many of the cells were replicating. All cells reached the lumen and none of the cells was short. On day 12 the epithelium remained simple columnar, but groups of DCG cells had become specialized to form neuroepithelial bodies (NEB), which were most obvious in the dorsal tracheal epithelium, the primordium of the pars membranacea. On day 13 the epithelium was pseudostratified (i. e., a few cells were short) and preciliated, and ciliated cells were present in the dorsal tracheal epithelium. About 92% of mitoses were in columnar cells and only 8% were in short cells on day 13.

The proliferating cells, both short (prebasal) and columnar (presecretory), appeared to be rather poorly differentiated until day 14. On this day primitive hemidesmosomes were seen anchoring the short cells to the basal lamina. As hemidesmosomes appear to be unique to basal cells (Kawanami et al. 1979; Tandler et al. 1981), we assume that the short cells had specialized to become basal cells by day 14. Moreover, RER had developed in the columnar cells, heralding the specialization of secretory cells. These cells were mucus producing by day 15.

It is clear from these studies that the poorly differentiated columnar cells were the primordial stem cells capable of division and giving rise to the specialized cells. The first specialized cells to be recognized were DCG cells on day 12, followed by ciliated cells on day 13. In human bronchi DCG cells become specialized before mucous cells (Cutz and Conen 1972) and DCG cells were the first cell type to specialize in rabbit bronchi (Sorokin et al. 1982). Basal cells and secretory cells did not become overtly specialized in fetal hamster trachea until gestational days 14 and 15 (hemidesmosomes in basal cells on day 14; mucous granules in secretory cells on day 15).

Proliferative activity in columnar cells (at first poorly differentiated and later secretory cells) dominated numerically over that in basal cells at all stages of fetal development. On the day of birth (day 16), about 80% of all mitotic activity was in the columnar secretory cells, yet the basal cells were well established at this time, accounting for about 30% of the epithelial cell population.

As the columnar secretory cells matured, so mucous granules were seen in their apices and a few mucous granules were seen in preciliated cells, as described earlier by Emura and Mohr (1975). Preciliated cells which are generated in adult tracheal epithelium following mechanical injury or vitamin A deficiency also contain mucous granules in their apical cytoplasm, suggestive of their lineage from secretory cells (see below).

Additional ongoing experiments in our laboratory are designed to determine the respective roles played by secretory cells and basal cells in the genesis of ciliated cells during postnatal tracheal development. The total mitotic rate (6 h colchicine blockade) was very low in tracheal epithelium of normal adult hamsters, at about 0.1% (Keenan et al. 1982a), but was considerably higher in young animals, being about 3% (2.4% secretory cells; 0.6% basal cells) at 5 weeks of age (McDowell et al. 1984a). Thus, about 80% of dividing cells were secretory cells in 5-week-old hamsters, a ratio similar to that seen on the day of birth. Preciliated cells, which contained a few mucous granules in the apical cytoplasm and which appeared morphologically similar to those in the developing fetal epithelium and in adult epithelium regenerating following injury, were also seen in the tracheas of juvenile hamsters (McDowell and Trump 1983; McDowell et al. 1984a). Preciliated cells with apical mucous granules have also been described during development of tracheal epithelium in chicks (Kalnins and Porter 1969), *Xenopus laevis* (Steinman 1968), and rats (Cireli 1966).

Regeneration Following Injury

Analysis of the regenerative process following mechanical and nutritional injury (vitamin A deficiency) has aided our understanding of the histogenesis of epidermoid metaplasia. Moreover, we have been able to study the kinetics of restoration of the pseudostratified mucociliary epithelium.

Mechanical Injury

The various stages of regeneration of adult hamster tracheal epithelium following mechanical injury have been characterized morphologically and quantified by cell type (Keenan et al. 1982a–c). Some of the changes that occur in the regenerating epithelium resemble early preneoplastic changes in TBE (Section: "Preneoplastic Lesions and Tumorigenesis") and other changes recapitulate certain aspects of fetal and neonatal development (see above). In our experimental model, all the epithelial cells were carefully scraped from the basal lamina at a focal area. Within a few hours the columnar secretory cells and short basal cells which were adjacent to the wound changed shape. The cells flattened to become squamous cells. Since secretory cells outnumbered basal cells in a ratio of about 6:1 in normal adult hamster trachea, most of the newly formed squamous cells were altered secretory cells. By 24 h after wounding about 30% of the cells in the wound were in mitosis. Thereafter they piled up and formed a typical multilayered epidermoid (keratin-producing) metaplastic epithelium. The keratinizing metaplastic cells also contained mucous granules, which is indicative of their secretory nature and consistent with their origin from secretory cells. Within 3 days some of the metaplastic cells sloughed from the epithelium, and division of metaplastic and columnar secretory cells gave rise to presecretory and preciliated cells. A nearly normal mucociliary epithelium was restored within 5–6 days following wounding.

Presecretory cells could not be readily distinguished from mature secretory cells with scant secretion granules, especially by light microscopy. However, the preciliated cells were quite distinctive. Preciliated cells were larger than other epithelial cells; they had an electron-lucent ctyoplasm and bore long slender apical microvilli. Ciliogenesis was evident at the cell apices as fibrogranular areas and/or basal bodies. Cell quantitation and labeling experiments with tritiated thymidine indicated that preciliated cells were the progeny of division of secretory cells, both normal columnar and polygonal keratinizing metaplastic secretory cells. Many preciliated cells contained mucous granules in their apical cytoplasm, but they lacked well-developed endoplasmic reticulum and Golgi apparatus, indicating that the mucous granules had been carried over from the parent secretory cell. Thus, presecretory cells and preciliated cells appeared to be the progeny of dividing secretory cells (Keenan et al. 1982c, 1983). Similar preciliated cells with a few mucous granules (formerly called "indifferent cells") were described in human bronchial epithelium (McDowell et al. 1978).

Vitamin A Deficiency

Vitamin A is a growth factor essential for maintaining the mucociliary state of TBE. Vitamin A deficiency results in epidermoid metaplasia of many epithelia (Wolbach and Howe 1925). In the vitamin A-deficient hamster tracheal epithelium a continuum of change was oberved, ranging from minimal morphologic change, through hyperplasia and stratification of secretory cells, to noncornifying and cornifying epidermoid metaplasia (McDowell et al. 1984a). Mitotic activity was greatly reduced in secretory cells in the minimally changed epithelium, but was increased in stratified lesions and at epidermoid foci. Epidermoid metaplastic cells were characterized by well-developed tonofilament bundles and desmosomes, yet they retained many features of secretory cells, i. e., abundant cytoplasm, well-developed RER and Golgi apparatus, and mucous granules, suggesting that the metaplastic cells were altered secretory cells. Epidermoid metaplasia also occurred in the tracheal glands during vitamin A deficiency (Chopra and Cooney 1983). Since basal cells are absent from the glandular epithelium, this observation further supports the premise that the metaplastic cells are altered secretory cells, not basal cells. The similarity of the metaplastic lesions induced by vitamin A deficiency and by mechanical injury is very striking.

The reconstitution of minimally changed epithelium in vitamin A-deficient hamsters to normal mucociliary epithelium following restoration of dietary vitamin A was studied quantitatively (McDowell et al. 1984b). In the vitamin A-deficient epithelium preciliated cells were virtually absent and mitotic activity of secretory cells was reduced 14-fold. The number of ciliated cells was also reduced. The basal cell mitotic rate, which was very low even in control hamsters, was reduced three-fold by deficiency of vitamin A. Within 2 days of restoration of vitamin A, mitotic activity began to increase in columnar secretory cells and a high mitotic rate was maintained over the next 5 days although the control mitotic rate was not exceeded. Mitotic activity of basal cells remained below the control level throughout the restorative period. Preciliated cells reappeared in the epithelium following increased proliferation of secretory cells. Many preciliated cells contained apical mucous granules but their cytoplasm lacked well-developed RER and Golgi apparatus, suggesting that the mucosubstances were carried over from the parent secretory cell, rather than synthesized by the preciliated cells themselves. The preciliated cells rapidly developed cilia and matured into ciliated cells. The normal mucociliary epithelium was restored over 7 days.

Boren and Paradise (1978) showed that columnar secretory cells gave rise to ciliated cells during normal maintenance of adult hamster trachea. Our results indicate that this

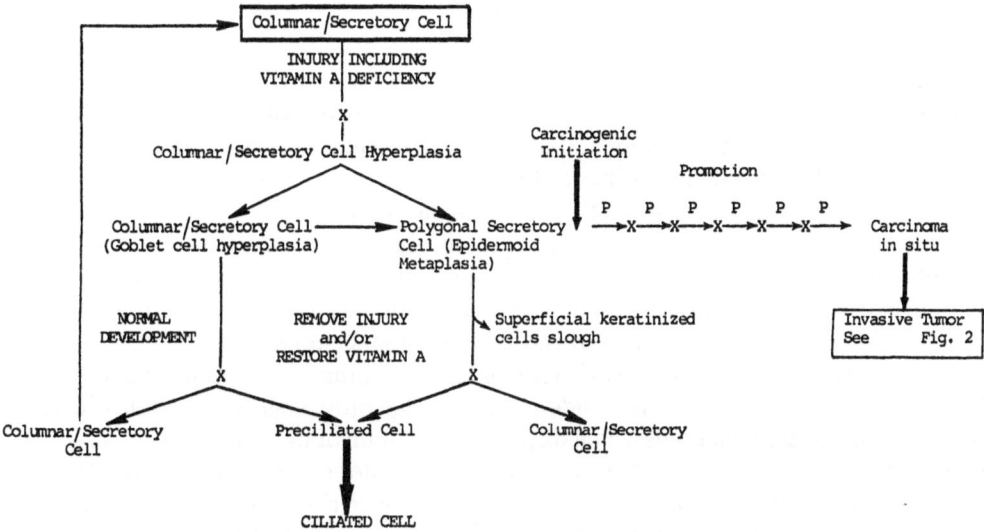

Fig. 1. Development, maintenance, regeneration and neoplasia of the mucociliary tracheobronchial epithelium. A working hypothesis. Secretory cell hyperplasia is a late change in vitamin A deficiency, associated with death and sloughing of epithelial cells (McDowell et al. 1984a). It is postulated that carcinogenic initiation may occur in either columnar or polygonal metaplastic secretory cells. *X*, mitotic division; *P*, promotion

process is greatly accelerated during fetal and neonatal development and during regeneration following injury. We predict that this phenomenon is common to many species and propose that columnar/secretory cells play a pivotal role in development, maintenance and regeneration of TBE (Fig. 1). It is already well established that Clara cells – another type of specialized secretory cell – are the progenitor cells for ciliated cells in the bronchiolar epithelium, where basal cells are normally absent (Evans et al. 1976, 1978).

Preneoplastic Lesions and Tumorigenesis

In the bronchus, as in other organs, neoplasia is believed to involve a series of steps following initiation by a carcinogenic agent. These steps result in the appearance of sequential populations of cells and may or may not involve subsequent mutagenic events. They require, almost certainly, the action of tumor promoters of various types which permit or even stimulate the selective growth of initiated cells. Tobacco smoke, for example, not only contains a number of known carcinogens and other mutagens but also contains a variety of agents which must be regarded as putative promoters (Wynder et al. 1975). It has been rather convincingly argued that the beneficial effects of smoking cessation can be best explained by the hypothesis that although the smoke contains initiators, a major factor in its association with lung cancer is that it acts as a tumor promoter (Hoffman et al. 1976).

The Cellular Origin(s) of Bronchogenic Carcinomas

Our interpretations of the histogenetic relationship between secretory cell hyperplasia and epidermoid metaplasia in TBE have been discussed above and are summarized in Fig. 1. Progression of the epithelial lesions to invasive carcinomas will now be described. The discussion will be limited to the genesis of four types of bronchogenic tumor: epidermoid carcinomas; adenocarcinomas; small cell carcinomas; and carcinoid tumors, all of which have been associated with the production of ectopic hormones in man.

Only cells that can be induced to divide are potentially capable of hyperplastic, metaplastic, and neoplastic change. In TBE, obvious candidates include secretory (mucous) cells and basal cells, both of which divide in adult epithelium and possibly DCG cells which divide in fetal TBE but rarely if ever divide in the normal adult epithelium (Section: "The Adult Epithelium"). It is now widely accepted that bronchogenic carcinomas derive from pluripotent epithelial cells which display a continuum of phenotypic expression, but there is no consensus regarding the nature of the progenitor cell. One hypothesis (McDowell et al. 1982) is that any one of these cell types may give rise to progeny which mature into *any* of the cell types of the mature epithelium, *irrespective of the nature of the parent cell*. Thus, cells arising from division of any one of these cell types could mature into basal cells, secretory cells, ciliated cells, DCG cells, or cells with mixed phenotypes, such as DCG-ciliated, DCG-mucous, or even more complicated mosaics, under the influence of presently ill-defined factors. Moreover, the hypothesis that any of the progeny has the capability to keratinize, thereby demonstrating an epidermoid phenotype, *in addition to and superimposed upon* the primary specialization. Thus, the hypothesis proposes that any cell capable of division has the potential to produce hyperplastic, metaplastic, and neoplastic lesions, composed of cells which may differ phenotypically from the parent cell(s). It follows that divisions of secretory cells could lead to hyperplastic, metaplastic, and neoplastic foci of DCG cells or that divisions of DCG cells could lead to foci of secretory cells, with or without keratinization, or to complicated phenotypic mosaics showing mixed patterns of mucus secretion, keratin production, and dense-cored granules.

Although the hypothesis does not exclude basal cells or DCG cells as progenitor cells, our bias, based on results of the experiments described above, is that columnar secretory cells are progenitor cells for the majority of bronchogenic neoplasms (Figs. 1 and 2).

Epidermoid Carcinomas

Foci of epidermoid metaplasia are often present in the bronchi of patients with invasive epidermoid carcinoma, and it may be significant that during regeneration of TBE following injury, preneoplastic traits such as epidermoid metaplasia and nuclear atypia are observerd. However, epidermoid metaplasia per se is not committed to neoplasia, and assuming that carcinogenic initiation has not occurred and that the adverse environment is removed, this lesion can reverse. Nevertheless, several studies have stressed that the development of invasive carcinoma is preceded by stratification and epidermoid metaplasia with ever-increasing degrees of atypia (Black and Ackerman 1952; Auerbach et al. 1961; Nasiell 1963; Nasiell et al. 1978). Cytological studies performed by Saccomanno et al. (1974) on the sputum of uranium miners yielded a considerable amount of data on the intermediate stages between initiation and invasive carcinoma. To a large extent these findings were corroborated by controlled studies in experimental animals (Stenbäck 1973; Schreiber et al. 1974; Becci et al. 1978). The stages are as follows: mucous cell hyperplasia

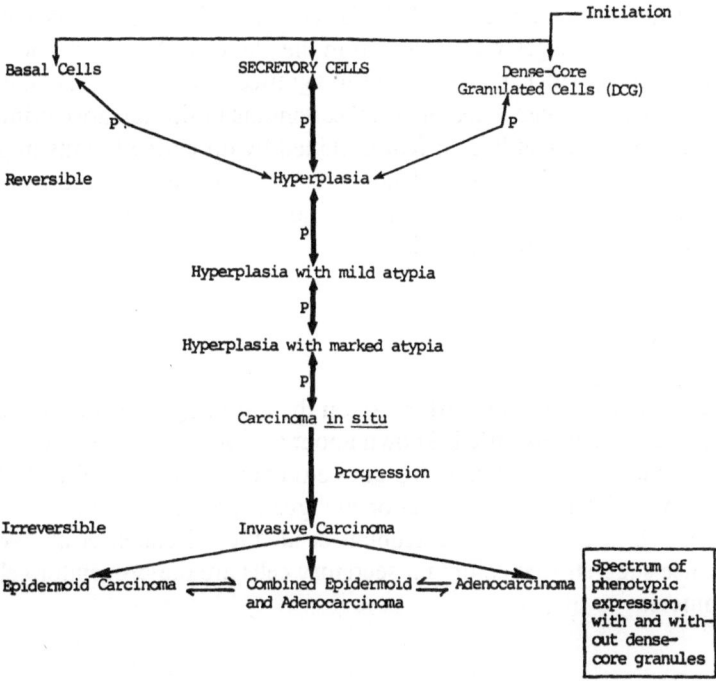

Fig. 2. Postulated progression of changes towards invasive bronchogenic carcinoma. Basal cells are absent from bronchial glandular epithelium. Secretory cells are of two types in adult human submucosal bronchial glandular epithelium, mucous cells and serous cells. Serous cells are absent from the adult surface epithelium. Although basal cells and DCG (endocrine) cells are included as progenitor cells in this scheme, our personal bias is that secretory cells are the major progenitor cell for the various preneoplastic and neoplastic states in the tracheobronchial epithelium (see text). *P,* promotion

and regular epidermoid (squamous) metaplasia; followed by epidermoid metaplasia with mild, moderate, and severe nuclear atypia; carcinoma in situ; and finally invasive cancer. The time of transition from moderate atypia to invasive carcinoma was estimated at around 5 years in humans, but this varied and may be somewhat longer (Saccomanno et al. 1974). It is unknown what proportion of persons demonstrating foci of atypia or even carcinoma in situ eventually develop invasive carcinoma.

Foci of atypical epidermoid metaplasia and carcinoma in situ may lie intercalated between more normal appearing epithelium. When this occurs the metaplastic/neoplastic cells and normal columnar cells interface with one another. Epidermoid metaplasia and carcinoma in situ involve the full epithelial thickness, and the appearance of columnar mucociliary epithelium overlying groups of atypical or neoplastic cells derives from fortuitous cuts through the periphery of a metaplastic/neoplastic focus at the interfacing zone, where the abnormal epithelial cells undermine the adjacent normal mucociliary epithelium.

Cells with combined epidermoid and mucus-secreting specializations are commonly found in areas of atypical epidermoid metaplasia and carcinoma in situ in man and experimental animals (Stenbäck 1973; Becci et al. 1978; Trump et al. 1978; Klein-Szanto et al. 1980 a, b), further strengthening the argument that these lesions develop from pathological changes in secretory cells. Furthermore, epidermoid metaplasia with atypia and carcinoma in situ are commonly found in submucosal glandular epithelium (Black and Ackerman 1952; Auerbach et al. 1967; Carter et al. 1976), where basal cells are normally absent. Neoplastic changes were reported in the glands of human bronchial specimens that had been treated with benzo(a)pyrene and grafted into nude mice (Ito et al. 1982). Some workers have questioned whether or not carcinoma in situ can arise primordially at this site, but evidence for a glandular origin is gained by observing lesions in glandular acini stained with a mucus stain such as Alcian blue-PAS. It is clear that the mucus-secreting cells of the gland are involved but that their secretions are diminished as epidermoid metaplasia with atypia progresses to neoplasia.

Adenocarcinomas

Few studies have been performed on the morphogenesis of bronchogenic adenocarcinomas, and relatively little is known about this process. In cases where the site of origin has been identified, the nonmalignant columnar epithelium of a small peripheral bronchus blended directly into atypical or malignant epithelium, from which the invasive cancer arose (Lisa et al. 1965; Rosenblatt et al. 1967; Bennett et al. 1969). The available evidence suggests that columnar secretory cells are the progenitor cells for these peripheral tumors.

Small Cell Carcinomas and Carcinoid Tumors

The cells of carcinoid tumors and of some small cell carcinomas (SCC) contain dense-cored granules similar to those seen in normal DCG cells. This has led to a long-held assumption that these tumors derive from DCG cells. However, several investigators are now questioning whether or not all (or any) of these tumors arise from pre-existing DCG cells (Yesner 1980; Gazdar et al. 1981; Sorokin et al. 1981).

An epithelial origin has been demonstrated for SCC, and focal areas of hyperplastic cells, which resemble cells of the invasive tumor, are seen within the basal layers of an otherwise rather normal-appearing mucociliary epithelium (Walter and Pryce 1955; Watson and Berg 1962; Lisa et al. 1965; Rosenblatt et al. 1967; McDowell and Trump 1981). Saccomanno et al. (1970) noted hyperplastic whorling of basally located small cells, below an otherwise normal mucociliary epithelium, in *non-smoking* uranium miners. This lesion predominated in the bronchi of patients who developed SCC. Hyperplastic foci at the base of an otherwise normal mucociliary epithelium contrast sharply with the form of carcinoma in situ that precedes epidermoid carcinoma, in that the latter is characterized by atypical metaplastic cells throughout the epithelial thickness and normal mucociliary differentiation is absent at the lesion (see above).

Carcinoid tumors also appear to arise from abnormal proliferations of cells at the base of the epithelium, overlaid by apparently normal mucociliary epithelium (Kay 1958; Gmelich et al. 1967; Salyer et al. 1975). We described a rather unusual carcinoid tumor which showed multiple continuities with clusters of abnormal DCG cells, situated at the base of bronchial surface and glandular epithelium (McDowell et al. 1981; Sorokin et al. 1981). Some secretory cells and ciliated cells within the clusters also contained putative DCG. In addition to continuities with the underlying tumor mass, some of the epithelial clusters were interconnected but a number of schmall clusters were clearly isolated from one another, as shown by serial sectioning (Sorokin et al. 1981). The population density of the intraepithelial clusters far exceeded the normal distribution of DCG cells in human bronchi and provided testimony favorable to the idea that these clusters, and hence tumors such as this one, may arise from phenotypic changes in pre-existing epithelial cells which do not contain DCG.

Conclusions and Hypothesis

It has been conventional to assume that bronchogenic epidermoid carcinomas arise from basal cells, adenocarcinomas from mucous cells (especially those in the submucosal glands), and SCC and carcinoid tumors from DCG cells. There are many reasons to believe that such a generalization is simplistic and erroneous. For example, tumors grown in rats from clones derived from single cells of combined epidermoid and adenocarcinomas showed phenotypic instability and the nascent tumors expressed different degrees of epidermoid and mucus-secreting specializations in an apparently uncontrolled manner (Steele and Nettesheim 1981). Furthermore, when a transplantable adenocarcinoma of the colon (an organ derived from endoderm, as is TBE) made up of mucous, DCG, columnar and undifferentiated cells, was cloned each of the nascent tumors derived from a single cell was composed of mucous, DCG, columnar, and undifferentiated elements (Cox and Pierce 1982).

There is no definitive evidence that basal cells are stem cells for the TBE. Clearly they are *not* analogous to basal cells in the skin. In hamster trachea, columnar/secretory cells are much more numerous than basal cells and play the major proliferative role during fetal and neonatal development and during epithelial regeneration following mechanical and nutritional injuries. Epidermoid metaplasia induced by vitamin A deficiency (Section: "Epithelial Development and Regeneration Following Injury") and carcinoma in situ and epidermoid carcinomas (Section: "Preneoplastic Lesions and Tumorigenesis") can arise in the submucosal glands, yet basal cells are absent from this anatomic site. In some other epithelial models of carcinogenesis, such as proximal tubules of the kidney, which are made up of simple cuboidal epithelium consisting of specialized epithelial cells without a basal or stem cell population, regeneration and tumorigenesis occur (Dees et al. 1980).

The hypothesis is presented that bronchogenic carcinomas can arise from any normal cell type in TBE that can divide. Such cells include secretory cells, basal cells, and also perhaps DCG cells, although no ability of these last cells to divide has been demonstrated in the adult. Thus, we do not claim that basal cells or DCG cells are incapable of giving rise to some bronchogenic carcinomas. However, on the basis of information gained primarily from studies of fetal development and regeneration of the adult TBE, we are of the opinion that the majority of bronchogenic carcinomas arise from pathologic alterations in the columnar secretory (mucous) cells of the TBE.

References

Auerbach O, Stout AP, Hammond EC, Garfinkel L (1961) Changes in bronchial epithelium in relation to cigarette smoking and in relation to lung cancer. N Engl J Med 265: 253–267
Auerbach O, Stout AP, Hammond EC, Garfinkel L (1967) Multiple primary bronchial carcinomas. Cancer 20: 699–705
Banks-Schlegel S, Schlegel R, Pinkus G (1981) Keratin protein domains within the human epidermis. Exp Cell Res 136: 465–469

Becci PJ, McDowell EM, Trump BF (1978) The respiratory epithelium: IV. Histogenesis of epidermoid metaplasia and carcinoma in situ in the hamster. J Natl Cancer Inst 61: 577–586

Bennett DE, Sasser WF, Ferguson TB (1969) Adenocarcinoma of the lung in men. A clinicopathologic study of 100 cases. Cancer 23: 431–439

Bensch KG, Gordon GB, Miller LR (1965) Studies on the bronchial counterpart of the Kultschitsky (argentaffin) cell and innervation of bronchial glands. J Ultrastruct Res 12: 668–686

Black H, Ackerman LV (1952) The importance of epidermoid carcinoma in situ in the histogenesis of carcinoma of the lung. Ann Surg 136: 44–55

Boren HG, Paradise LJ (1978) Cytokinetics of lung. In: Harris CC (ed) Pathogenesis and therapy of lung cancer. Dekker, New York, pp 369–418

Carter D, Marsh BR, Robinson Baker R, Erozan YS, Frost JK (1976) Relationships of morphology to clinical presentation in ten cases of early squamous cell carcinoma of the lung. Cancer 37: 1389–1396

Chopra DP, Cooney RA (1983) Squamous metaplasia of tracheal submucosal glands induced by benzo(a)pyrene and vitamin A deficiency. Carcinogenesis 4: 1345–1347

Cireli E (1966) Elektronenmikroskopische Analyse der prä- und postnatalen Differenzierung des Epithels der oberen Luftwege der Ratte. Z Zellforsch Mikrosk Anat 74: 132–178

Conen PE, Balis JU (1969) Electron microscopy in study of lung development. In: Emery J (ed) The anatomy of the developing lung. Heinemann, London, pp 18–48

Cox WF, Pierce GB (1982) The endodermal origin of the endocrine cells of an adenocarcinoma of the colon of the rat. Cancer 50: 1530–1538

Cutz E, Conen PE (1972) Endocrine-like cells in human fetal lungs. An electron microscopic study. Anat Rec 173: 115–122

Dees JH, Heatfield BM, Trump BF (1980) Adenocarcinoma of the kidney: IV. Electron microscopic study of the development of renal adenocarcinomas induced in rats by N-(4'-fluoro-4-biphenylyl)acetamide. J Natl Cancer Inst 64: 1547–1562

Emura M, Mohr U (1975) Morphological studies on the development of tracheal epithelium in the Syrian golden hamster: I. Light microscopy. Z Versuchstierkd 17: S 14–26

Evans MJ, Johnson LV, Stephens RJ, Freeman G (1976) Renewal of the terminal bronchiolar epithelium in the rat following exposure to No_2 or O_3. Lab Invest 35: 246–257

Evans MJ, Carbral-Anderson JJ, Freeman G (1978) Role of Clara cells in renewal of the bronchiolar epithelium. Lab Invest 38: 648–655

Gazdar AF, Carney DN, Guccion JC, Baylin SB (1981) Small Cell carcinoma of the lung. Cellular origin and relationship to other pulmonary cancers. In: Greco FA, Oldham RK, Bunn PA (eds) Small cell lung cancer. Grune and Stratton, New York, pp 145–175

Gmelich JT, Bensch KG, Liebow AA (1967) Cells of Kultschitsky type in bronchioles and their relation to the origin of peripheral carcinoid tumor. Lab Invest 17: 88–98

Harris CC, Frank A, Barrett LA, McDowell EM, Trump BF, Paradise LJ, Boren H (1975) Cytokinetics in the respiratory epithelium of hamster, cow and man. J Cell Biol 67: 158 a

Hoffman D, Hecht SS, Ornaf RM, Wynder EL (1976) Chemical studies on tobacco smoke: XXXIX. On the identification of carcinogens, promoters and cocarcinogens. In: Proceedings of the Third World conference on smoking and health. U.S. Government Printing Office, Washington, pp 125–145

Ito M, Tamada J, Aoki M (1982) Induction of squamous cell carcinoma with 3,4-benzo(a)pyrene in the human bronchus transplanted into nude mice. Gann 73: 141–146

Jeffery PK, Ayers M, Rogers D (1982) The mechanisms and control of bronchial mucous cell hyperplasia. In: Chantler EN, Elder JB, Elstein M (eds) Mucus in health and disease: II. Plenum, New York, pp 399–409

Kalnins VI, Porter KR (1969) Centriole replication during ciliogenesis in the chick tracheal epithelium. Z Zellforsch Mikrosk Anat 100: 1–30

Kawanami O, Ferrans VJ, Crystal RG (1979) Anchoring fibrils in the normal canine respiratory system. Am Rev Respir Dis 120: 595–611

Kay S (1958) Histologic and histogenetic observations on the peripheral adenoma of the lung. Arch Pathol 65: 395–402

Keenan KP, Combs JW, McDowell EM (1982a) Regeneration of hamster tracheal epithelium after mechanical injury: I. Focal lesions: Quantitative morphologic study of cell proliferation. Virchows Arch [Cell Pathol] 41: 193–214

Keenan KP, Combs JW, McDowell EM (1982b) Regeneration of hamster tracheal epithelium after mechanical injury: II. Multifocal lesions: Stathmokinetic and autoradiographic studies of cell proliferation. Virchows Arch [Cell Pathol] 41: 215–229

Keenan KP, Combs JW, McDowell EM (1982c) Regeneration of hamster tracheal epithelium after mechanical injury: III. Large and small lesions: Comparative stathmokinetic and single pulse and continous thymidine labeling autoradiographic studies. Virchows Arch [Cell Pathol] 41: 231–252

Keenan KP, Wilson TS, McDowell EM (1983) Regeneration of hamster tracheal epithelium after mechanical injury: IV. Histochemical, immunocytochemical and ultrastructural studies. Virchows Arch [Cell Pathol] 43: 213–240

Klein-Szanto AJP, Topping DC, Heckman CA, Nettesheim P (1980a) Ultrastructural characteristics of carcinogen-induced nondysplastic changes in tracheal epithelium. Am J Pathol 98: 61–82

Klein-Szanto AJP, Topping DC, Heckman CA, Nettesheim P (1980b) Ultrastructural characteristics of carcinogen-induced dysplastic changes in tracheal epithelium. Am J Pathol 98: 83–100

Leeson TS, Leeson CR (1964) A light and electron microscope study of developing respiratory tissue in the rat. J Anat 98: 183–193

Lisa JR, Salvador T, Rosenblatt MB (1965) Site of origin, histogenesis and cytostructure of bronchogenic carcinoma. Am J Clin Pathol 44: 375–384

McDougall J (1978) Endocrine-like cells in the terminal bronchioles and saccules of human fetal lung. An ultrastructural study. Thorax 33: 43–53

McDowell EM, Barrett LA, Glavin F, Harris CC, Trump BF (1978) The respiratory epithelium. I. Human bronchus. J Natl Cancer Inst 61: 539–549

McDowell EM, Sorokin SP, Hoyt RF, Trump BF (1981) An unusual bronchial carcinoid tumor. Light and electron microscopy. Hum Pathol 12: 338–348

McDowell EM, Trump BF (1981) Pulmonary small cell carcinoma showing tripartite differentiation in individual cells. Hum Pathol 12: 286–294

McDowell EM, Harris CC, Trump BF (1982) Histogenesis and morphogenesis of bronchogenic neoplasms. In: Shimosato Y, Melamed MR, Nettesheim P (eds) Morphogenesis of lung cancer, vol 1. CRC Press, Boca Raton, pp 1–36

McDowell EM, Trump BF (1983) Histogenesis of preneoplastic and neoplastic lesions in tracheobronchial epithelium. Surv Synth Pathol Res 2: 235–279

McDowell EM, Keenan KP, Huang M (1984a) Effects of vitamin A-deprivation on hamster tracheal epithelium: A quantitative morphologic study. Virchows Arch [Cell Pathol] 45: 197–219

McDowell EM, Keenan KP, Huang M (1984b) Restoration of mucociliary tracheal epithelium following deprivation of vitamin A: A quantitative morphologic study. Virchows Arch [Cell Pathol] 45: 221–240

Meyrick B (1977) Mucus-producing cells of the tracheobronchial tree. In: Elstein M, Parke DV (eds) Mucus in health and disease. Plenum, New York, pp 61–76

Meyrick B, Reid L (1970) Ultrastructure of cells in the human bronchial submucosal glands. J Anat 107: 281–299

Meyrick B, Sturgess JM, Reid L (1969) A reconstruction of the duct system and secretory tubules of the human bronchial submucosal gland. Thorax 24: 729–736

Nasiell M (1963) The general appearance of the bronchial epithelium in bronchial carcinoma: A histopathological study with some cytological viewpoints. Acta Cytol (Baltimore) 7: 97–106

Nasiell M, Kato H, Auer G, Zetterberg A, Roger V, Karlen L (1978) Cytomorphological grading and Feulgen DNA - analysis of metaplastic and neoplastic bronchial cells. Cancer 41: 1511–1521

Rosenblatt MB, Lisa JR, Collier F (1967) Criteria for the histologic diagnosis of bronchogenic carcinoma. Dis Chest 51: 587–595

Saccomanno G, Saunders RP, Archer VE, Auerbach O, Brennan L (1970) Metaplasia to neoplasia. In: Nettesheim P, Hanna MG, Deatherage JW (eds) Morphology of experimental respiratory carcinogenesis. US Atomic Energy Commission AEC Symposium Series #21, pp 63–80

Saccomanno G, Archer VE, Auerbach O, Saunders RP, Brennan LM (1974) Development of carcinoma of the lung as reflected in exfoliated cells. Cancer 33: 256–270

Said JW, Nash G, Banks-Schlegel S, Sassoon A, Murakami S, Shintaku IP (1983) Keratin in human lung tumors. Patterns of localization of different molecular-weight keratin proteins. Am J Pathol 113: 27–32

Salyer DC, Salyer WR, Eggleston JC (1975) Bronchial carcinoid tumors. Cancer 36: 1522–1537

Schreiber H, Saccommano G, Martin DH, Brennan L (1974) Sequential cytological changes during development of respiratory tract tumors induced in hamsters by benzo(a)pyrene-ferric oxide. Cancer Res 34: 689–698

Sorokin SP, Hoyt RF, McDowell EM (1981) An unusual bronchial carcinoid tumor analyzed by conjunctive staining. Hum Pathol 12: 302–313

Sorokin SP, Hoyt RF, Grant MM (1982) Development of neuroepithelial bodies in fetal rabbit lungs: I. Appearance and functional maturation as demonstrated by high-resolution light microscopy and fromaldehyde-induced fluorescence. Exp Lung Res 3: 237–259

Steele VE, Nettesheim P (1981) Unstable cellular differentiation in adenosquamous cell carcinoma. INCI 67: 149–154

Steinman RM (1968) An electron microscopic study of ciliogenesis in developing epidermis and trachea in the embryo of *Xenopus laevis*. Am J Anat 122: 19–52

Stenbäck F (1973) Morphologic characteristics of experimentally induced lung tumors and their precursors in hamsters. Acta Cytol (Baltimore) 17: 476–486

Tandler B, Sherman J, Boat TF (1981) EDTA-mediated separation of cat tracheal lining epithelium. Am Rev Respir Dis 124: 469–475

Thomas P, Said JW, Nash G, Banks-Schlegel S (1984) Profiles of keratin proteins in basal and squamous cell carcinomas of the skin. An immunohistochemical study. Lab Invest 50: 36–41

Trump BF, McDowell EM, Glavin F, Barrett LA, Becci PJ, Schürch W, Kaiser HE, Harris CC (1978) The respiratory epithelium: III. Histogenesis of epidermoid metaplasia and carcinoma in situ in the human. J Natl Cancer Inst 61: 563–575

Walter JB, Pryce DM (1955) The histology of lung cancer. Thorax 10: 107–116

Watson WL, Berg JW (1962) Oat cell lung cancer. Cancer 15: 759–768

Willis RA (1962) Regeneration and repair: Resumed embryonic growth. In: The borderland of embryology and pathology, 2nd edn Butterworths, Washington, pp 495–518

Wolbach S, Howe P (1925) Tissue changes following deprivation of fat-soluble A vitamin. J Exp Med 42: 753–778

Wynder EL, Hoffman D, Chan P, Reddy B (1975) Interdisciplinary and experimental approaches: Metabolic epidemiology. In: Fraumeni JF Jr (ed) Persons at high risk of cancer. Academic, New York, pp 485–501

Yesner R (1980) Are small-cell carcinomas of the lung derived from neural crest? N Engl J Med 303: 51

Peptide Hormone Production in Primary Lung Tumors*

K. Yamaguchi, K. Abe, I. Adachi, S. Kimura, M. Suzuki, A. Shimada,
T. Kodama, T. Kameya, and Y. Shimosato

Endocrinology Division, National Cancer Center Research Institute, 5-1-1, Tsukiji Chuo-ku,
Tokyo 104, Japan

Introduction

It is well established that the most common nonendocrine tumor associated with peptide hormone production is lung carcinoma. Many peptide hormones have been reported to be produced by lung tumors, and their biochemical and immunological characteristics have been described (Odell 1981; Orth 1981; Abe et al. 1984 a). Recent progress in peptide hormone research has revealed that many previously unknown peptides are present in vivo, and one of these newly discovered peptides, GRP, has been shown to be an important product of primary lung tumors (Moody et al. 1981; Wood et al. 1981; Erisman et al. 1982; Sorenson et al. 1983; Yamaguchi et al. 1983 a). We also found that small cell lung carcinoma can produce GRF (Yamaguchi et al. 1983 b). In the present study, a large number of primary lung tumors were examined for production of 18 peptides, including classic and new peptides.

Materials and Methods

Seven human fetal lungs were obtained from legally aborted fetuses with gestational ages of 16–34 weeks. Thirty-two macroscopically normal adult lungs used were parts of lobectomy or pneumonectomy specimens obtained at the time of operations for primary lung cancers. The 157 specimens of primary lung tumor tissue (44 squamous cell carcinomas, 45 adenocarcinomas, 11 large cell carcinomas, 40 small cell carcinomas and 17 carcinoids) examined were all obtained at surgery, except for those from 17 small cell carcinomas, which were obtained at autopsy. Immediately after removal, the tissue samples were stored at $-80\,°C$ until extraction. Pathological diagnoses were made according to the revised WHO classification.

* This investigation was supported in part by a Grant-in-Aid from the Ministry of Health and Welfare for Comprehensive 10-year strategy for Cancer Control, by Grants-in-Aid for Cancer Research (57-8, 57 S, 58-12, 59-1, 59 S-1, 60 S) and a Research Grant for Intractable Diseases from the Ministry of Health and Welfare, by Grants-in-Aid for Cancer Research from the Ministry of Education, Science and Culture, and by the Special Coordination Fund from the Science and Technology Agency, Japan

Radioimmunoassays

Tissue concentrations of 18 peptides were determined by the appropriate radioimmunoassays (RIAs). These were ACTH, β-MSH, ME-AGL, leumorphin, CT, CGRP, secretin, GRF, VIP, PP, NPY, PYY, GRP, SS, NT, ADH, big gastrin (1–15), and motilin.

The minimum concentration detectable by the individual RIAs varied from 2.9 to 170 fmol/tube. The RIAs used here were highly specific, and there was no crossreaction with the other peptides examined here up to 100 pmol/tube, except for the NPY-RIA, in which PYY and PP showed 5.7% and 0.013% crossreactivity, respectively.

Tissue Extraction

The frozen tissues (0.2 g) were extracted by the boiling water method and lyophilized, as reported previously (Yamaguchi et al. 1983 a). Before assay, the lyophilized samples were reconstituted to 1 ml with 1 N acetic acid. The solution was diluted 20-fold and serially diluted further by adding the assay buffer, and 0.1-ml quantities of each dilution were used for the RIAs. When significant displacement was observed in only one dilution the substance was considered undetectable. When 50- and 500-ng quantities of the synthetic peptides examined in this study were extracted with bovine muscle tissue that had been shown to contain no detectable peptides the average recovery rates were in the range of 65%–100%.

The minimum detectable concentrations of the peptides in tissues, with 0.2 g tissue used in each case as the starting material are shown in Table 1.

Table 1. Minimum tissue peptide concentration detectable in appropriate RIAs

Peptides	Minimum detectable conc. (pmol/g)
GRP	7.0
CT	1.8
SS	1.5
CGRP	5.2
GRF	1.9
NT	0.38
ACTH	2.8
β-MSH	4.7
ME-AGL	6.8
NPY	5.9
VIP	1.9
ADH	0.15
PP	1.5
Big gastrin	8.5
Secretin	3.2
Motilin	7.3
PYY	5.9
Leumorphin	1.8

Data Analysis

To establish a strict criterion for tumor production, only when the concentration of peptide detected was 10 pmol or more per g wet weight was the peptide considered to be produced by the tumor. There are two reasons why the concentration of 10 pmol or more per g wet weight was selected. First, no peptide was detectable at a higher concentration than this in normal adult lungs. This excluded the possibility that the peptide derived from normal lung mixed with the tumor tissue might mistakenly be thought to be of tumor origin. Second, in the RIA used, the minimum detectability varied from 0.15 to 8.5 pmol/g wet weight, and the above criterion covered the detectable range of every peptide examined when 0.2 g tissue was used as the starting material. However, there is one disadvantage: the frequency of peptide production by the tumor is underestimated.

Results

Frequency of Peptide Hormone Production by 157 Primary Lung Tumors

The frequency of production of each peptide by the 157 primary lung tumors is shown in Table 2. GRP was produced in 33 (21%) of the tumors. CT, SS, and CGRP were produced in about 10% of the tumors. GRF, NT, ACTH, ME-AGL, β-MSH, NPY, VIP, ADH, and PP were produced in 2%–7%. In four NPY-producing tumors neither PYY nor PP was detected. On the other hand, big gastrin, secretin, motilin, PYY, and leumorphin were not produced in any of these tumors.

Table 2. Frequency of peptide hormone production by 157 lung tumors

Peptides	Produced by	
	No. of tumors	(%)
GRP	33	(21)
CT	18	(11)
SS	17	(11)
CGRP	15	(10)
GRF	11	(7)
NT	10	(6)
ACTH	8	(5)
ME-AGL	8	(5)
β-MSH	6	(4)
NPY	4	(3)
VIP	4	(3)
ADH	3	(2)
PP	3	(2)
Big gastrin	0	(0)
Secretin	0	(0)
Motilin	0	(0)
PYY	0	(0)
Leumorphin	0	(0)

To determine the relationship between the frequency of peptide production and the type of lung tumor, the data were examined according to tumor type, as shown in Table 3. The data on β-MSH were deleted because this peptide is known to be produced from a common precursor molecule of proACTH. The frequency of production of GRP, CT, or NT was higher in carcinoids and small cell carcinomas than in adenocarcinomas and squamous cell carcinomas. On the other hand, SS and CGRP were produced in the adenocarcinomas with a frequency almost equal to that of the small cell carcinomas. GRF, ME-AGL, NPY, and VIP were produced only in the carcinoids and the small cell carcinomas, ACTH and ADH only in the small cell carcinomas, and PP only in the carcinoids.

The frequency of production of at least 1 of the 17 peptides in normal lung and lung tumors is shown in Table 4. The frequency was 100% in the fetal lungs, in which only GRP was detected. In adult lungs no peptide was detectable at a concentration of 10 pmol or more per g wet weight. In all, 65% of the bronchial carcinoids and 83% of the small cell

Table 3. Frequency (%) of peptide hormone production in various types of lung tumors

Peptides	Type of tumor				
	Carci-noid($n = 17$)[a]	Small cell Ca ($n = 40$)	Adeno-carcinoma ($n = 45$)	Squamous cell Ca ($n = 44$)	Large cell Ca ($n = 11$)
GRP	5 (29)[b]	23 (58)	3 (7)	2 (5)	0 (0)
CT	3 (18)	11 (28)	4 (9)	0 (0)	0 (0)
NT	2 (12)	6 (15)	2 (4)	0 (0)	0 (0)
SS	3 (18)	7 (18)	7 (16)	0 (0)	0 (0)
CGRP	5 (29)	4 (10)	5 (11)	1 (2)	0 (0)
GRF	1 (6)	10 (25)	0 (0)	0 (0)	0 (0)
ME-AGL	4 (24)	4 (10)	0 (0)	0 (0)	0 (0)
NPY	1 (6)	3 (8)	0 (0)	0 (0)	0 (0)
VIP	1 (6)	3 (8)	0 (0)	0 (0)	0 (0)
ACTH	0 (0)	8 (20)	0 (0)	0 (0)	0 (0)
ADH	0 (0)	3 (8)	0 (0)	0 (0)	0 (0)
PP	3 (18)	0 (0)	0 (0)	0 (0)	0 (0)

[a] Number of tumors examined
[b] Number of tumors producing peptides (%)

Table 4. Frequency of production of at least one peptide in normal lung and lung tumor tissues

Tissue	No. examined	No. positive[a] (%)
Fetal lung	7	7 (100)
Adult lung	32	0 (0)
Carcinoid	17	11 (65)
Small cell Ca	40	33 (83)
Adenocarcinoma	45	9 (20)
Squamous cell Ca	44	3 (7)
Large cell Ca	11	0 (0)

[a] At least one peptide produced

carcinomas produced at least one peptide. The frequency was 20% in adenocarcinomas and 7% in squamous cell carcinomas, but none of the peptides were found in large cell carcinomas.

Concentrations of Various Peptides Produced by Tumors

The tissue concentrations of various peptides produced by lung tumors are shown in Fig. 1. Most of the tumors with peptide concentrations of 10 pmol or more per g wet weight are small cell carcinomas or carcinoids, and the concentrations detected are mostly in the range of 10-1000 pmol/g wet weight. However, it is interesting to note that in the cases of GRP, CT, SS, and CGRP some adenocarcinomas produced concentrations roughly equal to those found in small cell carcinomas or carcinoids.

Multiple Hormone Production by Primary Lung Tumors

The frequency of multiple hormone production by lung tumors is summarized in Table 5. It was found that 8 carcinoids, 20 small cell carcinomas, and 6 adenocarcinomas produced two or more peptides at a concentration of 10 pmol or more per g wet weight. These results indicate that about half the carcnoids and the small cell carcinomas were multiple-hormone-producing tumors, and that some adenocarcinomas also produced multiple hormones. Neither the squamous cell carcinomas nor the large cell carcinomas examined here produced multiple hormones.

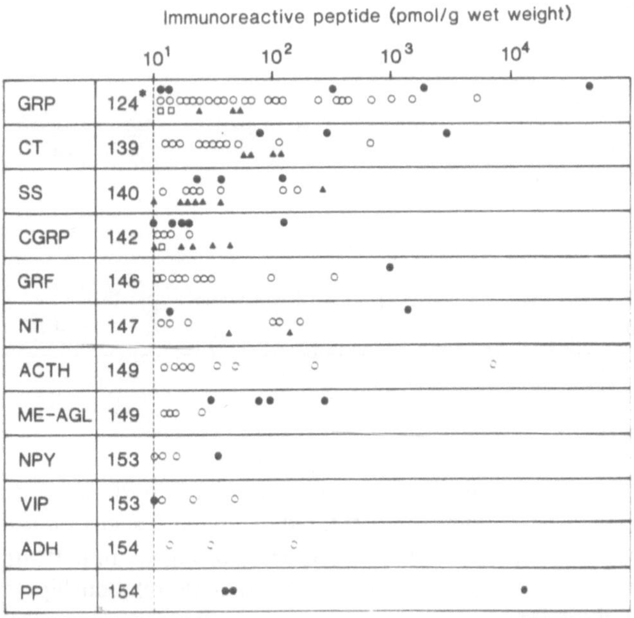

Fig. 1. Tissue concentrations of peptides in 157 primary lung tumors. An *asterisk* indicates the number of tumors with a peptide concentration of less than 10 pmol/g wet weight. *Symbols:* ●, bronchial carcinoid; ○, small cell carcinoma; ▲, adenocarcinoma; □, squamous cell carcinoma

Table 5. Multiple hormone production by primary lung tumors

Type of tumor	No. of peptides produced						
	0	1	2	3	4	5	7
Carcinoid	6[a]	3	2	4	2		
Small cell Ca	7	13	6	5	5	3	1
Adenocarcinoma	36	3	1	4	1		
Squamous cell Ca	41	3					
Large cell Ca	11						

[a] No. of tumors

Table 6. Combination of peptides produced by 20 multiple-hormone-producing small cell carcinomas

Case no.	Peptides produced										
	GRP	CT	ACTH	GRF	SS	CGRP	NT	ME-AGL	VIP	NPY	ADH
1	+	+	+	+	+	+				+	
2	+	+	+	+		+					
3	+	+	+			+					
4	+	+	+	+	+						
5	+	+				+	+	+			
6	+	+		+	+						
7	+	+					+				
8	+	+			+						
9	+	+									
10	+		+	+	+						
11	+		+	+						+	
12	+		+								
13	+			+							+
14	+						+	+			
15	+							+	+		
16	+							+			
17	+								+		
18	+										+
19			+	+				+		+	
20					+		+				

The combinations of peptides produced in 20 cases of multiple-hormone-producing small cell carcinoma are presented in tabular form in Table 6. GRP was produced in 18 tumors. CT, ACTH, GRF, and SS were produced in 6–9 tumors. The other peptides were produced in 4 or fewer tumors. No specific relationship was found among any of these combinations.

In Table 7, the combinations of peptides produced by six multiple-hormone-producing adenocarcinomas are summarized. More than two hormones from among GRP, CT, CGRP, SS, and NT were produced in each of six adenocarcinomas. It is interesting to

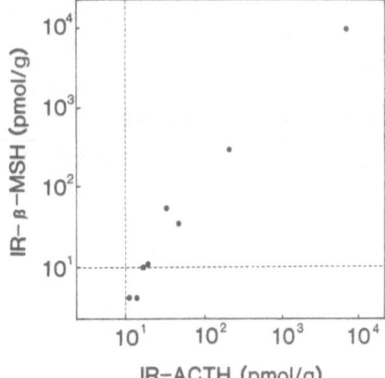

Fig. 2. Immunoreactive (IR)-ACTH and IR-β-MSH determined simultaneously in primary lung tumors

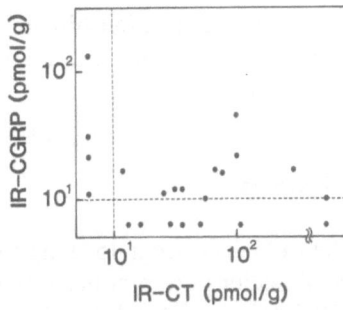

Fig. 3. IR-CT and IR-CGRP determined simultaneously in primary lung tumors

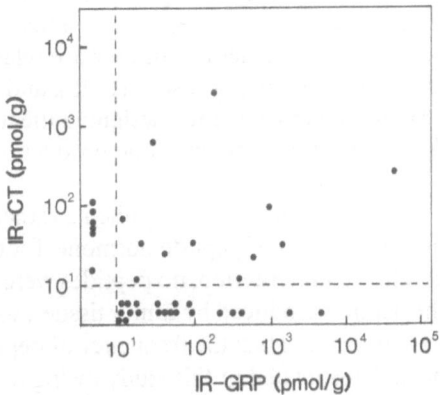

Fig. 4. IR-CT and IR-GRP determined simultaneously in primary lung tumors

note that SS was produced in all these tumors. To determine the relationship between the concentrations of the peptides produced by the tumor, the data were analyzed for the tumors in which at least one peptide was shown to be produced at a concentration of 10 pmol or more per g wet weight. The relationship between ACTH and β-MSH is shown in Fig. 2. A significant correlation was observed between the concentrations of these peptides. The relationship between CT and CGRP is shown in Fig. 3. In seven tumors CT alone was produced, and in four, CGRP alone; but in 11 tumors both these peptides were

Table 7. Combinations of peptides produced by six multiple hormone-producing adenocarcinomas

Case No.	Peptides produced				
	SS	CT	CGRP	GRP	NT
1	+			+	+
2	+			+	
3	+	+	+		+
4	+	+	+		
5	+	+	+		
6	+	+	+		

produced. The relationship between GRP and CT is shown in Fig. 4. In 21 tumors GRP alone was produced, and in six, CT alone. In 12 tumors, however, both these peptides were found.

Discussion

Data obtained from 157 lung tumors show that various peptide hormones in different combinations are produced by primary lung tumors. The production of GRP, CT, SS, GRF, NT, ACTH, β-MSH, VIP, ADH, and PP by primary lung tumors had already been shown, but the production of CGRP, ME-AGL, and NPY by these tumors has not been reported previously. The physiological role of these new peptides is not yet known, and further investigation is required to elucidate how they contribute to clinical features in the patients and whether or not they are related to tumor growth. It is worth noting that leumorphin, secretin, big gastrin (1–15), and motilin were not produced at a concentration of 10 pmol or more per g wet weight in the tumors examined in this study, indicating that the frequency of peptide hormone production by lung tumors is very different for each peptide.

In the present study we applied a rather strict criterion for analyzing the data on the tumor production of peptide hormone, for the reasons given in *Materials and Methods*. According to this criterion, no peptides were produced by normal adult lungs, indicating that the peptide produced by tumor tissue was truly of tumor cell origin. However, it needs to be kept in mind that the frequency of peptide production by primary lung tumors tends to be underestimated in this study owing to this strict criterion. We found that many tumor tissues contained peptides at concentrations lower than 10 pmol/g wet weight, and we regarded these cases as nonproducers.

The data obtained so far indicate that GRP is produced by primary lung tumors with the highest frequency. GRP is stated to have a stimulatory effect on the growth of small cell lung carcinoma cells (Minna 1984), and therefore it is reasonable to assume that some autocrine mechanism plays a role in the proliferation of these lung cancer cells. Although GRP is a very important product of lung cancers, as mentioned above, we have found that some medullary thyroid carcinomas and pancreatic endocrine tumors produce GRP in concentrations almost equal to those found in carcinoids and small cell carcinomas (Abe et al. 1984b; Yamaguchi et al. 1984). These results indicate that the production of GRP is not confined to lung tumors, and further studies are therefore, required to clarify the biological implications of GRP production in tumors other than lung cancers.

In the present study, all the tumors used were clinical samples obtained from patients, and this made it possible to analyze the data from the viewpoint of the relationship between routine pathological diagnosis of lung tumors and their hormone-producing activity. The data obtained so far show that 65% of bronchial carcinoids and 83% of small cell carcinomas produce at least one peptide. These results indicate that peptide hormone production is a very common phenomenon in carcinoids and small cell carcinomas, even though the data were analyzed with reference to a very strict criterion. As far as other types of lung carcinomas are concerned, our results show that non-small cell carcinomas can also produce peptide hormones, and it is interesting to note that CGRP and SS were produced in the adenocarcinomas at a frequency and concentration almost equal to those found in carcinoids and small cell carcinomas. The fact that not only small cell carcinoma and carcinoid, but also non-small cell carcinomas, of the lung can produce peptide hormones suggests that the endocrine nature of the tumor cells does not necessarily depend on morphological differentiation as determined by routine histological methods. Non-small cell lung carcinomas with an endocrine nature should be examined more carefully from the clinical as well as the pathological standpoint.

Another interesting point regarding peptide hormone production by lung tumors is that they frequently produce multiple hormones. Although the actual mechanism of multiple hormone production by the tumor has not been clarified, at least two mechanisms can be proposed. The first is the derivation of two or more peptides from a common precursor, and the second is alternative processing of messenger RNA. The concomitant production of ACTH and β-LPH is caused by the former mechanism (Nakanishi et al. 1979), and in general the concentrations of these peptides derived from a common precursor are in a ratio of almost 1:1. The second mechanism (Amara et al. 1982; Rosenfeld et al. 1983) could explain the concomitant production of CT and CGRP. Although the alternative messenger RNA processing in the CT gene is said to be tissue specific, our data on lung tumors indicate that considerable numbers of lung tumors can produce both these peptides. While these two mechanisms of multiple hormone production by tumors can be proposed, there are still many findings which could not be explained by either of them. As shown in Fig. 4, GRP and CT were produced in many lung carcinomas. We have reported that these two peptides are also concomitantly produced by medullary thyroid carcinomas, and that there is a significant relation between the concentrations of these peptides produced by medullary thyroid carcinomas (Yamaguchi et al. 1984). It is unlikely, however, that these to hormones are derived from a common precursor or generated by alternative processing of messenger RNA, because in medullary thyroid carcinomas the concentrations of CT were always 100–1000 times higher than those of GRP; because in lung tumors there is no correlation between them; and because the structure of the rat CT gene does not contain a coding sequence corresponding to GRP (Amara et al. 1982). Further studies are required to clarify the actual mechanisms of these associations.

Acknowledgments. The authors thank the physicians of the National Cancer Center Hospital for collecting the samples and K. Otsubo, C. Hani-uda, and K. Nagasaki for their excellent technical assistance.

Abbreviations. GRP, gastrin-releasing peptide; *ACTH,* adrenocorticotropic hormone; *β-LPH,* β-lipotropic hormone; *β-MSH,* β-melanocyte stimulating hormone; *ME-AGL,* methionine-enkephalin arginine-glysine-leucine; *CT,* calcitonin; *CGRP,* calcitonin gene-related peptide; *GRF,* growth hormone-releasing factor; *VIP,* vasoactive intestinal peptide; *PP,* pancreatic polypeptide; *NPY,* neuropeptide tyrosine; *PYY,* peptide tyrosine tyrosine; *SS,* somatostatin; *NT,* neurotensin; *ADH,* antidiuretic hormone.

References

Abe K, Kameya T, Yamaguchi K, Kikuchi K, Adachi I, Tanaka M, Kimura S, Kodama T, Shimosato Y, Ishikawa S (1984a) Hormone-producing lung cancers. In: Becker KL, Gazdar AF (eds) The endocrine lung in health and disease. Saunders, Philadelphia, pp 549-595

Abe K, Yamaguchi K, Adachi I, Suzuki M, Kimura S, Shimada A, Maruno K, Yanaihara N (1984b) Production of gastrin-releasing peptide (bombesin) by tumors. In: Endocrinology 1984. Elsevier/North-Holland Biomedical, Amsterdam, pp 77-80

Amara SG, Jonas V, Rosenfeld MG, Ong ES, Evans RM (1982) Alternative RNA processing in calcitonin gene expression generates mRNAs encoding different polypeptide products. Nature 298: 240-244

Erisman MD, Linnoila RI, Hernandez O, DiAugustine RP, Lazarus LH (1982) Human lung small-cell carcinoma contains bombesin. Proc Natl Acad Sci USA 79: 2379-2383

Minna JD (1984) Recent advances of potential clinical importance in the biology of lung cancer. Proceedings of the 75th Annual Meeting, American Association for Cancer Research, 9-12 May 1984, Toronto, pp 393-394 (abstract)

Moody TW, Pert CB, Gazdar AF, Carney DN, Minna JD (1981) High levels of intracellular bombesin characterize human small-cell lung carcinoma. Science 214: 1246-1248

Nakanishi S, Inoue A, Kita T, Nakamura M, Chang ACY, Cohen SN, Numa S (1979) Nucleotide sequence of cloned cDNA for bovine corticotropin-β-lipotropin precursor. Nature 278: 423-427

Odell WD (1981) Humoral manifestations of cancer. In: Williams RH (ed) Textbook of endocrinology, 6th edn. Saunders, Philadelphia, pp 1228-1241

Orth DN (1981) Ectopic hormone production. In: Felig P, Baxter JD, Broadus AE, Frohman LA (eds) Endocrinology and metabolism, McGraw-Hill, New York, pp 1191-1217

Rosenfeld MG, Mermod JJ, Amara SG, Swanson LW, Sawchenko PE, Rivier J, Vale WW, Evans RM (1983) Production of a novel neuropeptide encoded by the calcitonin gene via tissue-specific RNA processing. Nature 304: 129-135

Sorenson GD, Bloom SR, Ghatei MA, Del Prete SA, Cate CC, Pettengill OS (1983) Bombesin production by human small cell carcinoma of the lung. Regul Pept 4: 59-66

Wood SM, Wood JR, Ghatei MA, Lee YC, O'Shaughnessy D, Bloom SR (1981) Bombesin, somatostatin and neurotensin-like immunoreactivity in bronchial carcinoma. J Clin Endocrinol Metab 53: 1310-1312

Yamaguchi K, Abe K, Adachi I, Suzuki M, Kimura S, Kameya T, Yanaihara N (1984) Concomitant production of immunoreactive gastrin-releasing peptide and calcitonin in medullary carcinoma of the thyroid. Metablism 33: 724-727

Yamaguchi K, Abe K, Kameya T, Adachi T, Taguchi S, Otsubo K, Yanaihara N (1983a) Production and molecular size heterogeneity of immunoreactive gastrin-releasing peptide in fetal and adult lungs and primary lung tumors. Cancer Res 43: 3932-3939

Yamaguchi K, Abe K, Suzuki M, Adachi I, Kimura S, Shimada A, Ohno H, Kanai A, Kameya T (1983b) Production of immunoreactive pancreatic growth hormone-releasing factor in small cell carcinoma of the lung. Gann 74: 814-817

Peptide Hormone Production in Lung Cancer Cell Lines of Different Histopathological Types

W. Luster, C. Gropp, H. F. Kern, R. Wahl, H. D. Röher, and K. Havemann

Klinikum der Philipps-Universität, Zentrum für Innere Medizin, Medizinische Klinik und Poliklinik, Schwerpunkt Hämatologie, Onkologie, Immunologie, Baldingerstrasse, 3550 Marburg, FRG

Introduction

In patients with small cell lung cancer a number of peptide hormone-immunoreactive proteins have been demonstrated by radioimmunoassay (Silva et al. 1974; Berson and Yalow 1966; Horai et al. 1973; Ratcliffe et al. 1982). The small cell carcinoma of the lung is the most common non-endocrine tumor for which an association with the production of a variety of hormones and hormone-immunoreactive proteins has been described (Gropp et al. 1980). In some cases lung tumors of this histological type are associated with paraneoplastic syndromes caused by the peptide hormone production of the tumor. On the basis of these and some other biological characteristics of small cell lung tumors it has been suggested, in particular, that these tumor cells might be derived from pulmonary endocrine cells, the Kultschitzky-like cells in the bronchial submucosa (Skrabanek and Powell 1978; Gazdar et al. 1980). In vivo and in vitro results suggest that the secretion of calcitonin and bombesin is specific for small cell lung tumors (Roos et al. 1980; Silva et al. 1979; Moody et al. 1981). Immunohistological demonstrations of calcitonin and other peptide hormones in small cell as well as in non-small cell lung tumor cells by our group were contradory to this concept (Gropp et al. 1981; Luster et al. 1982).

In an attempto answer the question as to whether all lung tumors produce a variety of peptide hormones, 30 permanent and more than 60 primary tumor cell cultures were established from pleural and pericardial exudates or wedge biopsies from human bronchocarcinoma. To investigate the biological function of the peptide hormone synthesis, we examined the hormone production during several passages of long-term cell cultures and the influence of physiological human hormones on proliferation of tumor cell cultures.

Methods

Establishment of Cell Lines

Long-term cell lines were established from small cell, large cell, squamous and adeno lung tumors (WHO classification) (Kreyberg et al. 1967). Small cell lung tumor cell lines were derived from malignant effusions, while cell cultures of the other tumors originated from surgical biopsies (Fig. 1). Most of the cell lines were set up by direct cloning of disintegrated tumor tissue in MEM Dulbecco's or RPMI 1640 medium containing 16.6% fetal calf serum.

Recent Results in Cancer Research. Vol 99
© Springer-Verlag Berlin · Heidelberg 1985

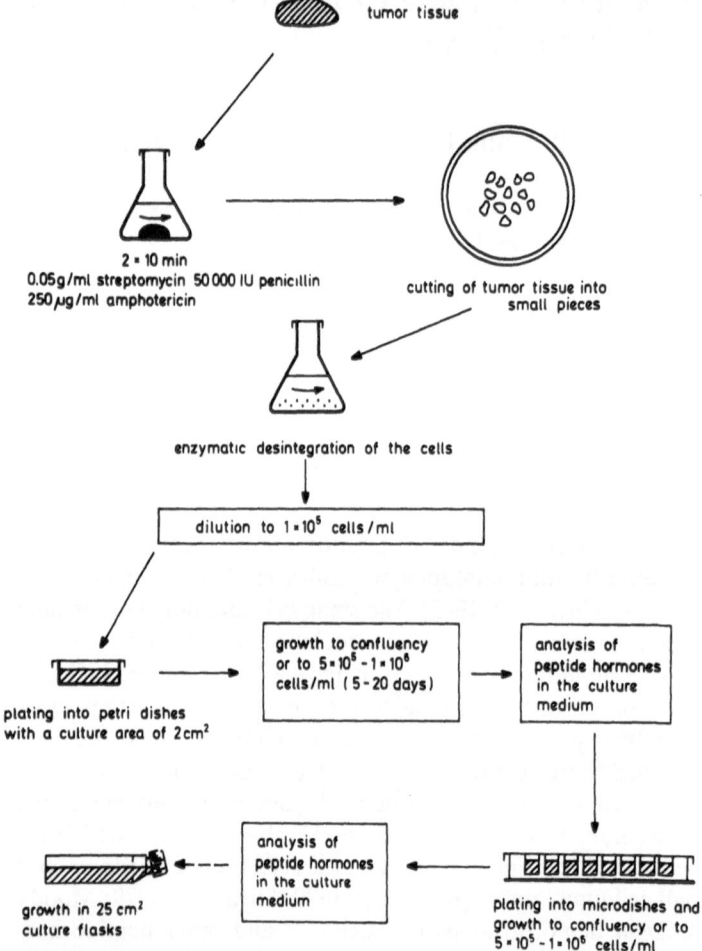

Fig. 1. Schematic documentation of the method of establishing tumor cell lines. Non-small cell lung tumor cell lines were established from surgically obtained fresh tumor tissue. Solid tissue specimens were washed with antibiotics and minced into small pieces. To obtain a cell suspension, the material was disintegrated by collagenase. In the case of an incomplete disintegration incubation in the presence of trypsin followed. After growth of the cells to confluency the culture medium was analyzed for peptide hormone content

Measurement of Peptide Hormones

ACTH, calcitonin, salmon-like calcitonin, bombesin, and neurotensin were determined by commercially available RIA.

Xenotransplantation in Athymic Mice

Cell suspensions from permanent cell lines were injected s.c. to athymic nude mice (NMRI). All handling of mice took place in a laminar flow hood. Mice were maintained in sterile plastic cages in a standard animal room (37 °C, 70% atmospheric humidity). Once tumors reached a size greater than 4 cm³ they were transplanted into new mice or prepared for cell culture and histological examinations (Shimosato et al. 1976).

Determination of Cell Proliferation by Incorporation of Radiolabeled Thymidine

Tumor cells 1×10^5/ml medium were cultured in microdishes. Once they had proliferated to 6 to 8×10^5 cells/ml medium, the cultures were incubated for 16 h in the presence of "stimulation" medium containing one of the following substances: 0.01%, 0.02%, or 0.1% EDTA; 6.2 mM, 12.5 mM, or 25 mM $CaCl_2$; insulin, transferrin, and selenium in concentrations of 5 µg/ml, 5 µg/ml, and 5 ng/ml, 50 µg/ml, 50 µg/ml and 50 ng/ml, or 200 µg/ ml, 200 µg/ml, and 200 ng/ml; 5 µg/ml, 50 µg/ml, or 200 µg/ml insulin; 0.05 ng/ml, 0.125 ng/ml, or 1 ng/ml bombesin; 0.01 ng/ml or 0.5 ng/ml or 1 ng/ml or 2 ng/ml ACTH; 0.1 ng/ml, 0.5 ng/ml, or 1 ng/ml calcitonin; 2.5 IU/ml, 5 IU/ml, 25 IU/ml or 50 IU/ml TSH; and 1 µg/ml, 2 µg/ml, 5 µg/ml or 10 µg/ml acetylcholine. This incubation was followed by addition of 1 µCi thymidine (methyl-^{3}H) to each culture plate and another incubation for 4 h. The cells were collected with an automatic cell harvester and the incorporation of the radiolabeled thymidine into the DNA was determined.

Determination of Cell Proliferation by Soft Agar Assay

Agar 0.5% in cell culture medium was poured into 22-mm dishes and allowed to set. Single cells (3×10^5) were suspended in 0.3% stimulation medium agar solution and added onto the base layer. Isolated colonies were analyzed 3 weeks later.

Cytodiagnostics

Cell suspensions were centrifuged directly on a microscopical slide. Cells were stained according to Pappenheim (Henning 1966) and analyzed with a light microscope (Takahasi 1981). Carbohydrates were demonstrated by the periodic acid-Schiff reaction (Barck 1982), not only in single cells but also in paraffin- or bouin-fixed tumor sections.

Electron Microscopy

Specimens of the original tumor biopsies, from xenografts and from cell lines, were fixed by immersion in 2.5% glutaraldehyde and 2% freshly prepared formaldehyde buffered in 0.1 M cacodylate at pH 7.4. After postfixation in 1% osmium tetroxide and dehydration in a graded series of alcohol, tissue samples were embedded in Epon. From each tumor six to eight blocks were analyzed, first by light microscopy using semithin sections (0.5 µm) stained with 1% Azur II; suitable areas were selected for thin sections, which were stained with 1% uranyl acetate and lead citrate (Reynolds 1963). Sections were examined in a Zeiss EM 9 S electron microscope.

Results and Discussion

The permanent cell lines established so far have their origin in six small cell, five large cell, nine squamous cell, and five adenocarcinomas of the lung (Sorenson et al. 1981; Luster et al. 1983). These cell lines have been stable for 12–24 months. In vivo secretion of ACTH, bombesin, calcitonin, and neurotensin has been demonstrated for lung tumor cells belonging to the four main different histological types (Fig. 2). When all cell cultures established were examined in long-term and short-term cultures, positive ACTH levels were found in 31% of the cell cultures deriving from small cell, 30% from large cell, 24% from squamous cell, and 20% from adenocarcinoma of the lung. Bombesin production was observed in 50% of small cell, 60% of large cell, 63% of squamous, and 46% of the adenocarcinoma cell cultures. Elevated calcitonin levels were demonstrated in culture media of 43% of small cell, 50% of large cell, 20% of squamous, and 39% of adenocarcinoma cell cultures. Neurotensin positivity was determined in 25% of the small cell, 40% of the large cell, 20% of the squamous and 30% of the adenocarcinoma cell cultures. Additionally, we were able to demonstrate salmon calcitonin-like activity in long-term cultures of all four histological cell types (Ziegler 1981).

In 5%–35% of all bronchogenic carcinomas, however, a mixed histology is present. It is not yet clear whether these tumors are stages of conversion from one tumor type to the other or whether originate from two different tumor stem cells. When light microscopical techniques alone are used the lack of specific markers makes it impossible to associate the various types of cells growing in cell culture with distinct types of tumors classified in pathology. The possibility might be considered that the demonstrated synthesis of peptide hormones in vivo derived from some small cells which were already present in the primary tumor or that small cells differentiated to non-small cells in vitro (Gazdar et al. 1981). To exclude these possibilities, a nude mouse can be useful system for the characterization of cultured cells (Shimosato et al. 1976).

Xenografts have been established from permanent cell cultures and compared with the original tumor specimens from which the cell culture was started (Pettengill et al. 1980). Xenografts which were established from small cell lung cancer cell lines grew within a few weeks to solid tumors (Fig. 3). A close histological identity of the xenotransplants with the

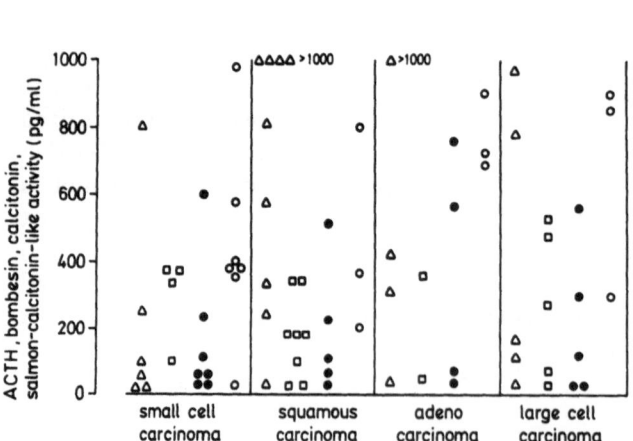

Fig. 2. Peptide hormones secreted into the culture medium by different tumor cell lines

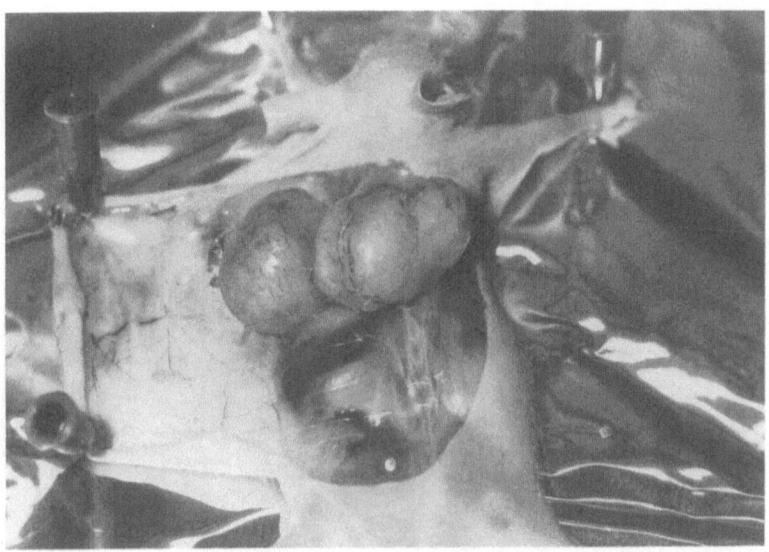

▲ Fig. 3. A xenograft established from a small cell lung cancer cell line 6 weeks after inoculation of 1×10^7 cells SC

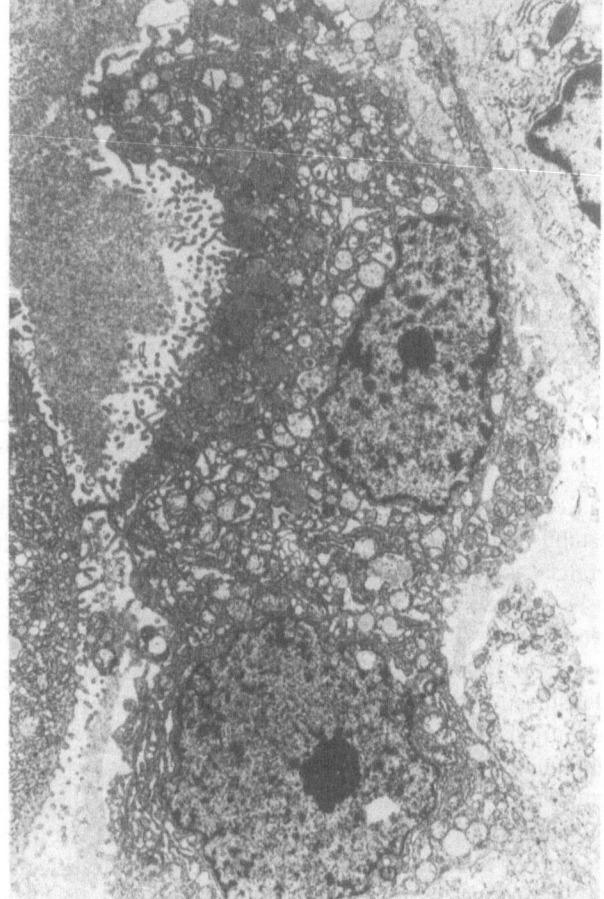

Fig. 4. Adenocarcinoma of the lung after xenotransplantation into nude mice. (2850 ×)

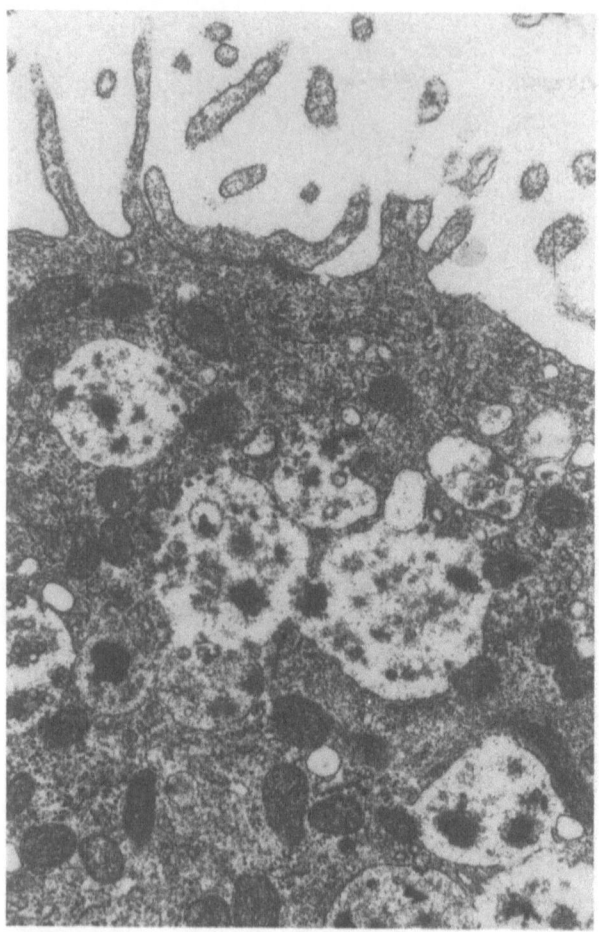

Fig. 5. Adenocarcinoma of
the lung grown in cell culture.
$(12\,000\times)$

primary tumor was demonstrated. This shows that the cell lines which synthesize peptide
hormone-immunoreactive proteins consist of cells belonging to the histological type of
the original tumor.

Representative fine structural results obtained in each type of non-small cell tumor as
established in xenografts and in cell culture show the histological identity of the xenografts established from cell culture and the primary tumor. The cells of a xenograft established from a peptide hormone producing long-term cell culture derived from adenocarcinoma cells grew as acini or tubules composed of epithelial cells containing abundant
rough endoplasmatic reticulum, elaborate Golgi complexes, and numerous electron-
translucent secretory granules (Fig. 4). At the light microscopical level these granules were
identified as mucin granules. On their luminal surface the tumor cells projected numerous
microvilli into the tubular lumen and revealed signs of exocytotic discharge of mucin
granules. Most of their characteristics, such as surface microvilli, rough endoplasmatic
reticulum, and mucin granules, were also preserved in cell culture (Fig. 5). Xenografts of
squamous cell carcinoma grew in the typical stratified pattern, with intracellular bridges
and occasional signs of keratinization (Fig. 6). In tissue culture the tumor cells grew partly
in epithelial contact, but formation of desmosomes and keratinization were minimal or

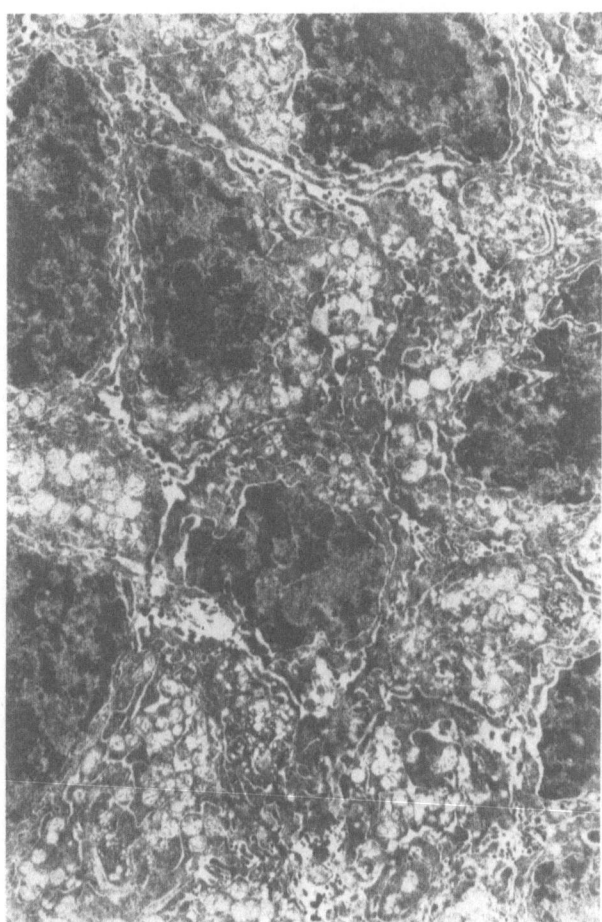

Fig. 6. Squamous cell carcinoma of the lung after xenotransplantation into nude mice. (2850×)

completely absent. These tumor cells contained predominantly free ribosomes, occasional profiles of rough endoplasmatic reticulum, and few dense-core vesicles (Fig. 7). The large cell carcinoma was characterized by tumor cells with large prominent nucleoli and a dense network of intermediate filaments in the cytoplasm. These features and the compact growth pattern of the tumor cells were preserved after xenotransplantation (Fig. 8). The data demonstrate the preservation of most tumor-type-specific structural criteria in the ex vivo systems.

Moreover, it has been shown that tumor cells not only from small cell carcinoma but also from other histological types are capable of synthesizing a broad spectrum of peptide hormone-immunoreactive proteins (Table 1). The wide difference observed in vivo between hormone production in small cell and in non-small cell bronchial carcinoma, which is in contrast to our in vitro results, may be explained by the different proliferation behavior of small cell and non-small cell lung tumors in vivo and in vitro. The non-small lung tumor cells in culture exhibited a higher proliferation than cells from small cell lung carcinoma. In contrast, the clinical course of the small cell lung tumor is characterized by the highest growth rate of all lung tumor types. Xenotransplants established from slow-proliferating small cell lung tumor cell lines again show a comparably fast development to a

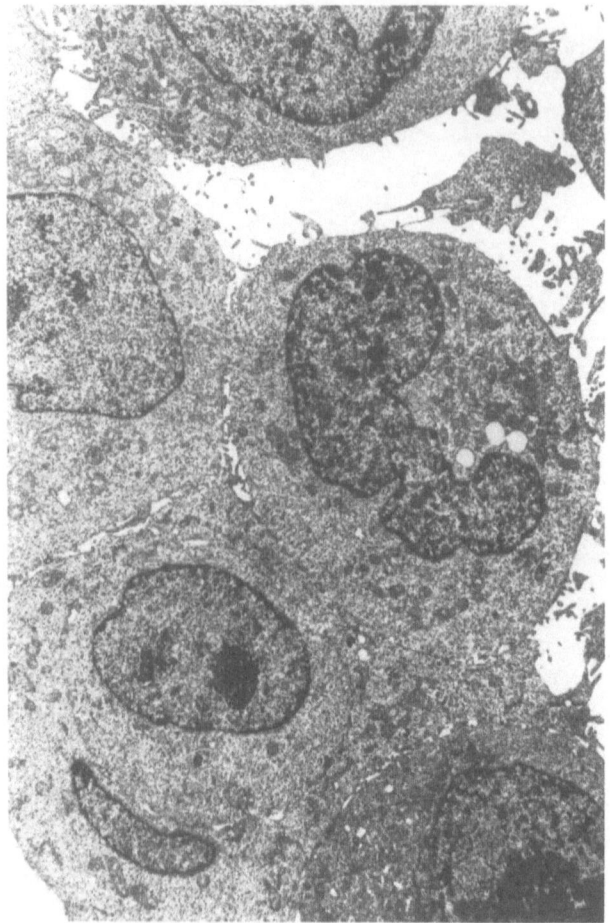

Fig. 7. Squamous cell carcinoma of the lung after xenotransplantation into nude mice. (2850 ×)

solid tumor. Another reason for the observed in vivo and in vitro differences of the hormone concentrations at the periphery of the tumor may be the variable degradation of the hormones. Bombesin, for instance, and some other small peptides are characterized by a very short half-life in the body.

To find whether the multiple hormone production is the result of the potency of one tumor cell or whether different tumor cells synthesize various hormones, *monoclonal* tumor cell lines were established (Fig. 9) and analyzed for their peptide hormone production.

Fig. 9. Establishment of monoclonal tumor cell lines: A 2- to 3-mm layer of 0.5% agar in MEM-Dul- ▶ becco's cell culture medium containing 16.6% fetal calf serum was poured into petri dishes with a gas-permeable membrane bottom. After hardening, another 2- to 3-mm layer of a 0.3% agar solution followed. When the agar had been cooled to 37° C, 1 ml small cell lung tumor cell suspension with 1×10^4 cells was added to the soft agar. Petri dishes containing not only single cells but also cell clusters were discarded. The single cells in soft agar were incubated for 3–8 weeks at 37° C in an atmosphere of 5% CO_2. Cell clones were transferred into microplates and cultivated until confluency or until the growth of 5×10^5 to 5×10^6 cell per ml. Peptide hormone content of the culture medium of the monoclonal cell cultures was determined by radioimmunoassay

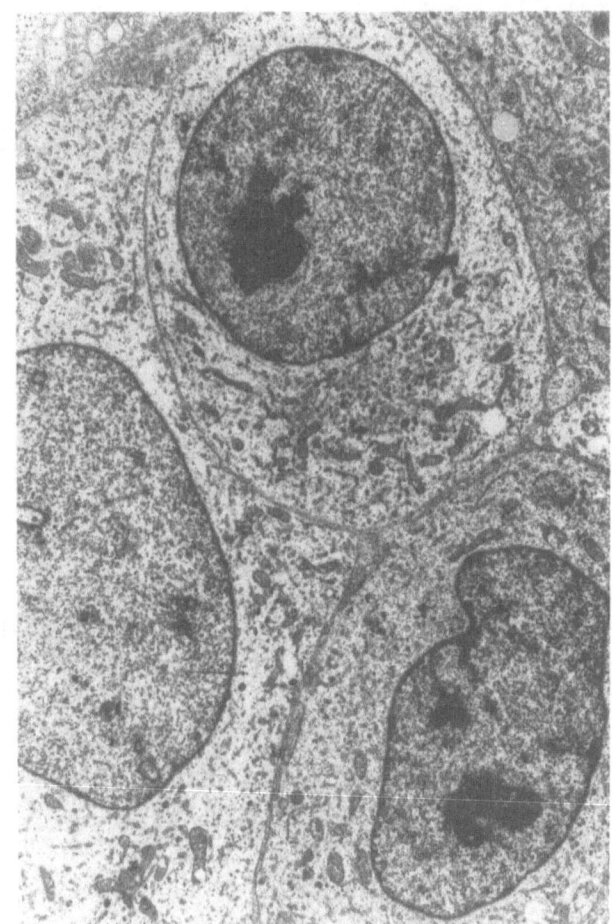

Fig. 8. Large cell carcinoma of the lung after xenotransplantation into nude mice. (2850 ×)

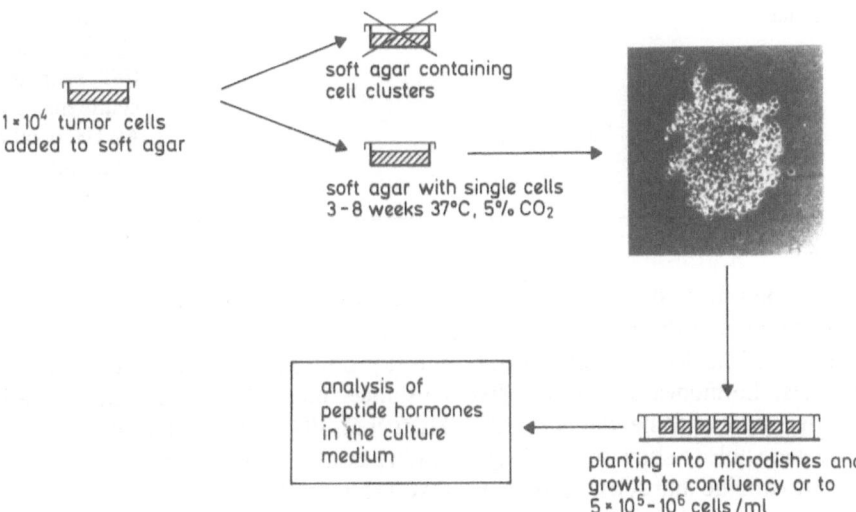

1×10^4 tumor cells added to soft agar

soft agar containing cell clusters

soft agar with single cells 3–8 weeks 37°C, 5% CO_2

analysis of peptide hormones in the culture medium

planting into microdishes and growth to confluency or to 5×10^5–10^6 cells/ml

Table 1. Characteristics of permanent lung tumor cell lines

Lung cancer cell line		Tumor growth, mouse 1×10^7 cells SC	Doubling time (days) in culture	Doubling time (days) in mouse	Hormones produced
Small cell carcinoma	MR-22	14	4	2– 7	Bombesin, calcitonin, salmon-like calcitonin, neurotensin, estriol
	MR-55		5		ACTH, bombesin, calcitonin, salmon-like calcitonin, neurotensin
	MR-86	7	4	1– 2	–
	MR-103	14–28	3–5	3– 5	Bombesin, calcitonin, salmon-like calcitonin, neurotensin
Squamous carcinoma	MR-9		1.5		ACTH, bombesin, neurotensin
	MR-25	80	1	20	ACTH, bombesin, calcitonin, neurotensin, substance P
	MR-32	70	1.5	6–12	Bombesin, calcitonin, salmon-like calcitonon, substance P
	MR-65	56	1	4– 7	Bombesin, calcitonin, salmon-like calcitonin, neurotensin
	MR-90	100	1	7–10	–
Adeno-carcinoma	MR-5	140	1–3	20	ACTH, bombesin, salmon-like calcitonin, β-lipotropin
	MR-13	./.	./.	6	ACTH, bombesin, calcitonin, salmon-like calcitonin, neurotensin
Large cell carcinoma	MR-8	35–42	3	6–10	ACTH, bombesin, calcitonin, salmon-like calcitonin, CSF, β-lipotropin, neurotensin, estriol
	MR-97	56–70	1	7–18	ACTH, bombesin, salmon-like calcitonin, neurotensin

These experiments showed that *one* tumor cell was able to synthesize all hormones found in the medium of the primary tumor cell culture. The next step was the search for an explanation of the described universal hormone production by lung tumor cells. We suggest that these hormones or hormone-like substances may play an important role in tumor growth regulation (Sherwin et al. 1981; Ham et al. 1981; Luster et al. 1984).

We were able to show by a soft agar cloning assay and by the incorporation of radiolabeled thymidine that substances such as ACTH, calcitonin, bombesin, and ITS influence the growth behavior of tumor cell colonies. Moreover, modification of the proliferation

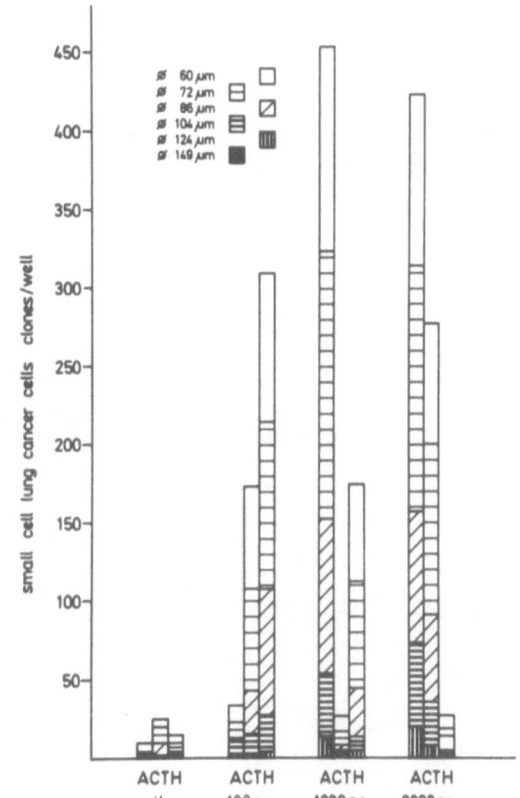

Fig. 10. Influence of ACTH on the proliferation of small cell lung tumor cell lines, tested in the soft agar colony assay

rate of the cell cultures was achieved by incubation of the cells in the presence of peptide hormones and other biological active substances. Cell lines from different tumors were influenced in a different manner by various substances (Fig. 9–11).

In conclusion, it can be said that the importance of the universal characteristic of hormone production by lung tumor cells of all four histological types in the autocrine or paracrine regulation of growth or differentiation of these tumors (Roth et al. 1982) is established.

References

Barck HC (1982) Histologische Technik. Thieme, New York, p 148
Berson SA, Yalow RS (1966) Parathyroid hormone in plasma in adenomatous hyperparathyroidism, uremia and bronchogenic carcinoma. Science 153: 907–909
Gazdar AF, Carney DN, Russell EK, et al (1980) Establishment of continuous clonable cultures of small cell carcinoma of the lung which have amine precursor uptake and decarboxylation cell properties. Cancer Res 50: 3502–3507
Gazdar AF, Carney DN, Guccion JG, et al (1981) Small cell carcinoma of the lung: cellular origin and relationship to other pulmonary tumors. In: Greco A, Bunn PA, Oldham R (eds) Small cell lung cancer. Grune and Stratton, New York, pp 145–175
Gropp C, Havemann K, Scheuer A (1980) Ectopic hormones in lung cancer patients at diagnosis and during therapy. Cancer 46: 347–354

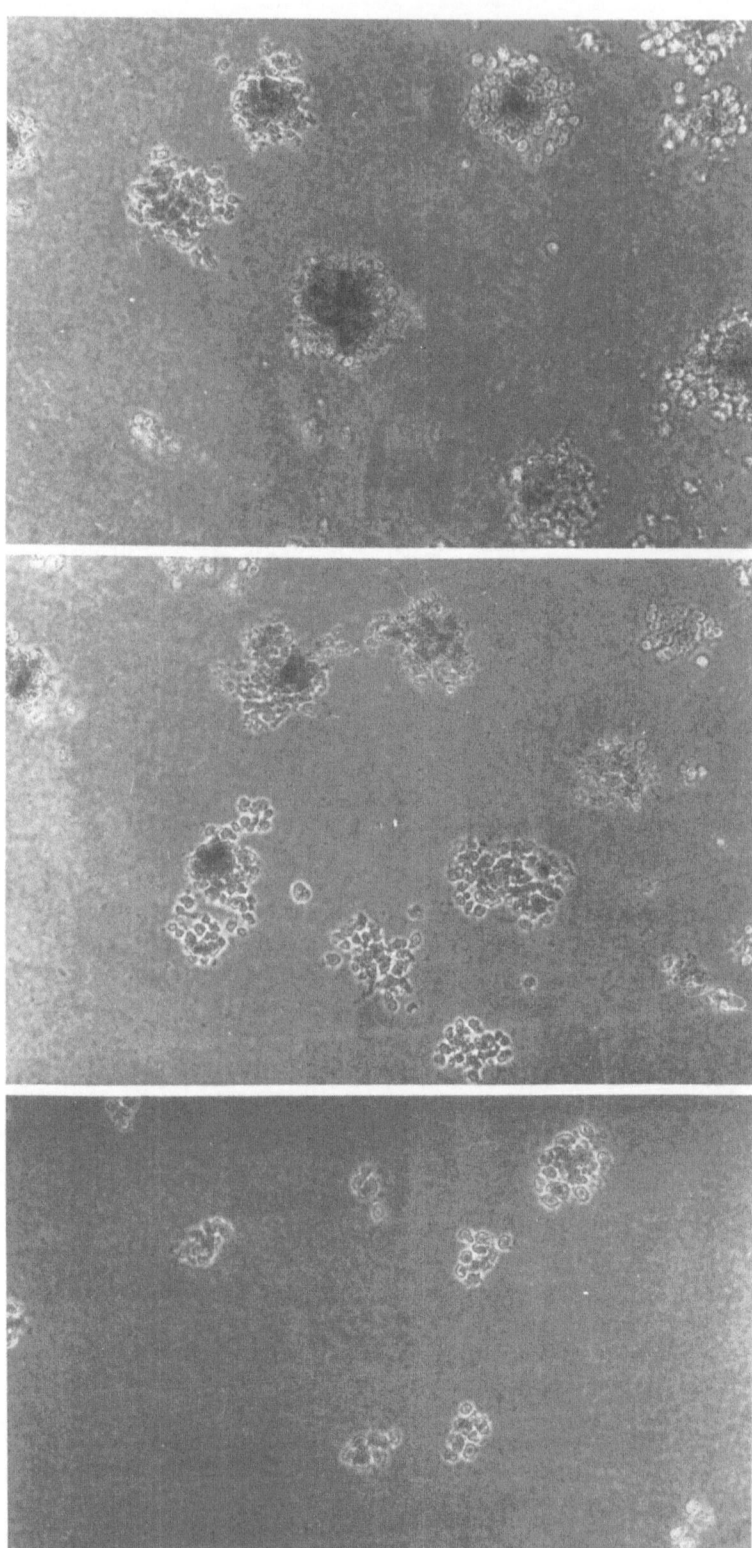

Gropp C, Sostmann H, Luster W, Kalbfleisch H, Lehmann FG, Havemann K (1981) ACTH, β-Lipotropin, β-Endorphin, β-HCG, calcitonin and CEA in lung tumor tissues. In: Uhlenbruck G, Wintzer G (eds) CEA und andere Tumormarker. Tumor-Diagnostik, Leonberg, pp 217–226

Ham J, Ellison ML, Lumsden J (1980) Tumor calcitonin: Interaction with specific calcitonin receptor. Biochem J 190: 545–550

Henning N (1966) Klinische Laboratoriumsdiagnostik. Urban and Schwarzenberg, Munich, pp 149

Horai T, Nishihara H, Tateishi R, Matsuda M, Hattori S (1973) Oat-cell carcinoma of the lung simultaneously producing ACTH and serotonin. J Clin Endocrinol Metab 27: 212–219

Kreyberg L, Liebow AA, Uehlinger EA (1967) Histological typing of lung tumors. World Health Organization, Geneva

Luster W, Gropp C, Sostmann H, Kalbfleisch H, Havemann K (1982) Demonstration of immunoreactive calcitonin in sera and tissues of lung cancer patients. Eur J Cancer Clin Oncol 18: 1275–1283

Luster W, Gropp C, Havemann K (1983) Peptide hormone synthesizing lung tumor cell lines: Establishment and first characterization of biosynthetic products. Acta Endocrinol (Copenh) [Suppl] 253: 24–25

Luster W, Gropp C, Loeck MR, Havemann K (1984) Modification of tumor cell proliferation and peptide hormone secretion. In: Peeters H (ed) Protides of the biological fluids. XXXI. annual colloquium. Pergamon, London, pp 727–730

Moody TW, Pert CB, Gazdar AF, et al (1981) High levels of intracellular bombesin characterize human small cell lung carcinoma. Science 214: 1246–1248

Pettengill OS, Curphey TJ, Cate CC, Flint CF, Maurer LH, Sorenson GD (1980) Animal model for small cell carcinoma of the lung; effect of immunosuppression and sex of mouse on tumor growth in athymic nude mice. Exp Cell Biol 48: 279–297

Ratcliffe JG, Podmore J, Stack BHR, Spilg WGS, Gropp C (1982) Circulating ACTH and related peptide in lung cancer. Br J Cancer 45: 230

Reynolds ES (1963) The use of lead citrate at high pH as an electron opaque stain in electron microscopy. J Cell Biol 17: 208–212

Roth J, LeRoith D, Shiloach J, et al (1982) The evolutionary origins by hormones, neurotransmitters, and other extracellular chemical messengers. N Engl J Med 306: 523–527

Roos BA, Lindall AW, Baylin SB, O'Neill JA, Frelinger AL, Birnbaum RS, Lambert PW (1980) Plasma immunoreactive calcitonin in lung cancer. J Clin Endocrinol Metab 50: 659–666

Sherwin SA, Minna JD, Gazdar AF, Todaro GJ (1981) Expression of epidermal and nerve growth factor receptors and soft agar growth factor production by human lung cancer cells. Cancer Res 41: 3538–3542

Shimosato Y, Kayema T, Tvagi K, et al (1976) Transplantation of human tumors into nude mice. J Natl Cancer Inst 56: 1251–1260

Silva OL, Becker KL, Primack A, Doppman I, Snider RH (1974) Ectopic secretion of calcitonin by oat-cell carcinoma. N Engl J Med 290: 1122–1124

Silva OL, Broder LE, Doppman JL, Snider RH, Moore CF, Cohen MH, Becker KL (1979) Calcitonin as a marker for bronchogenic cancer. Cancer 44: 680–684

Skrabanek P, Powell D (1978) Unifying concept of non-pituitary ACTH secreting tumors. Evidence of common origin of neutral chest tumors, carcinoids, and oat-cell carcinomas. Cancer 42: 1263–1269

Sorenson GD, Pettengill OS, Brinck-Johnson T, Cate CC, Maurer LH (1981) Hormone production by cultures of small cell carcinoma of the lung. Cancer 47: 1289–1296

Takahasi M (1981) Color atlas of cancer cytology. Thieme, Stuttgart

Ziegler R (1981) Calcitonin. Sandoz, Basle (Kurzmonographie no. 27)

◄ **Fig. 11.** Small cell lung tumor cell colonies in soft agar: *above* 2000 pg/ACTH; *middle* 1000 pg/ACTH; *below* control with no hormone additive

The Endocrine Lung Tumor Cell: Morphological Aspects

Morphological Growth Characteristics and Hormone Secretion of Small Cell Carcinoma of the Lung In Vitro

O. S. Pettengill, D. H. Wurster-Hill, C. C. Cate, and G. D. Sorenson

Department of Pathology, Dartmouth Medical School, Hanover, NH 03756, USA

Introduction

Since 1974 a group at Dartmouth Medical School has been actively engaged in in vitro studies of small cell carcinoma of the lung (SCCL), utilizing tumors which have been obtained from patients with a known diagnosis of this disease. Continuous cell lines have been established from biopsy or autopsy specimens from patients with a known pathologic or cytologic diagnosis of SCCL.

Prior to 1974, the first continuous SCCL cell line to be isolated was described by Oboshi et al. (1971). Since then, and concurrently with our own efforts, additional cell lines have been described in Japan (O'Hara and Okamoto 1977), England (Ellison et al. 1976), USA (Fisher and Paulson 1978; Luk et al. 1981), and Sweden (Bergh et al. 1982), and a major group of cell lines from the laboratories of Gazdar, Carney, and Minna, at the National Cancer Insitute in the United States. Their work is described in more detail elsewhere in this volume (see chapters by Carney et al. and Gazdar et al.).

The properties of the Dartmouth (DMS) cell lines have been characterized with regard to growth properties (Pettengill et al. 1977; 1980b; 1984a; 1984b; Pettengill and Sorenson 1981), hormone secretion (Bertagna et al. 1978a, 1978b; 1979; Brinck-Johnsen et al. 1979; Sorenson et al. 1982; Sorenson and Pettengill 1980; Szabo et al. 1980), factors which regulate hormone secretion (Sorenson et al. 1983; 1984b), ultrastructural properties (Sorenson and Pettengill 1980), membrane antigens (Bergh et al. 1983; Ball et al. 1984), biogenic amine content (Pettengill et al. 1982), membrane electrical activity (McCann et al. 1981; 1984), karyotype (Wurster-Hill et al. 1984), and growth in vivo when inoculated in nude athymic mide (Pettengill et al. 1980a; Sorenson et al. 1981b; Chambers et al. 1981; Cate et al. 1984).

In this presentation we have accumulated data collected under similar conditions for five cell lines, to relate their hormone secretory activity to morphological correlates in the in vitro milieu. We have looked for evidence of alteration, i. e., gain or loss of hormone secretory activity and its relationship to morphological characteristics over a long period (3–5 years) of in vitro growth.

Materials and Methods

Cell Lines

Continuous cell lines were isolated from biopsy or autopsy specimens taken from patients. Cell lines DMS 44, DMS 53, DMS 79, DMS 153, and DMS 235 have been examined for these changes. All but DMS 79 has been propagated in Waymouth's MB 752/1 medium with 20% added heat-inactivated fetal calf serum. DMS 79 has been propagated in RPMI 1640 with similar serum additive. Cells have been routinely plated at 5×10^5 cells/ml, medium changed twice weekly, and subcultured using 0.25% trypsin containing 0.02% EDTA at intervals ranging from 10 days to 3 weeks, depending on the cell line. DMS 79 cells do not require trypsinization, and are counted and resuspended at the appropriate density.

Liquid Nitrogen Storage

Approximately 2×10^6 cells in 1 ml growth medium containing 10% DMSO, in plastic Nunc vials, are frozen at $1°$ C/min in a Linde controlled-rate freezer and stored in liquid nitrogen. When thawed, cells are brought to $37°$ C as rapidly as possible, diluted slowly into 20 ml growth medium, and allowed to sit for 20 min to allow DMSO to diffuse out of the cells. Cells are centrifuged and placed in fresh complete growth medium for incubation.

Hormone Assays

Radioimmunoassays for ACTH, BN, and CT have been utilized. ACTH has been assayed as described by Orth (1979), utilizing his antisera. Bombesin (BN) has been assayed in the laboratory of S. Bloom, Hammersmith Hospital, London, as previously described (Bloom and Long 1982; Sorenson et al. 1983), and CT was assayed as previously described (Sorenson et al. 1983).

Results and Discussion

Since the cell of origin of SCCL tumors has yet to be identified and mixed tumors occur in vivo, we were interested in determining the degree of heterogeneity in a given cell line both at a given time, by isolation of clones, but also over a span of months or years of in vitro growth of individual cell lines. Most of the data to be presented here have been accumulated from control data of a variety of experiments. The cell lines have not been regularly tested or monitored to document changes which may occur in subsequent passages of in vitro culture. We have been particularly observant of alterations which would be characteristic of variant cell lines as reported by Carney et al. and Gazdar et al. (this volume) from their laboratory, in which secretion of some peptides and other APUD properties associated with SCCL tumor cells are lost along with changes in growth pattern and growth rate.

A change in growth morphology in vitro is the most direct means by which cell lines can be monitored in the laboratory to provide clues for other physiological and biochemical alterations, which may then be examined. We have described three varieties of growth patterns among the DMS SCCL cultures (Pettengill et al. 1980b; Pettengill and Sorenson 1981).

1. Attached cell population. This is most pronounced in the DMS 53 cell line, which is growing as an attached population and has continued to do so for 5 years or more of in vitro culture.
2. Attached and suspended cell lines. In these cultures, cells attach to the substrate after trypsinization and spread out to some degree. This is different from cells which only stick to the substrate but do not spread out. As cell density increases, clumped areas of the monolayer detach and continue to grow as clumps in the medium.
3. Cells which grow only as clumps in suspension, but show no indication of attaching to the substrate. DMS 79 is an example of this type of growth.

The cultures have appeared to be stable in their growth pattern in vitro, and only after several years were any changes observed. Alterations in growth patterns, karyotype and hormone secretion appear to be associated with the storage of cell lines in liquid nitrogen, and have been observed when cells have been thawed and placed into culture again.

The term "in vitro age" has been used to describe the number of months a given cell line has been in culture. It is defined as the length of time a cell line has been propagated in vitro, excluding intervals of liquid nitrogen storage. For example, DMS 53, first isolated in 1975, has been cultured and stored in liquid nitrogen, and the highest passage level in vitro has an in vitro age of over 60 months. The subculture or passage level number has been used to indicate the use of trypsin, but some cell lines, notably DMS 79, have been passaged more frequently without the aid of trypsin. Intervals between passage by trypsinization vary widely for different cell lines. In this discussion, only the term in vitro age will be used, and not passage level, to indicate the length of in vitro growth of a cell line.

The morphological growth pattern history of DMS 235 is tabulated in Table 1, where it is seen that there is a gradual trend from an attached cell line to a suspension culture with successive freezes in liquid nitrogen. In this case, there is variation in the ability of the cells to attach in two instances following freezing in liquid nitrogen. But in the most recent return to in vitro culture, at 28 months, the cells are completely in suspension after three pas-

Table 1. In vitro characteristics of DMS 235 with time in culture

In vitro age (mo)	Liquid N_2 freezes	Growth pattern	ACTH[a] (pg/10^6 cells)	LH[a] (IU/µg protein)	CT[a] (pg/µg)	BN[a] (fmol/ml)	Chromosome[b] mode(s)
1–30	0	Attached, pavement-like epithelium	13.5	8.9	192	58	44/80 (17 mo)
16–25	1	Attached					
22–30	2	Did not attach immediately			15	100	45–46 (25 mo) 45–47 (26 mo) 44–46 (27 mo)
16–23	1	Did not attach					
18–23	2	Attached					
28–	3	Never attached					

[a] *ACTH*, adrenocorticotropin; *LH*, luteinizing hormone; *CT*, calcitonin; *BN*, bombesin
[b] Wurster-Hill et al. (1984)

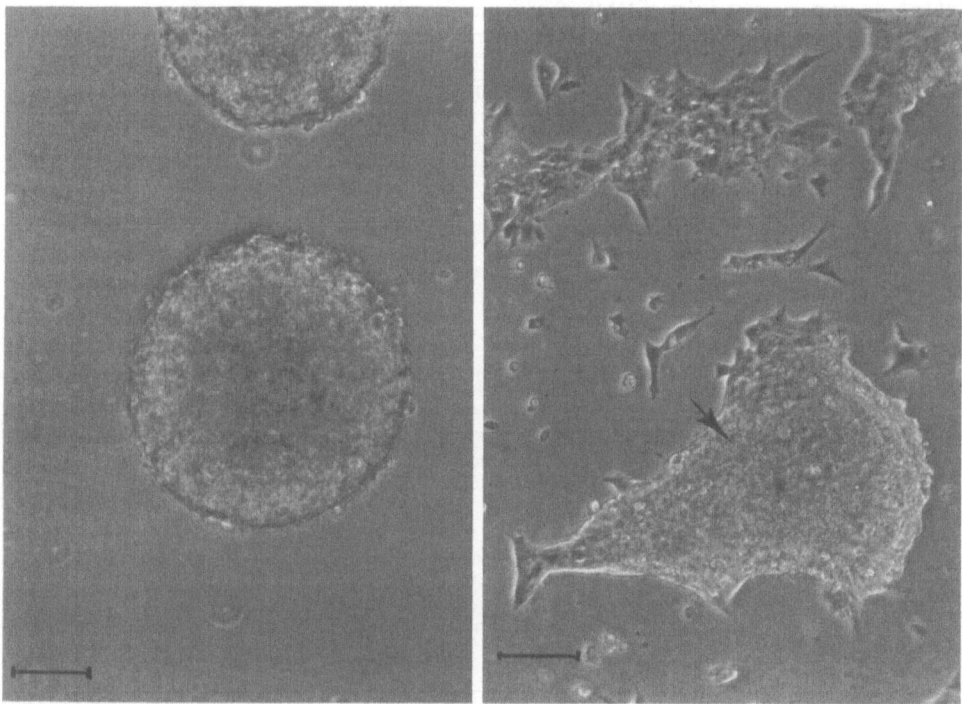

Fig. 1 *(left)*. Phase micrograph of culture of DMS 44$_{34}$ cells, which are growing in tight clumps in suspension. *Bar* = 100 μm

Fig. 2 *(right)*. DMS 153 at 65 months of in vitro age. The culture attaches to the substrate when subcultured, but as cell density increases areas *(arrow)* detach from the substrate and grow as clumps in suspension. *Bar* = 100 μm

sages through liquid nitrogen. This is in contrast to the original culture, which was propagated continuously for 30 months and remained an attached cell line throughout this time.

DMS 44 grew as an attached population for 16 months under continuous culture (Pettengill et al. 1977). When a passage at a similar age level was retrieved from liquid nitrogen it was not as firmly attached to the substrate. Clumps of cells may have stuck to the plastic, but there was little of the spreading of attached cells that had been observed in the earlier culture. In a third retrieval from liquid nitrogen, after three cycles of freezing, a culture grew (Fig. 1) which had no tendency to stick to the culture flask, but instead grew as tight clumps of cells in suspension. This is similar to the characteristic, typical pattern of growth as described by Gazdar et al. (1980) and by Baylin's Group for SCCL in vitro (Luk et al. 1981). These two cell lines (DMS 44 and DMS 235) have exhibited altered growth patterns. In DMS 235, there is circumstantial evidence that this may be related to storage in liquid nitrogen. In the other, DMS 44, the alteration was observed after liquid nitrogen storage, but there was not an uninterrupted culture of the same age to serve as a control. This remains to be tested specifically, by freezing and thawing a cell line while simultaneously maintaining its duplicate in continuous culture.

An attempt has been made to identify alterations which may have occurred in those cell lines which secrete high levels of hormone and which have been used extensively in exper-

Table 2. In vitro characteristics of DMS 79 with time in culture

In vitro age[a]	Doubling time (days)	ACTH (pg/μg protein)	BN (pg/mg protein)	Chromosome mode(s)[b]
4		8.5		
6.5				63–66/76–80**
$\underline{7}$				40–42**/76–80
$\underline{8}$		27.5	TLTM***	40–41**/76
$\underline{12}$				37–42**/74
$\underline{13}$				39–46**/78–81
$\underline{15}$				37–40**/77–78
$\underline{16}$		10.22		
$\underline{17}$			799	
$\underline{21}$	2.5	7.5		
$\underline{26}$		13.9		
$\underline{29}$		24.5		
$\underline{31}$	3.9			
$\underline{\underline{46}}$	5.8	31.65		
$\underline{\underline{49}}$		34.0		
$\underline{\underline{60}}$		8.9		
$\underline{\underline{66}}$		18.4		
$\underline{\underline{67}}$			1627	

[a] The number of lines under a number indicating in vitro age shows how many times the particular culture had been frozen in liquid nitrogen
[b] Wurster-Hill et al. (1984)
** Predominant mode
*** Too low to measure

imentation. These are DMS 53, which secretes nanogram quantities of CT and of 11 other peptides, including significant quantities of BN, and DMS 79, which secretes picogram quantities of ACTH. DMS 153, a moderate producer of ACTH, BN, and CT has also been used. DMS 406, a producer of nanomolar amounts of BN, has been used extensively, but sufficient retrospective data for this more recently developed cell line are not yet available for comparison.

The changes in ACTH secretion, doubling time, and chromosomal modal numbers observed in DMS 79 cells during 5 years, propagation in vitro are summarized in Table 2. The lower chromosomal modal number of this bimodal cell line drops from approximately 64 to 41 after one passage through liquid nitrogen and remains at this level during subsequent months. The upper mode was predominant prior to freezing and the lower mode became prominent following freezing (Wurster-Hill et al. 1984). No changes in growth pattern have been observed, but there is a possible increase in ACTH secretion associated in time with this chromosomal alteration.

The retrospective data for DMS 53 have been summarized in Table 3. This is a relatively stable picture with respect to growth morphology, karyotype, and hormone secretion, where there is no significant change over a long period of time. Levels of BN and CT secreted into the medium do fluctuate. The chromosomal modal numbers of three spearate cell lineages, all of which had been frozen prior to chromosome analysis, are listed in this table, to illustrate the level of variation in chromosomal alteration.

Table 3. In vitro eharacteristics of DMS 53 with time in culture

In vitro age[a]	Doubling time (days)	CT (pg/µg protein)	ACTH (pg/µg protein)	BN (pg/mg protein)	SRIF[b] (ng/ml)	Chromo-some (1)	modal (2)	number[a] (3)
16				942				
20						60–69		
26	4.0							
27		63.6						
31					8.8			
32				522				
36	5.7							48–57
37		31.9						
38					22.3			
39				1155				
41							50–58	
42								54–58
43							52/61	
44.5	3.0							
45			0.44	616		59		
49		37.7						
54	4.9		0.024					
59			0.94					
64					7.8			
66				1745				

[a] The number of lines under a number indicating in vitro, age shows how many times the particular culture had been frozen in liquid nitrogen
[b] *SRIF*, Somatostatin
[c] Wurster-Hill et al. (1984)

DMS 153 cells have been used as they secrete moderate amounts of three peptides, ACTH, BN, and CT. Although the levels of peptides measured during the 1 year of in vitro growth were low, since that time, the growth morphology and hormone secretion have remained constant through two passages in liquid nitrogen with minor fluctuations, and the chromosome mode has dropped slightly (Table 4). The growth morphology illustrated in Fig. 2 is very similar to that of the original cell line; cells attach and form colonies, which detach from the substrate and grow in suspension as cell density increases.
Some of these cell lines have been analyzed karyotypically at several in vitro ages, and the change in modal number of chromosomes associated with liquid nitrogen storage has been documented (Wurster-Hill et al. 1984). It was found in some cases that cells at higher poloidy (i. e., triploid to tetraploid) did not survive as well as cells with a paradiploid chromosome mode when thawed from liquid nitrogen and returned to in vitro conditions. In addition, a gradual but marked shift to lower ploidy levels of the percentage of cells in each chromosomal mode was observed in successive karyotypic analysis. This alteration appears to correlate with growth morphology in the cell lines with which it has been possible to make the comparison. There may also be changes in the level of hormone secreted, but it does not appear to be the same for all peptides. It has previously been shown that bombesin levels may increase with time in vitro (Sorenson et al. 1982), whereas other hormone levels may decrease.

Table 4. In vitro characteristics of DMS 153 with time in culture

In vitro age[a]	Doubling time	ACTH (pg/µg protein)	BN (pg/mg protein)	CT (pg/µg protein)	Chromosome modal number[b]
2					41–44
12		0.076		None detectable	
17		0.966		2.80	
23		0.683		1.81	
30				+	
38			2147		
42	5.7	0.487			
43			2146	0.73	
46			2076		
47					39
48					39–40
49				2.8	
50			5254		
51	5.9				37–41
65				1.5	

[a] The number of lines under a number indicating in vitro age shows how many times the particular culture had been frozen in liquid nitrogen
[b] Wurster-Hill et al. (1984)

A number of other factors present in in vitro conditions may be responsible for inducing such changes in cultured cells. We have examined a few of them. The cell lines DMS 53 and DMS 79 have been used in numerous experiments. In these instances, where these cell lines were used for specific experimental studies, considerable variation was observed in baseline secretion of ACTH by DMS 79 cells, and BN or CT secretion by DMS 53 cells. The range of CT secretion is approximately five-fold in these data; however, CT secretion by DMS 53 cells has remained at a level three times greater than that observed in any other cell line and ten-fold higher than in most DMS SCCL cell lines. Similarly, DMS 79 retains its ten-fold greater level of secretion of ACTH compared with other DMS cell lines.

The variation in CT and BN secretion by DMS 53 cells, where the hormones were measured under standardized culture conditions, is illustrated in Fig. 3. These represent the control values in a large series of experiments, where the hormone secretions were measured over 2-h periods. In another experiment, ACTH secretion in DMS 79 cells was measured twice weekly over a 3-week period (Fig. 4). The cells were adjusted to 5×10^5 cells/ml at weekly intervals. (These times are indicated by the arrows on the ordinate of this graph.) DMS 79 cells do not require trypsinization before counting in a hemacytometer, and the adjustment of cells to 5×10^5 cells/ml was accomplished by resuspending a pellet of counted cells in fresh medium. A lag period during which cells reattach to a substrate is not present in this cell line, as these cells grow in suspension. From these data, where hormone secretion is expressed per microgram of protein as a measure of cell number, it is apparent that more hormone is secreted when cell density is higher, i.e., approximately 10^6 cells/ml at day 4 compared with 2×10^6 cells/ml at day 7.

The effects of fetal calf serum from two manufacturers, newborn calf serum (NBCS) from two manufacturers, and horse serum (HoS) in various concentrations and combina-

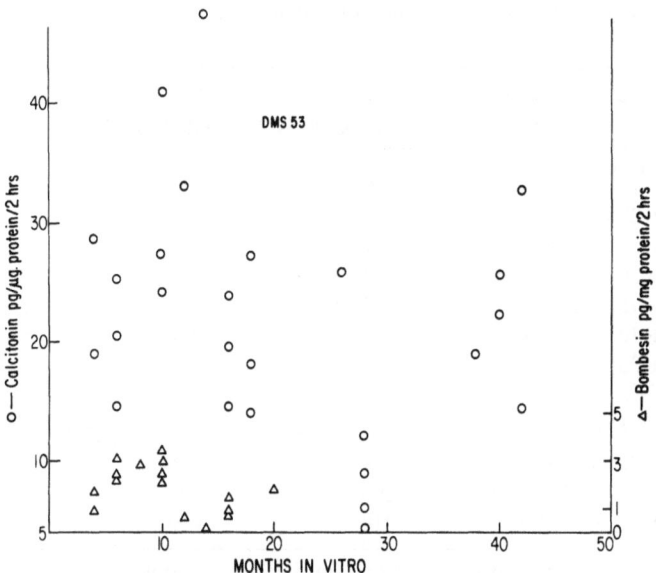

Fig.3. Variation in CT and BN secretion by DMS 53 cells over a period of 50 months in culture

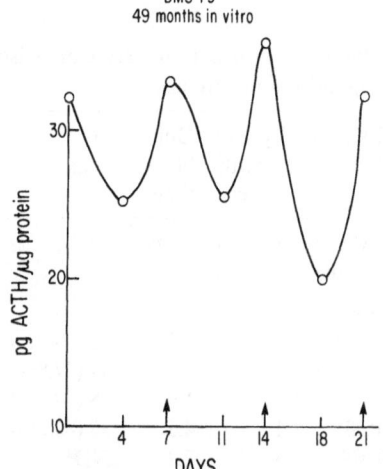

Fig.4. Variation in ACTH secretion by DMS 79 cells over a period of 21 days. At 0, 7, 14, and 21 days, cell density was adjusted to 5×10^5 cells/ml. Culture medium was completely changed 4 days later, at which time it was harvested for ACTH RIA. Cells were washed and assayed for protein. Duplicate flasks were harvested at 7 days, before adjustment of cell number

tions on the secretion of ACTH and CT from two cell lines, DMS 53 and DMS 79 are tabulated in Table 5. DMS 79 is routinely cultured in RPMI 1640 with 20% heat-inactivated fetal calf serum, and DMS 53 is routinely cultured in Waymouth's MB 752/1 medium with 20% heat-inactivated fetal calf serum (Table 5).

The major point to be made on the basis of these data is that there is again wide variation in the secretion of these hormones, depending on the composition of the medium. These data reflect an apparent stimulation of ACTH secretion by DMS 79 and DMS 53 cells in the presence of NBCS. In the absence of complete characterization of the various serum lots for factors which they may contain which stimulate secretion, one cannot draw any conclusions.

Table 5. Effect of serum source and concentration on hormone secretion by DMS 79[a] and DMS 53[b] cells

Serum source and concentration[c]	CT pg/µg protein	ACTH pg/µg protein
DMS 79		
R 20 FCS (R 511)	0.207	21.9
R 20 FCS (SS)	0.945	20.3
R 20 NBCS (GB)	1.14	59.0
R 15 FCS	1.11	15.0
R 10 HoS/5 FCS	1.27	23.6
R 20 NBCS (KC)	–	71.7
DMS 53		
W 20 NBCS (KC)	170	1.24
W 20 FCS (R 511)	290	0.94
W 20 FCS (SS)	310	0.59

[a] Frozen and thawed three times; in vitro age 48 months
[b] Frozen and thawed once; in vitro age 56 months
[c] R 20 FCS, RPMI 1640 with 20% fetal calf serum; W 20, Waymouth's MB 752/1 medium with 20% fetal calf serum; letters and numbers in parenthesis are codes for various manufacturers or serum lot numbers

Table 6. In vitro data for six clones isolated from and compared with their parent, DMS 53, at 45 months of in vitro age

DMS 53 clone	Type of growth pattern	Dou-bling time	G_1 peak	C.V.[b]	G_2 peak	C.V.	% Cells in S (calc'd)	DDC[c] (units/mg protein)	CT (pg/µg protein)
DMS 53 (parent)	1	4 days	29 (77.3%)	4.9	63 (10.1%)	4.3	12.6	203	37.7
Clone									
29/E 5	3	2.2	39 (78.9%)	5.3	74 (7.4%)	3.9	13.7	272	78.9
29/E 6	3	1.8	34 (67.8%)	5.6	68 (4.2%)	4.2	20.9	455	35.5
29/F 4	2	1.9	38 (84.0%)	4.9	69 (6.3%)	3.6	9.7	432	21.8
30/B 9	1	4.5	35 (83.2%)	5.4	69 (6.3%)	4.3	10.5	416	24.1
30 F 9	3	2.2	ND[a]	ND	ND	132	48.4		
30/H 11	1	3.0	35 (81.0%)	5.0	72 (8.6%)	3.9	10.4	299	27.6

[a] Not done
[b] Coefficient of variation
[c] Dopa-decarboxylase activity

In addition to these effects, another source of variation lies in the phenotypic composition of the cell line itself. A series of 19 clones has been isolated from DMS 53 cells at 45 months of in vitro age (Table 6). Because it was apparent that some clones were closely similar on the basis of preliminary examination, only six were selected for more complete examination. The selection was based on their in vitro growth patterns and their initial rate of growth. In a preliminary assay, all appeared to be secreting CT. Three different growth patterns were observed, which are illustrated in Fig. 5.

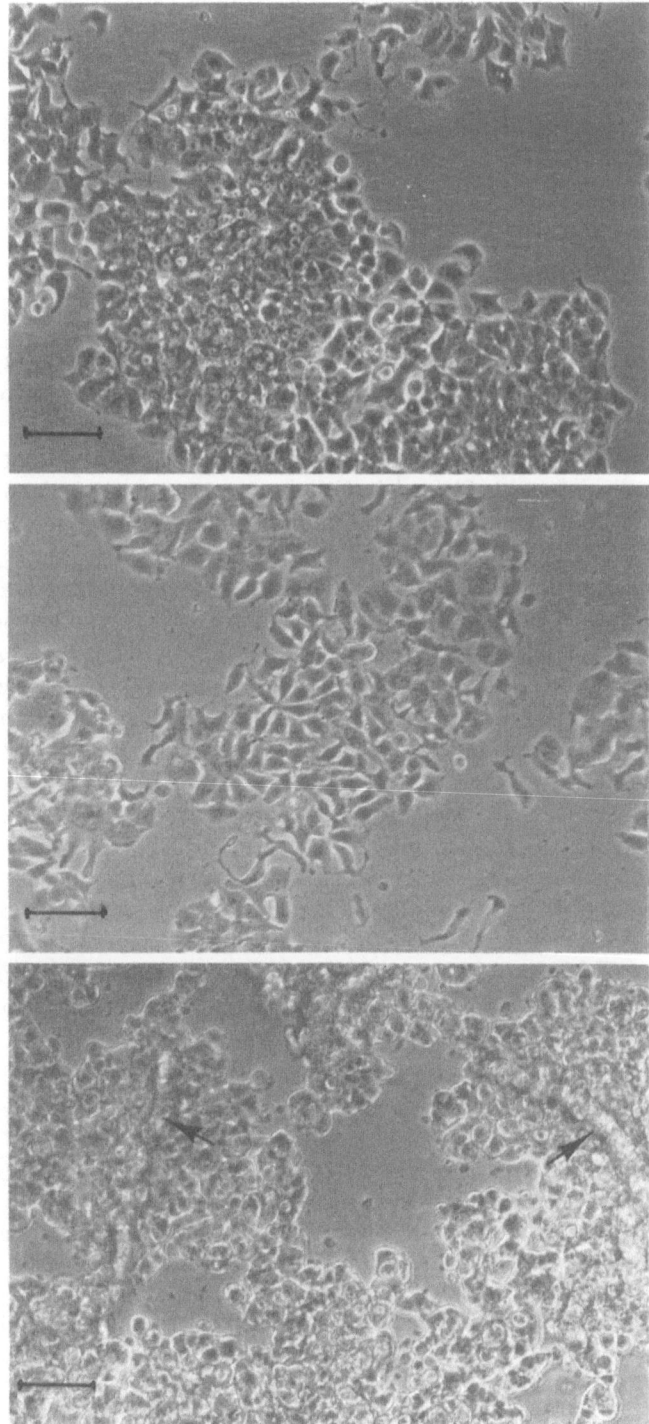

Fig. 5. In vitro growth patterns observed in DMS 53 clones. *Above:* Type 1 growth pattern (described in text); *middle:* Type 2 growth pattern; *below:* Type 3 growth pattern (*arrows* indicate clumped cell pattern)

Table 7. In vitro data for clones isolated from DMS 79 at 66 months of in vitro age

Clone	ACTH (pg/µg protein)	CT (pg/µg protein)
Parent DMS 79	18.4	0.415
Cline C7	29.9	0.415
D3	23.2	0.476
H5	39.6	0.381

Type 1 is very similar to DMS 53 parent cells. There is a very closely packed monolayer, displaying small amounts of process formation at the periphery of each colony (Fig. 5, above).

Type 2 has a much flatter morphology and is more pleomorphic, containing areas suggestive of rosette formation. Cellular processes are prominent (Fig. 5, middle).

Type 3 is characterized by colonies which, although attached and not floating, do not flatten and exhibit almost no cellular processes. When these populations become denser, ridges of clumped cells appear in the central part of the colony. They show a tendency to become floating spheres when the culture reaches high cell density (Fig. 5, below).

The doubling times of the clones vary from 1.9 to 4.5 days, as against 4 days for the parent culture (Table 6). No karyotypic data are available for these clones, but their relative ploidy levels have been assessed by flow cytometry using the method of Vindelor et al. (1983). There is minimal evidence for a difference in ploidy between clones 29/E5 and 29/E6, where the G1 and G2 peaks vary slightly more than 5%. At the time the clones were isolated, the parent population exhibited only one modal chromosome range. CT secretion varies six-fold among the clones, and dopa decarboxylase (DDC) varies three-fold among the clones. All levels of DDC are higher than that seen in DMS 53, while CT production ranges above and below that of the parent DMS 53 cells.

A smaller series of clones has been isolated from DMS 79 cells, at 66 months of vitro age (Table 7). Although it appears that the clones may be producing more ACTH than the parent cells, this is probably within the limits of normal fluctuation in this cell line. CT secretion is closely similar to the parent cell line.

References

Ball ED, Graziano RF, Pettengill OS, Sorenson GD (1984) Monoclonal antibodies reactive with small cell carcinoma of the lung. JNCI 72: 593–598

Bergh J, Larsson E, Zech L, Nilsson K (1982) Establishment and characterization of two neoplastic cell lines (U-1285 and U-1568) derived from small cell carcinoma of the lung. Acta Pathol Microbiol Immunol Scand [A] 90: 149

Bergh J, Nilsson K, Cate CC, Pettengill OS, Sorenson GD (1983) Cell surface glycoprotein patterns of cell lines derived from human small cell carcinoma of the lung. Acta Pathol Microbiol Immunol Scand [A] 91: 91–95

Bertagna XY, Nicholson WE, Pettengill OS, Sorenson GD, Mount CD, Orth DN (1978) Extopic production of high molecular weight calcitonin and corticotropin by human small cell carcinoma cells in tissue culture: evidence for separate precursors. J Clin Endocrinol Metab 47: 1390–1393

Bertagna XY, Nicholson WE, Sorenson GD, Pettengill OS, Mount CD, Orth DN (1978) Corticotropin, lipotropin, and B-endorphin production by a human non-pituitary tumor in culture: evidence for a common precursor. Proc Natl Acad Sci USA 75: 5160–5164

Bertagna XY, Nicholson WE, Tanaka K, Mount CD, Sorenson GD, Pettengill OS, Orth DN (1979) Ectopic production of ACTH, lipotropin ind B-endorphin by human cancer cells. Structurally related markers. Recent Results Cancer Res 67: 16–25

Bloom SR, Long RG (1982) Radioimmunoassay of gut regulatory peptides. Saunders, London

Brinck-Johnsen T, Pettengill OS, Brinck-Johnsen K, Sorenson GD, Maurer LH (1979) Estrogen production by an established cell line from pulmonary small cell anaplastic carcinoma. J Steroid Biochem 10: 330–340

Cate CC, Douple EB, Andrews KM, Pettengill OS, Curphey TJ, Sorenson GD, Maurer LN (1984) Calcitonin as an indicator of the response of human small cell carcinoma of the lung cells to drug and radiation. Cancer Res 44: 949–954

Chambers WF, Pettengill OS, Sorenson GD (1981) Intracranial growth of pulmonary small cell carcinoma cells in nude athymic mice. Exp Cell Biol 49: 90–97

Ellison ML, Hillyard CJ, Bloomfield GA, Rees LH, Coombes RC, Neville AM (1976) Ectopic hormone production by bronchial carcinomas in culture. Clin Endocrinology [Suppl] 5: 397s

Fisher ER, Paulson JD (1978) A new in vitro cell line established from human large cell variant of lung cell cancer. Cancer Res 38: 3830

Gazdar AF, Carney DN, Russell EK, Sims HL, Baylin SB, Bunn PA Jr, Guccion JG, Minna JD (1980) Establishment of continuous, clonable cultures of small-cell carcinoma of the lung which have amine precursor uptake and decarboxylation cell properties. Cancer Res 40: 3502–3507

Luk GD, Goodwin G, Marton LJ, Baylin SB (1981) Polyamines are necessary for the survival of human small cell carcinoma in culture. Proc Natl Acad Sci USA 78: 2355–2358

McCann FV, Pettengill OS, Cole JJ, Russell JA, Sorenson GD (1981) Calcium spike electrogenesis and other electrical activity in continuously cultured small cell carcinoma of the lung. Science 212: 1155–1157

McCann FV, Cole JJ, Pettengill OS, Russell JAG, Sorenson GD (1985) Electrical properties of human pulmonary endocrine-like cells: small cell carcinoma of the lung. In: Poisner A, Trifaro J (eds) Electrophysiology of the secretory cell, chap 10. Elsevier, Amsterdam (in press)

Oboshi S, Tsugawa S, Seido T, Shimosato Y, Koide T, Ishikawa S (1971) A new floating cell line derived from human pulmonary carcinoma of oat cell type. GANN 62: 505–516

O'hara H, Okamoto T (1977) A new in vitro cell line established from human oat cell carcinoma of the lung. Cancer Res 37: 3088–3095

Orth DN (1979) Adrenocorticotropic hormone. In: Jaffe BM, Behrman H (eds) Methods of hormone radioimmunoassay. Academic, New York, pp 245–284

Pettengill OS, Sorenson GD (1981) Tissue culture and in vitro characteristics. In: Greco FA, Oldham RK, Bunn PA (eds) Small cell lung cancer. Grune and Stratton, New York, pp 51–77

Pettengill OS, Faulkner CS, Wurster-Hill DH, Maurer LH, Sorenson GD, Robinson AG, Zimmerman EA (1977) Isolation and characterization of a hormone producing cell line from human small cell anaplastic carcinoma of the lung. J Natl Cancer Inst 58: 511–518

Pettengill OS, Curphey TJ, Cate CC, Flint CF, Maurer LH, Sorenson GD (1980a) Animal model for small cell carcinoma of the lung: effect of immunosuppression and sex of mouse on tumor growth in nude athymic mice. Exp Cell Biol 48: 279–297

Pettengill OS, Sorenson GD, Wurster-Hill DH, Curphey TJ, Noll WW, Cate CC, Maurer LH (1980b) Isolation and growth characteristics of continuous cell lines from small cell carcinoma of the lung. Cancer 45: 906–918

Pettengill OS, Bacopoulos NG, Sorenson GD (1982) Biogenic amine metabolites in human lung tumor cells: Histochemical and mass-spectrographic demonstration. Life Sci 30: 1355–1360

Pettengill OS, Carney DN, Sorenson GD, Gazdar AF (1984a) Establishment and characterization of cell lines from human small cell carcinoma of the lung: a comparative study from two institutions. In: Becker AF, Gazdar AF (eds) The endocrine lung. Saunders, Philadelphia, pp 460–468

Pettengill OS, Cate CC, Flint CF, Sorenson GD (1984b) In vitro isolation of malignant cells from small cell carcinomas. In: Prelow TG, Pretlow TP (eds) Cell separation: methods and applications, vol IV. Academic, New York, pp 123–137

Sorenson GD, Pettengill OS (1980) Structural and fuctional characteristics of small cell carcinoma of the lung, in vitro and in vivo. Biologie Cellulaire 39: 277–280

Sorenson GD, Pettengill OS, Brinck-Johnsen T, Cate CC, Maurer LH (1981 a) Hormone production by cultures of small cell carcinoma of the lung. Cancer 47: 1289–1296

Sorenson GD, Pettengill OS, Cate CC (1981 b) Studies of xenografts of small cell carcinoma of the lung. In: Greco FA, Oldham RK, Bunn PA (eds) Small cell lung cancer. Grune and Stratton, New York, pp 95–121

Sorenson GD, Bloom SR, Ghatei MS, Cate CC, Pettengill OS (1982) Bombesin production by small cell carcinoma of the lung. Regul Pept 4: 59–66

Sorenson GD, Pettengill OS, Cate CC, Ghatei MA, Molyneux KE, Gosselin EJ, Bloom SR (1983) Bombesin and calcitonin secretion by pulmonary carcinoma is modulated by cholinergic receptors. Life Sci 33: 1939–1944

Sorenson GD, Pettengill OS, Cate CC, del Prete SA (1984 a) Biomarkers in small cell carcinoma of the lung. In: Aisner J (ed) Lung cancer. Churchill, Livingstone, New York

Sorenson GD, Pettengill OS, Cate CC, Ghatei MA, Bloom SR (1984 b) Modulation of bombesin and calcitonin secretion in cultures of small cell carcinoma of the lung. In: Becker AF, Gazdar AF (eds) The endocrine lung. Saunders, Philadelphia, pp 596–602

Szabo M, Berelowitz M, Pettengill OS, Sorenson GD, Frohman LA (1980) Ectopic production of somatostatin-like immuno- and bioactivity by cultured human pulmonary small cell carcinoma. J Clin Endocrinol Metab 5: 978–987

Vindelov LL, Christensen IJ, Jensen G, Nisson NI (1983) A detergent trypsin method for the preparation of nuclei for flow cytometric DNA analysis. Cytometry 3: 323–327

Wurster-Hill DH, Cannizzaro LA, Pettengill OS, Sorenson GD, Cate CC, Maurer LH (1984) Cytogenetics of small cell carcinoma of the lung. Cancer Genet Cytogenet 13: 303–330

Regulation of Hormone Production in Small Cell Carcinoma of the Lung

G. D. Sorenson, C. C. Cate, and O. S. Pettengill

Department of Pathology, Dartmouth Medical School, Hanover, NH 03756, USA

Introduction

Patients with pulmonary carcinoma frequently have elevated levels of one or more circulating tumor-produced hormones. This is particularly characteristic of small cell carcinoma of the lung (SCCL), where an increased frequency of hormone production is seen more commonly than in other types of lung cancer. The most frequently elevated hormones in such patients are calcitonin (CT), ACTH and neurophysins (Sorenson et al. 1984a). In addition to the noteworthy frequency of hormone production by tumors in these patients, there is also a wide variety of hormones produced by individual SCCL tumors. The reasons for the common occurrence and the great variety of hormones produced by SCCL are unknown, as are the molecular mechanisms involved.

As a background for considering the possible mechanisms for increased hormone production, it is important to consider how hormone production in SCCL compares with hormone production in the cell of origin for these tumors. Unfortunately it is not possible to answer this question definitively, because of the uncertainties concerning the precise cell of origin, so that direct comparisons between hormone production by these cells and SCCL cannot be made. CT, bombesin/gastrin-releasing peptide (BN/GRP), and leu-enkephalin have been demonstrated in a very limited population of cells in human bronchial mucosa by immunocytochemistry (Wharton et al. 1978; Becker et al. 1980; Cutz et al. 1981). Positively staining cells are more readily identifiable in lungs from fetuses and newborns than in the adult. Cells related to these hormone-containing cells in the bronchial mucosa (probably precursor cells to them) are the presumed cell or origin for SCCL.

The multiple production of hormones by these tumors in the same patient has been well documented (Abe et al. 1984; Suzuki et al. 1984), and this multiplicity of hormone production by single tumors has been particularly well demonstrated in cell cultures (Sorenson et al. 1981). Some cultures produce few hormones, while others will produce many, e.g., as many as 13 different peptide hormones. Cultures derived from single-cell clones will also produce multiple hormones that are present in the parent culture. Individual normal endocrine cells produce only one or two hormones, so that in this context, neoplastic cells producing many hormones are abnormal. This can be extended to suggest that those tumors producing many hormones are more abnormal in this regard than those producing few hormones. This increase in hormone production could be due to either a decrease in negative (down) regulation and/or an increase in positive (up) regulation of hormone gene expression.

Recent Results in Cancer Research. Vol. 99
© Springer-Verlag Berlin · Heidelberg 1985

Another important characteristic of hormone production by SCCL is the heterogeneity of the secreted product. There is a variety of molecular forms of any individual hormone secreted, including apparently the prohormone as well as partially processed products. This heterogeneity may be due in part to the presence of multiple forms of messenger RNA (mRNA) and therefore to multiple forms of translation product. It may also be due to defective processing, storage, and release mechanisms of the prohormone or a combination of these events. Thus, a tumor-secreted hormone may vary markedly in size from that of the prohormone to that of the normally produced finally secreted form (Becker et al. 1983, 1984). What is involved in causing this heterogeneity of secreted molecular forms by SCCL has not been studied in much detail.

Materials and Methods

Cell Lines and Culture Conditions

Continuous cell lines were established from biopsy or autopsy specimens from patients with confirmed tissue or cytologic diagnosis of SCCL (Pettengill et al. 1977, 1980b; Pettengill and Sorenson 1981; Sorenson and Pettengill 1980; Sorenson et al. 1981).

Bethanechol Experiment

SCCL cell culture (DMS 53) was established at 5×10^5 cells/ml in 100-mm petri dishes and grown for 4 days prior to the start of the experiment. At this time the medium was removed and replaced with Waymouth's medium with 20% fetal calf serum (Way 20/FCS) and $10^{-3} M$ bethanechol. Cells and medium were harvested at the end of 4 h; the cells were rinsed in phosphate-buffered saline (PBS); and then both specimens were frozen and stored at $-60 °C$.

Chloroquine Experiment

This was performed as described previously after the cultures had been established for 4 days. At this time the medium was replaced with Way 20/FCS which contained 50, 100, 200, or 400 M chloroquine. After 1 h this medium was replaced with a medium containing the same concentrations of chloroquine. Four hours later cells and medium were harvested as above.

Specimen Treatment

Washed cell pellets were homogenized in 1 ml 63 mM phosphate buffer, pH 7.4, containing 13 mM EDTA, 0.02% NaN$_3$, 15 ml aprotinin, and 0.2% Triton X-100, and then centrifuged for 15 min at 12000 g. The clear supernates were denatured, reduced, and alkylated by heating in a final concentration of 4 M urea and 10 mM dithiothreitol, followed by incubation for 30 min in the dark at room temperature in 50 mM iodoacetamide. Media samples were concentrated four-fold by lypohilization and reconstituted in phosphate buffer and the resultant concentrate denatured, reduced, and alkylated, as were the homogenate supernates.

Column Chromatography

Aliquots (1 ml) of denatured, reduced DMS 53 cell homogenates and media from the control and experimental cultures were applied to a 0.9×60 cm Sephadex G50 fine (Pharmacia) column equilibrated in elution buffer containing 0.9% NaCl, 5 mM HCl, 0.1% Triton X-100, at pH 2.5–2.8. Samples of 0.5 ml were collected for radioimmunoassay.

Radioimmunoassay

The calcitonin (CT) RIA is a two-step procedure involving an initial 48-h incubation of sample and goat anti-CT in phosphate/EDTA buffer, pH 7.2, containing 0.1% Triton X-100 and 2% CT-free normal human plasma, followed by a second, 24-h incubation in the presence of ^{125}I-labeled hCT. Separation of bound tracer is accomplished using rabbit anti-goat gamma globulin antiserum mixed with diluted goat serum and 3% polyethylene glycol 8000, permitting final globulin precipitation in 1–2 h. Intra- and interassay variabilities are less than 3% and 10%, respectively (Cate et al. 1984).

RNA Isolation

Total cellular RNA was isolated from seven SCCL cell lines, other human tissues, mouse 3T3 cells and *Drosophila* by standard procedures (Maniatis et al. 1982). In brief, this includes lysis of the cells, separation of the cell debris through a sucrose gradient, proteinase K digestion, and extraction with phenol:chloroform:isoamyl alcohol (P:C:IA). The nucleic acids were precipitated with ethanol, redissolved, and digested with DNase I. Subsequently, the RNA was extracted with P:C:IA and then precipitated and washed with ethanol.

Total RNA (0.5–3 µg in a volume of 200 µl) was applied to a Gene Screen membrane (New England Nuclear) using a Schleicher and Schuell vacuum manifold. The membrane was baked, prehybridized, and hybridized by standard methods (Stellway and Dahlbert 1980; Bittner et al. 1980) with a cDNA probe specific for human calcitonin (B24/22), provided by Dr. Roger Craig, Courtauld Institute of Biochemistry, Middlesex Hospital Medical School, London.

Results

The usual measure of hormone production by lung tumor cells in patients and in cultures has been quantification of the hormone secreted in the circulation or in the medium, respectively. Factors which will influence the secretion of two different hormones, i.e., CT and BN/GRP in cultures of small cell carcinoma, have been studied in detail (Sorenson et al. 1983a, b, 1984b). Substances influencing hormone secretion are listed in Table 1. Cholinergic, adrenergic, dopaminergic, and histaminergic ligands will stimulate secretion, which indicates that some of these cells have receptors for these ligands. Cholinergic stimulatory effects have been studied extensively, and a series of pharmacologic studies and also binding and competitive inhibition studies have revealed that at least some small cell cultures have cholinergic receptors of the muscarinic type (Sorenson et al. 1983a, b). Similarly, there is pharmacologic evidence indicating the presence of H1 histaminergic recep-

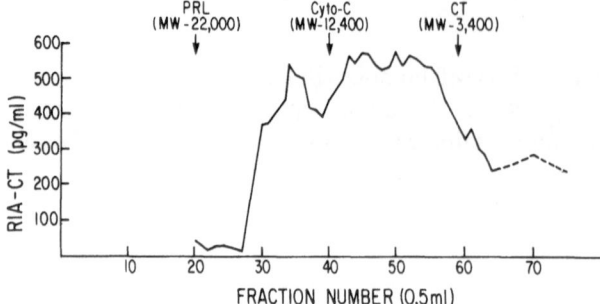

Fig. 1. Heterogeneity of plasma CT in SCCL. Gel filtration pattern on Sephadex G-50. Marker proteins were ^{125}I-labeled prolactin, catochrome C, and [^{125}I]h-calcitonin

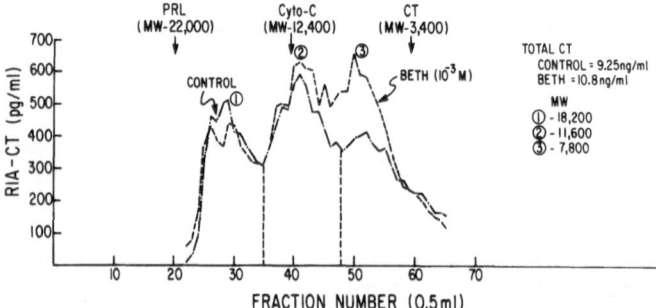

Fig. 2. Effect of bethanechol on CT secretion in DMS 53 cells. Gel filtration pattern on Sephadex G-50 of 4-h media. Marker proteins were ^{125}I-labeled prolactin, cytochrome C, and [^{125}I]h-calcitonin. Molecular size range (daltons) of the three major control peaks was 15000–20500, 9000–15000, and 2200–9000, respectively, with the average molecular weights of each peak indicated

tors on these cells (Delprete et al. 1985). It has also been determined that secretion is increased by increasing cAMP levels, thus indicating that this is part of the metabolic mechanism that is involved in the secretory process of the neoplastic cells, as has been observed in many normal endocrine cells. On the other hand, experiments have been done to evaluate the effect of a variety of hormones on secretion by these cells, and these have had no effect at the concentrations used (Table 2).

The heterogeneity of the molecular forms of CT in patients with SCCL and in media from cultures of SCCL has been clearly documented (Becker et al. 1984). We have extended these investigations and studied the molecular forms of CT which are secreted by the DMS 53 SCCL cell line and the extent to which the size of the secreted CT could be influenced. The molecular size of immunoreactive CT in the serum of patients with SCCL producing CT can vary from a molecular size of approximately 17000 daltons to one of about 3400 daltons (Fig. 1). In cultures there is also marked molecular size heterogeneity of the secreted CT. This is well illustrated by the G50 chromatographic profile of the immunoreactive calcitonin, which is secreted by the DMS 53 cell line in vitro (control profile Fig. 2). This profile shows a similar series of molecular forms, ranging from approximately 18000 to 3400 daltons. There are three major peaks, at about 18000, 12000, and 8000 daltons.

Furthermore, the molecular size of the secreted immunoreactive calcitonin after cholinergic stimulation has also been investigated. In these studies 10^{-3} M bethanechol was added to the culture and the medium was harvested after 4 h. This resulted in an increase in the total CT secreted and a shift of the sizes of the secreted molecules, resulting in an approximately 30% increase in the secretion of the smaller molecules of about 8000 molecular weight (Fig. 2, Table 3).

These results are reminiscent of studies of ACTH production in mouse pituitary AtT20 cells (Gumbiner and Kelly 1982). These investigations have defined two pathways of secretion in these tumor cells. One is referred to as a constitutive pathway by which relatively high-molecular-weight forms of ACTH and membrane glycoproteins are secreted. The second is termed as stimulatory pathway, and it is by this means that the smaller molecular weight forms, including the normal small form of ACTH, are secreted. The latter is referred to as a stimulatory pathway because it is responsive to stimulation with 8Br-cAMP, which would increase the level of the cAMP in the cell. Thus the effect of increasing cAMP is to increase secretion of the smaller molecular forms without affecting secretion of the large molecular forms. Subsequent studies by the same group have demonstrated

Table 1. Substances modulating BN/GRP and CT secretion by SCCL

Dibutyryl cyclic AMP	Adrenergic agonists
Theophylline	Epinephrine
Cations: Ca^{2+}, K^+	Norepinephrine
Chelators	Cholinergic agonists
EDTA, EGTA	Acetylcholine
Histamine	Bethanechol
Dopamine	Carbachol

Table 2. Hormones with no effect on bombesin secretion in DMS 53 over 2 h

Hormone	Final concentration
Cholecystokinin	$10^{-3} M$
Gastrin	$5 \times 10^{-6}M, 5 \times 10^{-7}M$
Glucagon	$10^{-6}M, 10^{-7}M$
Secretin	10 units/liter, 20 units/liter
Somatostatin	$10^{-6}M, 5 \times 10^{-5}M$
Thyroid releasing hormones	$2 \times 10^{-5}M, 5 \times 10^{-5}M$
Vasoactive intestinal peptide	$10^{-6}M, 10^{-7}M$
Arginine vasopressin	$1.8 \times 10^{-5}M, 3.6 \times 10^{-5}M$

Table 3. Effect of bethanechol on CT secretion by DMS 53 cells

	Total CT (ng/ml)	Percent of total recovery in chromatographic peaks		
		1	2	3
Control	9.2	31	39	30
Behanechol ($10^{-3}M$)	10.8	26	34	41

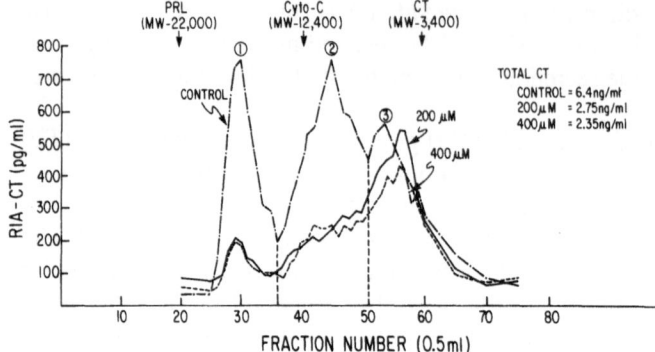

Fig. 3. Effect of chloroquine on CT secretion in DMS 53 cells. Gel filtration patterns on Sephadex G-50 of 4-h media. Marker proteins were ^{125}I-labeled prolactin, cytochrome C, and [^{125}I]h-calcitonin. Ranges of molecular sizes in the three major control peaks 15000–20500, 9000–15000, and 2200–9000 daltons, respectively

Table 4. Effect of chloroquine on CT secretion in DMS 53 cells

	Total CT (ng/ml)	Percent of total recovery in chromatographic peaks		
		1	2	3
Control	6.4	27	42	31
50 μM Chlorquine	2.0	24	37	39
100 μM Chloroquine	1.8	22	27	51
200 μM Chloroquine	2.8	14	29	58
400 μM Chloroquine	2.4	17	20	64

that the opposite effect occurs when the AtT20 cells are exposed to chlorquine (Moore et al. 1983). In these experiments, chloroquine induces an increase in secretion of the larger molecular weight forms by the constitutive pathway but has no effect on the secretion of the smaller forms.

We have studied the effect of chloroquine on calcitonin secretion with the DMS 53 cell line. Exposure to chloroquine for 4 h leads to alterations in secretion, but the alterations are different from those observed in the AtT20 cell (Fig. 3). The results indicate a decrease in secretion of the larger molecular weight forms during a 4-h period, with little effect on the small forms of CT. Moreover, this response is dose related (Table 4). With increasing concentrations of chloroquine the amount of the larger molecular forms secreted decreases and the relative percentage of small molecules secreted increases. However, as demonstrated in Fig. 3, there is no absolute increase in secretion of the smaller form. Intracellular CT is not significantly altered.

Hormone production can be controlled at multiple levels in the biosynthetic pathway, but it is clear that a central role in the regulation of hormone secretion is at the level of translation of mRNA. Depending on whether or not different regulatory steps or processes have occurred in the nucleus there may not be mRNA present. If there is no mRNA this can be attributed to one of the nuclear processes and no hormone secretion will be expected. Because of this central role for mRNA we have been interested in assessing our

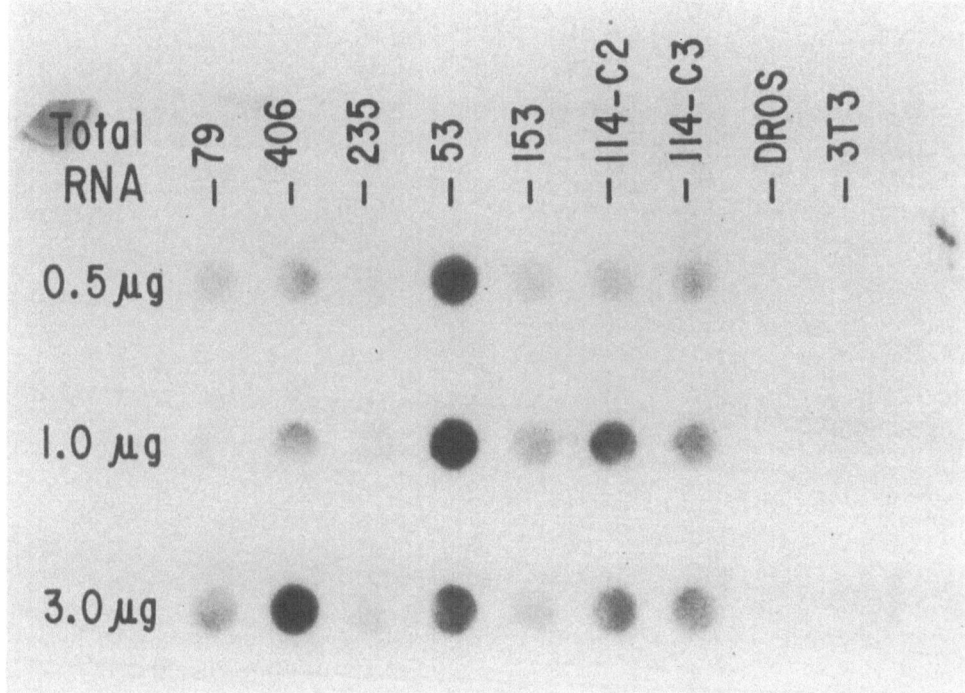

Fig. 4. Autoradiography of dot blots of total cellular RNA from a series of SCCL cultures (DMS 79, 406, 235, 53, 153, 114-C2, 114-C3), *Drosophila (Dros)* and mouse fibroblastic cell lines *(3T3).* Hybridization is apparent between the RNA from all the SCCL lines and a cDNA probe (B24/22) specific for calcitonin, which indicates the presence of CT mRNA in these tumors

SCCL cultures (and some other tumors) to see whether or not they contain mRNA for certain hormones. If it were present in all cell lines it would indicate that the genes are active in all cell lines, and if it was variably present it would indicate that one or another of the nuclear processes was apparently interfering with its production. We are also interested in assessing how close a correlation there is between the presence of mRNA for a certain hormone and secretion of the same hormone, since the latter is an end stage result and obviously may reflect changes at any one of multiple levels between the chromatin and the cell membrane.

We have looked for the presence of hormone mRNA by making dot blots of total cellular RNA. An autoradiograph of the blots of RNA from a series of SCCL cell lines is illustrated in Fig. 4. Three different quartities of each RNA are used, viz. 0.5, 1.0, and 3.0 µg. The negative controls are *Drosophila* RNA and 3T3 RNA (a mouse fibroblast cell line). Positive hybridization reactions occur between the CT probe (B24/22) and RNA from all the cell lines, including DMS 114-C2 and DMS 114-C3 (2 clones) which do not secrete detectable amounts of CT as evaluated by RIA.

Another similar series of dot blots of RNA from two SCCl lines, viz. DMS 153 and DMS 406 as positive controls with *Drosophila* and 3T3 RNAs as negative controls is illustrated in Fig. 5. The splenic RNA was obtained from a splenectomy specimen from a patient with idiopathic thrombocytopenic purpura. This RNA and the RNA from a colon carcinoma show no evidence of hybridization with the CT probe. However, RNAs from

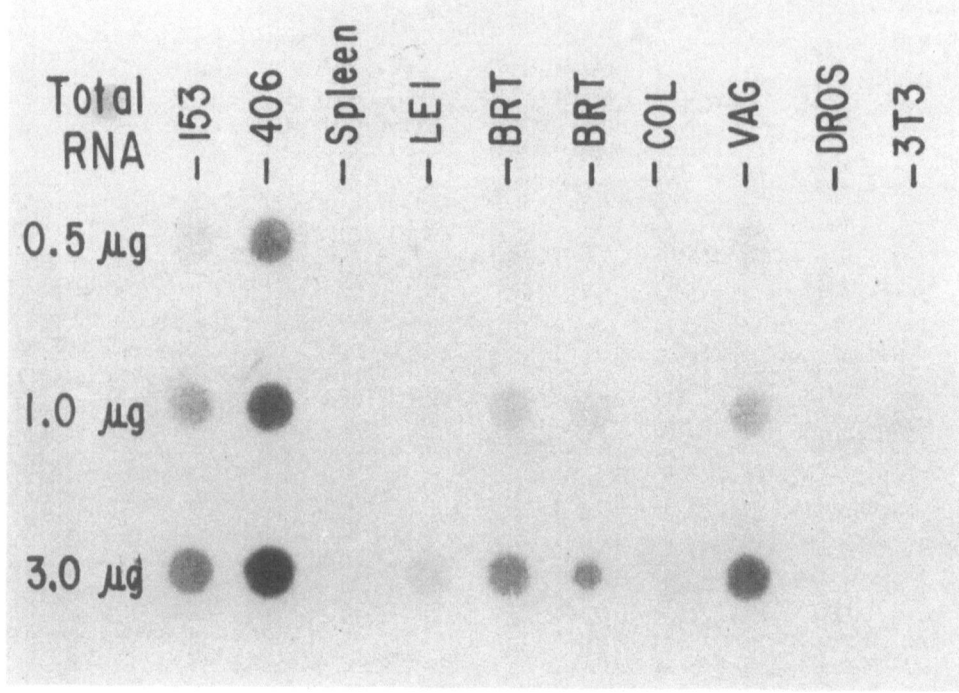

Fig. 5. Autoradiography of dot blots of total cellular RNA from SCCL cultures (DMS 153, 406), *Drosophila (Dros)*, 3T3 mouse fibroblastic cell line (3T3), non-neoplastic human spleen, gastric leiomyoblastoma *(Lei)*, breast carcinoma *(Brt)*, colon carcinoma *(Col)* and vaginal carcinoma *(Vag)*. Hybridization is apparent between the RNA from the SCCl cell lines, the two brest carcinomas and the vaginal carcinoma and a cDNA probe (B24/22) specific for calcitonin, which indicates the presence of CT mRNA in these tumors

two breast carcinomas, a vaginal squamous cell carcinoma, and a gastric leiomyoblastoma appear to be positive. The results of hybridization with RNA from a gastric leiomyoblastoma are equivocal and difficult to interpret definitively, since there appears to be slight homology between the cDNA probe and 3T3 RNA.

These are preliminary results although they are reproducible. They suggest that CT genes (and possibly other hormone genes) may be activated in a wide variety of tumors.

Discussion

The usual measure of hormone production by lung tumor cells in patients has been circulating levels, and that in culture has been the evaluation of the secreted hormone in the medium. Our studies have shown that secretion by SCCL may be influenced by cAMP, cation concentration as well as by the reactivity of multiple receptors. Secreted hormone is clear evidence of hormone gene activation. However, low or absent levels of the secreted product may be due to alterations at multiple levels of hormone biosynthesis as well as to extracellular catabolism of the hormone. In the latter instances, gene activation is not precluded even if secretion is not apparent.

Other studies have shown that the hormone levels in the medium from cultures may vary from time to time (Pettengill et al., this volume). Presumably this is due to variation in the amounts secreted by these cultures, although variable rates of catabolism have not been excluded. Most cells increase protein secretion by increasing the amount of mRNA available for translation, but there may also be an increase in the rate of translation, which contributes to the same result (Robbins et al. 1984).

There is also evidence that hormone production may develop in tumors which were previously nonproducers, and that tumors that are nonproducers may later in the clinical course become hormone producers (Havemann et al., this volume). Considering the heterogeneity of tumors and the selective pressures that may be brought to bear from therapeutic measures, it is not surprising that these alterations in hormone production occur.

Hormone secretion may be controlled in SCCL cells at variable and possibly multiple levels. Normal gene regulation may occur in mammalian cells in a variety of different ways (Lewin 1983). Regulation of hormone production may occur in the nucleus, in the cytoplasm, and at the cell membrane (Table 5). In the nucleus control may occur at the level of chromatin (Mathis et al. 1980; Razin and Friedman 1981; Weisbrod 1982). Evidence indicates that active genes such as the globin gene in erythrocytes and the ovalbumin gene in oviduct epithelium are susceptible to DNase 1 digestion, whereas inactive genes are not (Groudine and Weintraub 1982; Garel and Axel 1976). The implication is that since the inactive genes are protected from digestion by DNase 1 they are packaged differently.

Regulation may be at the level of DNA where it has been determined that in most cases inactive genes are methylated and active genes are hypomethylated (Felsenfeld and McGhee 1982; Busslinger et al. 1983; Cate et al. 1983; Feinberg and Vogelstein 1983). Although this correlation is frequently observed there are exceptions, and it is apparent that although hypomethylation can usually be correlated with gene activation it is not enough to activate genes by itself.

Control may be localized at the level of DNA transcription, and here promoter sequences and also possibly suppressor sequences can influence whether or not transcription takes place. Finally, within the nucleus, whether or not a cell produces a particular hormone may be related to RNA processing. Certain genes when they are transcribed will result in one or another form of mRNA, depending upon the type of RNA processing or splicing that has occurred. Thus, this may lead to the production of one hormone in one cell type and the production of another hormone or both hormones in a second cell type (Amara et al. 1982). The prototype for this is the formation of either CT or calcitonin gene-related peptide (CGRP), dependent upon how the mRNA is spliced after transcription of the calcitonin gene.

In the cytoplasm there are also multiple ways in which hormone production may be influenced (Table 5). It is possible that mRNA may not be present, because of regulation at the nuclear level. Even if the mRNA is present there are at least theoretical reasons for its

Table 5. Levels of cellular regulation of hormone biosynthesis and secretion

Nucleus	Cytoplasm	Cell membrane
Chromatin	Translation	Receptors
DNA	Peptide processing	
Transcription	Pathways of secretion	
RNA processing		

not being translated. After the mRNA is translated, hormones and many other proteins are typically processed, and this depends upon the processing enzymes within the cytoplasm. It has recently become apparent that certain peptides may be secreted by more than one pathway, and in particular, it has been shown that some larger molecular weight forms of specific hormones may be secreted by a different pathway than that of the smaller molecular weight forms (Gumbiner and Kelly 1982). The effects of cholinergic agonists and chloroquine on the molecular forms of CT secreted by the DMS 53 SCCL cell line suggest that at least two different pathways of secretion also occur in lung tumors. Secretion can also be regulated at the cell membrane. For example, stimulation of acetylcholine and histamine receptors will increase hormone secretion.

In addition to the steps in the normal hormone biosynthetic pathway, as reviewed above, which may differ between hormone producer and hormone nonproducer, there are other possible ways in which this abnormal hormone production may occur in these neoplastic cells. Hormone production by pulmonary tumors is usually explained by indicating that it is due to loss of hormone gene repression. This is probably true in a general sense, but there are multiple ways in which this might occur and the time has come and the techniques are available to sort this out more precisely.

It is important to do this not only as a means of obtaining a better understanding of normal hormone production, but also because these hormone genes in pulmonary tumors may be considered as models of genes activated as part of the malignant transformation process. Thus, understanding the regulatory mechanisms involved in the activation of these genes may provide clues as to factors involved in the activation and/or regulation of other important genes, such as genes involved in regulating proliferation in pulmonary tumors.

With this as background, let us consider some possible molecular mechanisms through which multiple hormone genes could be activated in a tumor cell. In considering how this might occur, we have to keep in mind that gene mapping studies indicate that human hormone genes are located on multiple different chromosomes. In Table 6 there is a compilation of this data, indicating that the ten hormone genes listed are located on eight different chromosomes (Shows et al. 1982; Naylor et al. 1983 a, b; Przepiorka et al. 1984).

Table 6. Hormone genes mapped to human chromosomes

	Chromosome
Calcitonin (CT)	11
Chorionic gonadotropin (CG)	10 and 18
	6 and 19
Chorionic somatomammatropin (CS)	17
(placental lactogen)	
Glucagon (GCG)	2
Growth hormone (GH)	17
Insulin (INS)	11
Parathormone (PTH)	11
Prolactin (PRL)	6
Pro-opiomelanocortin (POMC)	2
Somatostatin (SST)	3

What are some of the ways in which multiple hormone genes could be deregulated in SCCL? One possibility would be the occurrence of a point mutation in a repressor sequence or in a gene producing a repressor protein. However, a single mutation would activate only one hormone gene if the repressor was unique, or all the hormone genes uniformly if there was a common hormone gene repressor. Thus, this does not seem to be a likely explanation since multiple hormone genes are variably expressed in SCCL.

A second possible explanation for quantitative increases in hormone production would be amplification of the hormone gene. It is known that gene amplification (DNA amplification) occurs in tumors (Schimke, RT 1982; Fox 1984) and amplification of hormone genes probably occurs in SCCL.

A third possibility is translocation of DNA – transposition of positive or negative regulatory sequences and/or hormone genes could lead to hormone gene expression or increased degrees of expression. For example: (a) There might be activator sequences translocated to the sites of hormone genes; or (b) hormone genes might be translocated to the site of activator sequences; or (c) repressor sequences could be translocated away from the site of hormone genes, e.g., deletions. All these variable forms of DNA translocations could influence hormone production.

A fourth possibility would be that an oncogene-coded protein would increase hormone production. At least four different oncogenes have been identified in SCCL. These are c-*myc*, n-*myc*, c-*myb* and c-*raf* (Little et al. 1983; Griffen and Baylin and Rapp et al., this volume). There are some interesting although disconnected observations as background for considering this possibility. These are: (a) The epidermal growth factor (EGF) receptor resembles the v-*erb*-B oncogene product (Downward et al. 1984), and EGF increases transcription of the prolactin gene ten-fold in the GH_3 pituitary cell line, thus increasing prolactin secretion (Murdock et al. 1982a); and (b) cAMP increases transcription of the prolactin gene in GH_3 cells and prolactic secretion by increasing protein kinase (Murdock et al. 1982b) and protein kinases are encoded by at least several viral oncogenes (Bishop 1983).

A fifth possible mechanism for multiple hormone gene activation in SCCL is that this is due to multigenic activation associated with genetic instability as part of malignant transformation. Multigenic activation has been described in cultured embryonic chicken cells transformed by tumor viruses (Groudine and Weintraub 1980) as well as common multigenic activation in different human neoplasms (Hanania et al. 1983). Can it be that there are multiple types of genes activated in SCCL (and other tumors) and we recognize and focus on hormones because of clinical manifestation and/or because we have very sensitive RIAs which allow us to measure them? Widespread genomic activation as a possible mechanism for multiple hormone expression is worthy of much more investigation. Widespread genomic activation is a possible interpretation of the dot blot data illustrated, which suggested that CT genes may be activated in a variety of different types of tumor.

In summary, what we have done is to make a list of at least some of the ways in which, at various levels of hormone biosynthesis, cells which express hormone genes and produce hormones may differ from cells which do not. A list of ways in which genes may be activated in a more or less abnormal way in association with malignant transformation has been added. What the differences really are between hormone-producing and non-hormone-producing lung tumor cells is the question that needs to be addressed. We have just started.

References

Abe K, Kameya T, Yamguchi K, Kikuchi K, Adachi I, Tanaka M, Kimura S, Kodama T, Shimosato Y, Ishikawa S (1984) Hormone-producing lung cancers. Endocrinologic and morphologic studies. In: Becker KL, Gazdar AF (eds) The endocrine lung in health and disease. Saunders, Philadelphia, pp 549–595

Amara SG, Jonas V, Rosenfeld MG, Ong ES, Evans RM (1982) Alternative RNA processing in calcitonin gene expression generates mRNAs encoding different polypeptide products. Nature 298: 240–244

Becker KL, Managhan KG, Silva OL (1980) Immunocytochemical localization of calcitonin in Kulchitsky cells of human lung. Arch Pathol Lab Med 104: 196–198

Becker KL, Gazdar AF, Carney DN, Snider RH, Moore CF, Silva OL (1983) Calcitonin secretion by continuous cultures of small cell carcinoma of the lung: incidence and immuno-heterogeneity studies. Cancer Lett 18: 174–185

Becker KL, Silva OL, Gazdar AF, Snider RH, Moore CT (1984) Calcitonin and small cell cancer of the lung. In: Becker KL, Gazdar AF (eds) The endocrine lung in health and disease. Saunders, Philadelphia, pp 528–548

Bishop JM (1983) Cellular oncogenes and retroviruses. Ann Rev Biochem 52: 301–354

Bittner M, Kupferer P, Morris CF (1980) Electrophoretic transfer of proteins and nucleic acids from slab gels to diazobenzyloxymethyl cellulose or nitrocellulose sheets. Anal Biochem 102: 459–471

Busslinger M, Hurst J, Flavell RA (1983) DNA methylation and the regulation of globin gene expression. Cell 34: 197–206

Cate RL, Chick W, Gilbert W (1983) Comparison of the methylation patterns of the two rat insulin genes. J Biol Chem 258: 6645–6652

Cate CC, Douple EB, Andrews KM, Pettengill OS, Curphey TJ, Sorenson GD, Maurer LH (1984) Calcitonin as an indicator of the response of human small cell carcinoma of the lung cells to drugs and radiation. Cancer Res 44: 949–954

Cutz E, Chan W, Track, NS (1981) Bombesin, calcitonin and leu-enkephalin immunoreactivity in endocrine cells of the human lung. Experientia 37: 765–767

Debold CR, Schworer ME, Connor TB, Bird RE, Orth DN (1983) Ectopic pro-opiomelanocortin: sequence of cDNA coding for beta-melanocyte stimulating hormone and beta-endorphin. Science 220: 721–723

Delprete S, Pettengill OP, Cate CC, Ghatei M, Bloom SB, Sorenson GD (1985) Bombesin and calcitonin secretion by pulmonary carcinoma is stimulated by histaminic agonists (submitted for publication)

Downward J, Yarden Y, Mayes E, Scrace G, Totty N, Stockwell P, Ullrich A, Schlessinger J, Waterfield MD (1984) Close similarity of epidermal growth factor receptor and V-erb-B oncogene protein sequences. Nature 307: 521–527

Feinberg AP, Vogelstein B (1983) Hypomethylation distinguishes genes of some human cancers from their normal counterparts. Nature 301: 89–92

Felsenfeld G, McGhee J (1982) Methylation and gene control. Nature 196: 602–603

Fox M (1984) Gene amplification and drug resistance. Nature 307: 212–213

Garel A, Axel R (1976) Selective digestion of transcriptionally active ovalbumin genes from oviduct nuclei. Proc Natl Acad Sci USA 73: 3966–3970

Groudine M, Weintraub H (1980) Activation of cellular genes by avian RNA tumor viruses. Proc Natl Acad Sci USA 77: 5351–5354

Groudine M, Weintraub H (1982) Propagation of globin DNAse I hyposensitive sites in absence of factors required for induction: A possible mechanism for determination. Cell 30: 131–137

Gumbiner B, Kelly RB (1982) Two distinct intracellular pathways transport secretory and membrane glycoproteins to the surface of pituitary tumor cells. Cell 28: 51–59

Hanamia N, Shaool D, Harel J, Wiels J, Tursz T (1983) Common multigenic activation in different human neoplasias. EMBO 2: 1621–1624

Lehrach H, Diamond D, Wozney JM, Boedtker H (1979) RNA molecular weight determinations by gel electrophoresis under denaturing conditions, a critical reexamination. Biochemistry 16: 4743-4751

Lewin B (1983) Genes. Wiley, New York

Little CD, Nau MM, Carney DN, Gazdar AF, Minna JD (1983) Amplication and expression of the c-myc oncogene in human lung cancer cell lines. Nature 306: 194-196

Maniatis T, Fritsch EF, Sambrook J (1982) Molecular cloning: A laboratory Manual. Cold Spring Harbor Laboratory, Cold Spring Harbor, NY

Mathis D, Oudet P, Chambon P (1980) Structure of transcribing chromatin. Prog in Nucleic Acid Res Mol Biol 24: 1-55

Moore HP, Gumbiner B, Kelly RB (1983) Chloroquine diverts ACTH from a regulated to a constitutive secretory pathway in AtT-20 cells. Nature 302: 434-436

Murdock GH, Potter E, Nicolaisen AK, Evans RM, Rosenfeld MG (1982a) Epidermal growth factor rapidly stimulates prolactin gene transcription. Nature 300: 192-194

Murdock GH, Rosenfeld MG, Evans RM (1982b) Eukaryotic transcriptional regulation and chromatin associated protein phosphorylation by cyclic AMP. Science 218: 1315-1317

Naylor SL, Sakuguchi AY, Szoka P, Hendy GN, Kronenberg HM, Shows TB (1983a) Human parathyroid hormone gene (pth) is on the short arm of chromosome 11. Somatic Cell Genet 9: 609-616

Naylor SL, Sakaguchi AY, Shen L-P, Bell GI, Rutler WJ, Shows TB (1983b) Polymorphic human somotostatin gene is located in chromosome 3. Proc Natl Acad Sci USA 80: 2686-2689

Pettengill OS, Sorenson GD (1981) Tissue culture and in vitro characteristics, In: Greco FA, Oldham RK, Bunn PA (eds) Small cell lung cancer. Grune and Stratton, New York, pp 51-77

Pettengill OS, Faulkner CS, Wurster-Hill DH, Maurer LH, Sorenson GD, Robinson A, Zimmerman EA (1977) Isolation and characterization of a hormone producing cell line from human small cell anaplastic carcinoma of the lung. J Natl Cancer Inst 58: 511-518

Pettengill OS, Curphey TJ, Cate CC, Flint CT, Maurer LH, Sorenson GD (1980a) Animal model for small cell carcinoma of the lung: effect of immunosupression and sex of mouse on tumor growth in nude athymic mice. Exp Cell Biol 48: 270-297

Pettengill OS, Sorenson GD, Wurster-Hill DH, Curphey TJ, Noll WW, Cate CC, Maurer LH (1980b) Isolation and growth characteristics of continuous cell lines from small cell carcinoma. Cancer 45: 906-918

Przepiorka D, Baylin SB, McBride OW, Testa JR, deBustros A, Nelkin BD (1984) The human calcitonin gene is located on the short arm of chromosome 11. Biochem Biophys Res Commun 120: 493-499

Razin A, Friedman J (1981) DNA methylation and its possible biological roles. Prog Nucleic Acid Res Mol Biol 25: 33-52

Robbins DC, Tager HS, Rubenstein AH (1984) Biologic and clinical importance of proinsulin. N Engl J Med 310: 1165-1175

Schimke RT (ed) (1982) Gene amplification. Cold Spring Harbor Laboratory, Cold Spring Harbor, NY

Shows TB, Sakaguichi AY, Naylor SL (1982) Mapping the human genome, cloned genes, DNA polymorphisms and inherited disease. Adv Human Genet 12: 341-451

Sorenson GD, Pettengill OS (1980) Structural and functional characteristics of small cell carcinoma of the lung, in vitro and in vivo. Biol Cell 39: 277-280

Sorenson GD, Pettengill OS, Brinck-Johnsen T, Cate CC, Maurer LH (1981) Hormone production by cultures of small cell carcinoma of the lung. Cancer 47: 1289-1296

Sorenson GD, Bloom SR, Ghatei MA, Cate CC, Pettengill OS (1982) Bombesin production by human small cell carcinoma of the lung. Regul Pept 4: 59-66

Sorensen GD, Pettengill OS, Brown SJ, Bacopoulos NG (1983a) Evidence for muscarinic cholinergic receptors in cultured human small cell carcinoma of the lung. Fed Proc 42: 388

Sorenson GD, Pettengill OS, Cate CC, Gosselin EJ, Ghatei MA, Bloom SR (1983b) Bombesin and calcitonin secretion by pulmonary carcinoma is modulated by cholinergic receptors. Life Sci 33: 1939-1944

Sorensen GD, Pettengill OS, Cate CC, Delprete SA (1984a) Biomarkers in small cell carcinoma of the lung. In: Aisner J (ed) Lung cancer. Livingstone, New York

Sorenson GD, Pettengill OS, Cate CC, Ghatei MA, Bloom SR (1984b) Modulation of bombesin and calcitonin secretion in cultures of small cell carcinoma of the lung. In: Bedker KL, Gazdar AF (eds) The endocrine lung in health and disease. Saunders, Philadelphia, pp 596-602

Stellway ES, Dahlberg AE (1980) Electrophoretic transfer of DNA, RNA and protein onto diazo-benzyloxymethyl (DBM) paper. Nucleic Acid Res 8: 299-317

Suzuki H, Tsutsumi Y, Yamaguchi K, Abe K, Yokoyama T (1984) Small cell lung carcinoma with ectopic adrenocorticotropic hormone and antidiuretic hormone syndromes. J Clin Oncol 14: 129-137

Weisbrod S (1982) Active Chromatin. Nature 297: 289-295

Wharton J, Polak JM, Bloom SR, Ghatei MA, Solcia E, Brown MR, Pearse AGE (1978) Bombesin-like immunoreactivity in the lung. Nature 273: 769-770

The Serum-Free Establishment and in Vitro Growth Properties of Classic and Variant Small Cell Lung Cancer Cell Lines

D. N. Carney, G. Bepler, and A. F. Gazdar

Mater Misericordiae Hospital, Eccles Street, Dublin 1, Ireland

Introduction

Small cell lung cancer (SCLC) accounts for 20%–25% of all new cases of primary lung cancer in United States (Minna et al. 1981; Carney and Minna 1982). In 1984 it is anticipated that there will be in excess of 135000 new cases of lung cancer. Although major advances have been made in the past decade in our understanding of the important prognostic factors of SCLC, such as extent of disease, performance status, and response to therapy, clinical responses to cytotoxic therapy have remained unchanged over the past 3–5 years. With intensive combination chemotherapy, with or without radiation therapy, clinical responses can be achieved in 75%–80% of all patients, with complete responses (disappearance of all known tumor) occurring in 40%–50% of patients. However, in spite of these excellent initial responses to therapy, response duration is short and the median survival for all patients is 10–11 months. Approximately 10% of patients are cured of their disease.

To improve understanding of the biological properties of SCLC tumors, including the properties of tumors from newly diagnosed, untreated patients and from patients who have relapsed from prior intensive chemotherapy, and the properties of tumor cells in different metastatic sites (e. g., bone marrow, liver, etc.), there is a need to develop in vitro culture methods which will reproducibly support the establishment and long-term growth of cell lines of SCLC. In this chapter, the techniques used to establish cell lines of SCLC will be reviewed. In addition, the in vitro growth properties of cell lines established and maintained in both serum-supplemented medium and serum-free medium will be compared. Finally, the growth properties of two distinct subclasses of SCLC cell lines - classic and variant lines - will be discussed and results compared to oncogene expressions in these cell lines.

Establishment of SCLC Cell Lines:
Use of Serum-Supplemented and Serum-Free Defined Medium

Over the past 10 or more years, several different laboratories have reported on their attempts and success in the establishment of SCLC cell lines (Oboshi et al. 1971; Ohara et al. 1977; Pettengill et al. 1980; Gazdar et al. 1980; Bergh et al. 1982). In initial attempts a basal medium (RPMI 1640 or Weymouth's medium) supplemented with 10%–20% fetal

Recent Results in Cancer Research. Vol. 99
© Springer-Verlag Berlin · Heidelberg 1985

bovine serum was used. Although SCLC cell lines were successfully established using these media, the overall success rate in the establishment of SCLC cell lines from clinical specimens was low, ranging from 10% to 20% (Gazdar et al. 1980).

In an attempt to improve the rate of cell line establishment, and following the lead of Sato and coworkers (Barnes and Sato 1980), who suggested that each tumor type had specific growth factor(s) requirements, we have developed a defined serum-free medium for SCLC. The use of defined media for human tumor growth has several advantages over the use of serum-supplemented medium. In serum-supplemented medium the growth of cells is nonselective, so that proliferation of both malignant cells and admixed normal stromal cells may occur. In many instances the growth of the normal cells occurs at a much faster rate than the malignant cells, and the tumor cell population is frequently "outgrown" by the normal cells and is lost. Since the added factors in defined medium are known, it is easier to evaluate the secretory products from growing tumor cells, because of the many unknown proteins and other compounds in serum. In tests of in vitro cytotoxicity the presence of the many proteins in serum may bind, inactivate, or destroy the agents being tested, and thus lead to a misrepresentation of the activity of the test compounds. Finally, as it seems likely that the specific in vitro growth requirements of human tumors is different for each histologic type (i.e., breast cancer, colon cancer, etc.), it is possible that the use of defined medium in the initial growth of human tumors may aid in the typing of human tumors and in the detection of "occult" tumor cells in specimens otherwise pathologically negative for these cells.

The approach used in identifying the specific growth factor requirements for SCLC was as follows: (a) Determine the specific factors which would support the continuous growth of established SCLC cell lines; (b) evaluate the usefulness of this medium in supporting the growth of clinical (fresh) specimens of tumor cells of the same histologic type (i.e., SCLC); (c) evaluate the specificity and selectivity of the medium for SCLC growth by testing its ability to support the continuous growth of both normal, nonmalignant cells and the tumor cells of specimens of a different histologic type.

In the initial studies using established SCLC cell lines, a large number of different hormones and growth factors were tested for their mitogenic effects on SCLC growth (Simms et al. 1980) (Table 1). Although individual factors had only a slight stimulatory effect, the

Table 1. SERUM-free chemically defined HITES medium for SCLC

Factor	Concentration
Hydrocortisone	$10\,nM$
Insulin	$5\,\mu g/ml$
Transferrin	$10\,\mu g/ml$
17β-Estradiol	$10\,nM$
Selenium	$30\,nM$
Basal medium	RPMI 1640

Table 2. Use of serum-free defined HITES medium for the initial in vitro growth of SCLC specimens

Culture medium	Fraction with tumor growth
SCLC-positive	
HITES	39/53
Serum-supplemented	24/53
SCLC-negative	
HITES	0/243
Serum-supplemented	3/243[a]
Non-SCLC tumors	
HITES	3/40
Serum-supplemented	10/40

[a] Three B lymphoblastoid cell lines

combination of hydrocortisone (10^{-8} M), insulin (5 µg/ml), transferrin (10 µg/ml), 17-beta estradiol (10^{-8} M), and selenium (3×10^{-8} M) added to RPMI-1640 medium (HITES medium) was shown to support the long-term growth of cell lines previously established in RPMI-1640 medium supplemented with 10%-20% fetal bovine serum (serum-supplemented medium, SSM). After the transfer of cells from SSM to HITES, and following a short lag period, the growth rate of cells in HITES was comparable to that observed in SSM.

Following these studies, and because our success rate in establishing SCLC lines from fresh clinical specimens using SSM was only 10%-20% (Gazdar et al. 1980), a new approach using both serum-free HITES medium and SSM was designed to establish cell lines of SCLC (Carney et al. 1981a, 1984a).

Over the past 3 years, all specimens obtained from patients with SCLC have been tested for growth in both HITES medium and SSM. Results are indicated in Table 2. From a total of 300 specimens obtained from patients with a confirmed histological diagnosis of SCLC, 243 specimens were cytopathologically negative for tumor cells. When these negative specimens were plated in either HITES medium or SSM without added growth factors to support bone marrow proliferation, no growth of tumor cells was observed. In SSM, growth of adherent stromal cells was observed, whereas in HITES medium, death and lysis of all cells were usual by 7 days after plating. In addition, B lymphoblastoid cell lines were established from three specimens in SSM.

When specimens were plated in HITES medium tumor cell proliferation was observed in 72% of all specimens containing tumor cells. The outgrowth of tumor cells was observed within 2-3 days of plating (Table 2). In contrast, for the same specimens in SSM, tumor cell proliveration was observed in only 45% of these specimens. These data confirm the value of HITES medium in the initial outgrowth of SCLC cells from fresh clinical specimens. In addition, in tests of 40 human tumors of other cell types in HITES medium, tumor cell lines were established in three specimens, while ten of the same samples were established as cell lines in SSM. These latter data suggest that HITES is selective for SCLC growth in vitro.

Although initial studies with HITES medium indicated that after 6-24 weeks' culture SCLC proliferation either decreased or ceased in this medium, with minor modifications continuous cell lines can now be established with relative ease in HITES medium. The maintenance of tumor cells at a high cell density in HITES and the use of 50% "conditioned medium" at passage have supported the establishment and continual growth of SCLC cell lines in defined medium. Of 31 cell lines of SCLC established over a 2-year period, 22 cell lines (66%) were established in HITES medium. In addition, 9 lines which had failed to proliferate in HITES medium were established only in SSM. These data suggest that further additions to HITES medium are essential if all cell lines of SCLC can be cultivated and established in serum-free medium. With HITES medium, cell lines have been established with equal efficiency from treated and untreated patients and from a variety of different organ sites, including bone marrow, pleural effusions, lymph node, lung, and other surgically resected tumor masses. No differences have been observed in the biological properties of SCLC cell lines established from these different metastatic sites.

Growth Properties of SCLC Cell Lines in Serum-Free HITES
and Serum-Supplemented Medium

While initial studies using cell lines previously established in SSM had demonstrated no changes in the biological properties of the cells when transferred to and cultured in HITES, to further determine whether SCLC tumor cells cultured in HITES medium were similar to cells cultured in SSM we have compared and contrasted the biologic properties of lines established from the same clinical specimen in both SSM and HITES medium. Among a panel of lines cultured in both SSM and HITES, no differences have been observed in morphology, DNA content, nude mouse tumorigenicity, or the expression of a variety of APUD cell markers which we have previously shown to be elevated in SCLC, including L-dopa decarboxylase, bombesin, neuron-specific enolase, and the BB isozyme of creatine kinase (Baylin et al. 1980; Moody et al. 1981; Marangos et al. 1982; Carney et al. 1982; Gazdar et al. 1981; Carney et al. 1983 a). Results from a pair of cell lines are indicated in Table 3.

For further comparison of the growth properties of SCLC in SSM and HITES, doubling time studies for lines cultured in both media were indicated. As for the expression of the biomarkers, the growth rate of lines in HITES medium is equal to or more rapid than that observed in SSM (Fig. 1).

Soft agarose cloning studies have been carried out on cell lines isolated and established in HITES medium, using methods already described elsewhere (Carney et al. 1981b, 1984b). It is of interest that although SCLC cells could be maintained in mass culture in HITES medium, the colony-forming efficiency (CFE) of cell lines in HITES medium was very low, usually 10-fold less than the CFE in either SSM or HITES medium supplemented with 0.3%–5% bovine serum albumin. However, the addition of 10%–25% conditioned medium from lines cultured in HITES stimulated the soft agarose cloning of cell lines to an efficiency equal to that observed in SSM (Carney et al. 1983b).

Thus, the observation that the CFE of SCLC cell lines can be significantly improved by the addition of conditioned medium, and the observation that SCLC cell lines can be readily established in HITES with the use of "autologous" conditioned medium suggest that these cell lines synthesize and secrete growth-regulatory factors. The identification and characterization of these factors may provide a new way of controlling SCLC growth in vivo.

Table 3. Comparison of biological characteristics of SCLC cell line NCI-N592 cultured in serum-supplemented medium (SSM) and in serum-free chemically defined HITES medium

Biologic characteristic	SSM	HITES
Morphology	Tight aggregates	Tight aggregates
Doubling time	32 h	36 h
DNA index (flow cytometry)	1.6	1.6
L-Dopa decarboxylase[a]	204	320
Bombesin[b]	0.6	0.3
Neuron specific enolase[c]	1285	1138
Creatine kinase BB[c]	17555	9530

[a] nmol CO_2 h^{-1} mg^{-1}
[b] pmol/mg
[c] ng/mg

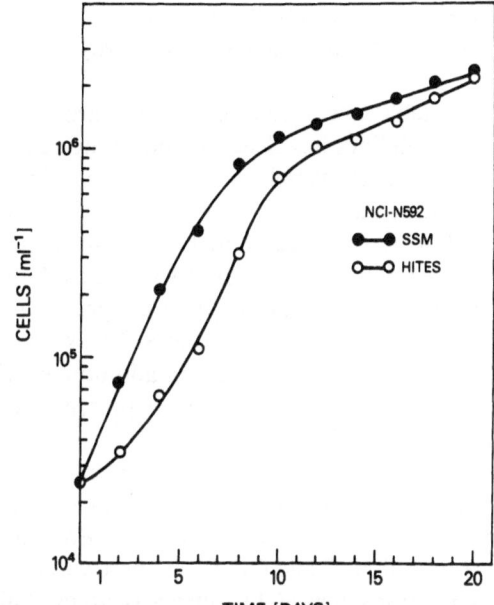

Fig. 1. Doubling time in vitro of SCLC cell line NCI-N592 cultured in serum-supplemented medium *(SSM)* and in serum-free HITES medium *(HITES)*. It can be seen that in log phase growth the doubling times of the cell line are almost identical in both culture conditions

Additions to HITES Medium

Although SCLC cell lines can now be established from 70%–75% of all clinical specimens, up to one-third of lines can only be established in serum-supplemented medium. These lines fail to demonstrate continued proliferation in HITES medium alone. Thus, it is clear that further additions to HITES medium are essential if (a) all specimens of SCLC can be successfully established as continuous SCLC cell lines; and (b) all SCLC cell lines can be established in defined medium.

In ongoing studies, the influence of a variety of hormones and growth factors on the growth of SCLC cells has been carried out using a panel of cell lines established in HITES medium. The hormones under study are those for which synthesis and secretion have been demonstrated in SCLC cell lines and/or which specific-binding receptors have been identified on the membrane of SCLC cells (Minna et al. 1982; Moody et al. 1983; Oie et al. 1984).

Bombesin/GRP (gastrin-releasing peptide), a 14/27 amino acid peptide, has been identified in most SCLC cell lines, is secreted by the cells into culture medium, and is bound by specific membrane receptors to many SCLC cell lines (Moody et al. 1983). These data suggest that BLI (bombesin-like immunoreactivity) may have an important physiologic role in SCLC. Using a clonogenic assay, we have evaluated the effect of bombesin (BN) on the growth of a panel of lung cancer cell lines. Preliminary data suggest that BN is a potent mitogen for SCLC but does not have any effect on the growth of a variety of other lung cancer cell lines (Carney et al. 1983b). Thus, BN may function as an "autocrine" growth factor for SCLC. Similar in vitro data have been observed with arginine vasopressin (AVP), but not with other hormones known to be produced by SCLC cells, including calcitonin and neurotensin.

Thus it appears that some hormones secreted by SCLC may function as autocrine growth factors. If these data are confirmed the development of monoclonal antibodies to these hormones, or to their specific receptors, may also provide an alternate means of in vivo growth control of SCLC.

Classic and Variant SCLC Cell Lines –
Growth Properties, Radiation Sensitivity, and Oncogene Expression

The establishment and detailed in vitro biological characterization of a large panel of SCLC cell lines has clearly demonstrated that these cell lines can be subdivided into two distinct classes, namely classic SCLC cell lines and variant SCLC cell lines. Classic SCLC lines, which account for 70% of the cell lines, express all four of a panel of biomarkers which we have evaluated in these cell lines; namely L-dopa decarboxylase, the key APUD enzyme (DDC); BLI; neuron-specific enolase (NSE); and the BB isozyme of creatine kinase (CK-BB). All four markers are present in significant amounts in classic lines, and in most fresh biopsy specimens of SCLC. None of these markers are elevated in non-SCLC (NSCLC) lung cancer lines. In addition, on electron microscopy examination in classic cell lines, dense core granules are present and easily identified either as clusters or single granules.

In contrast, variant cell lines of SCLC do not have detectable levels of the key APUD enzyme DDC. In addition, BLI is undetectable; NSE levels are significantly lower than those in classic cell lines; but CK-BB levels are the same as in classic SCLC cell lines. The presence of NSE and CK-BB in these variant lines distinguishes them from NSCLC lines and confirms their SCLC lineage (Carney et al. 1984b).

In addition to the biochemical differences which separate classic and variant lines, several other important biologic characteristics can distinguish these classes. Although some variant cell lines morphologically (in culture) and cytologically resemble classic SCLC lines, many variant lines grow as quite loose, floating aggregates of cells compared with the tight aggregates of classic SCLC lines. Cytologic examination of these lines reveal cells with more cytoplasm and more prominent nucleoli than are usually observed in classic cell lines (Figs. 2 and 3).

We have compared and contrasted the in vitro growth properties, including colony-forming efficiency in agarose and doubling time in mass culture, on a panel of classic and variant cell lines using assays already described elsewhere. Results demonstrated in Fig. 4 and Table 4 indicate that compared with the classic lines, variant SCLC lines have a significantly higher colony-forming efficiency (CFE) in agarose (1%–5% vs 5%–25%); a shorter doubling time (32 h vs 50 h) (Fig. 4); and a shorter latent period to tumor formation when inoculated sc to athymic nude mice.

Thus, variant cell lines, in addition to the biochemical differences, have a much more "malignant" and "aggressive" behavior than classic SCLC. These data suggest that variant lines may be more resistant to cytotoxic therapy than classic cell lines. Standard techniques and the clonogenic assay were used to study the in vitro radiobiologic properties of classic and variant cell lines (Carney et al. 1983c; Morstyn et al. 1984). All of nine classic SCLC cell lines were characterized by Do's ranging from 51 to 140 cGy, and a low extrapolation number (ñ) ranging from 1 to 3. In contrast, among variant cell lines, while similar Do's were observed (90–91 cGy), the extrapolation numbers were significantly greater (5.6–11.1) than those of the classic cell lines. The high extrapolation number of each of these variant lines indicates a greater ability of these cell lines to accumulate sublethal radiation damage, and thus an increased radiation resistance for variant cell lines compared with classic lines. In addition, the percentage of cells surviving a 2-Gy fraction was much greater for variant than for classic cell lines.

The recognition that variant lines have a much more aggressive in vitro growth behavior than classic cell lines has also prompted a study of the expression of oncogenes, including the c-*myc* oncogene, in a large panel of SCLC lines (Little et al. 1983; Nau et al. 1984). The

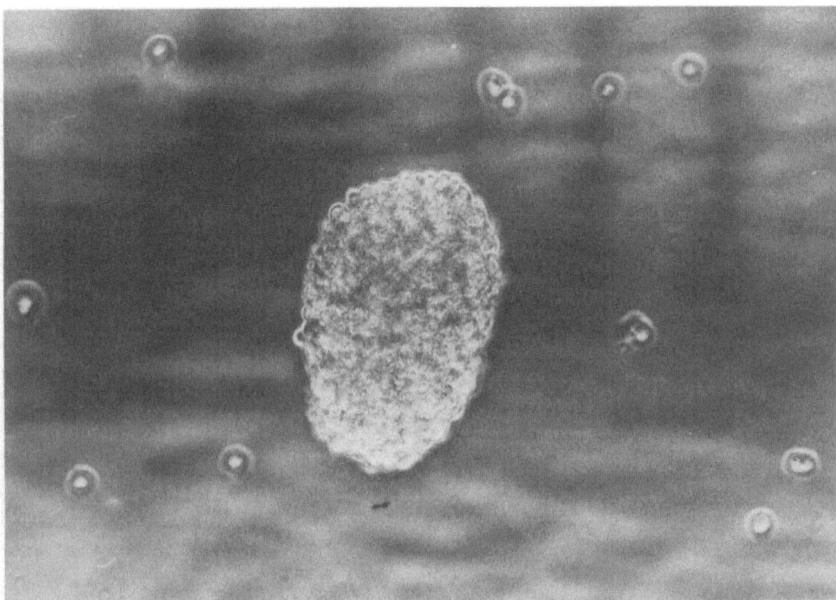

Fig. 2. Morphological examination of a classic SCLC cell line reveals tightly packed aggregates of floating cells

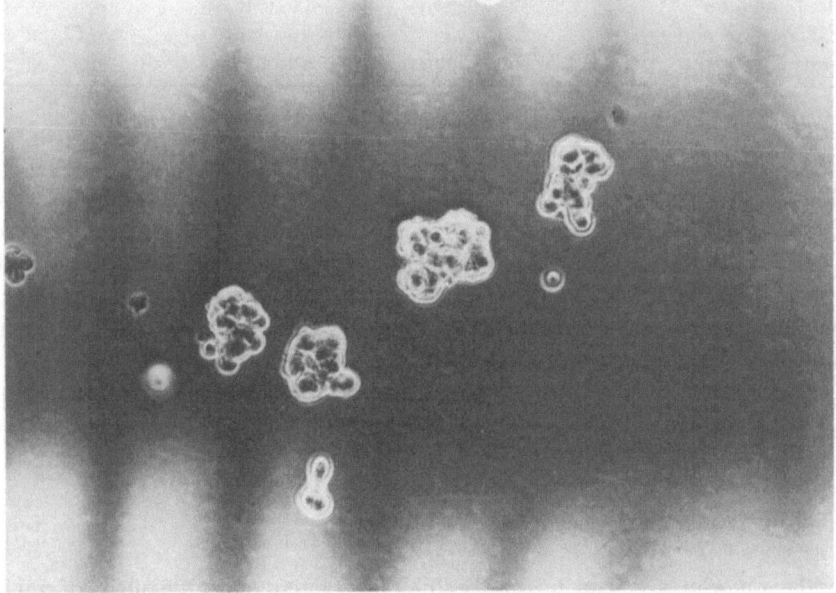

Fig. 3. Morphological examination of a variant SCLC cell line reveals much looser aggregates of floating cells than in classic SCLC cell lines

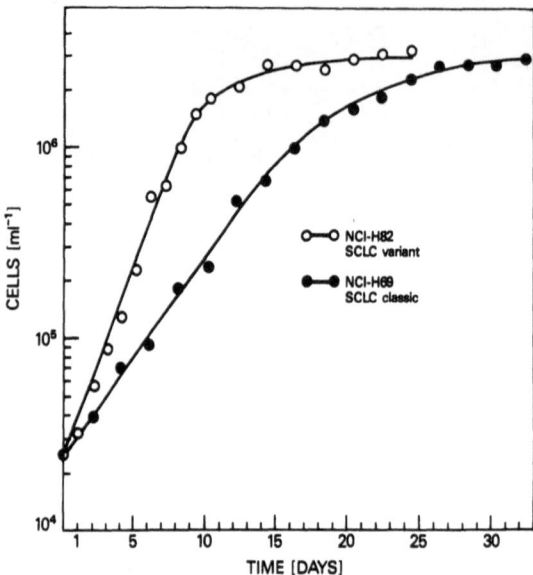

Fig. 4. Doubling times of a classic (NCI-H69) and of a variant (NCI-H82) cell line, showing that doubling time is significantly shorter for the variant cell line

Table 4. In vitro growth properties, radiation sensitivity, and c-*myc* oncogene amplification in classic and variant SCLC cell lines

Biologic characteristic	Classic	Variant
Colony-forming efficiency	1%–5%	5%–25%
Doubling time	32–132 h	18–36 h
Radiation sensitivity	Sensitive	Resistant
c-*myc* Oncogene	Not amplified	Amplified

results of these studies showed that in classic cell lines, minimal amplification of c-*myc* DNA was observed in only 1 of 12 cell lines. In contrast, significant amplification of c-*myc* DNA (20- to 75-fold) was observed in 5 of 6 variant cell lines. In addition, in these 5 lines c-*myc* RNA expression was also greatly increased. Cytogenetic studies of three variant lines have revealed one or more homogeneous-staining regions (HSRs) in each line, while none have been observed in classic cell lines.

Thus, variant cell lines are associated with an increased growth rate in vitro, are radioresistant, and are associated with amplification of the c-*myc* oncogene.

Clinical Correlates of Variant SCLC Cell Lines

As the majority of variant cell lines exhibit this phenotype at the outset of cell culture, and can be recognized as variant lines within 2–3 days of plating by their rapid growth rate and loose morphology, it is likely that clinical correlates of this subclass of SCLC cell lines exist.

Although histologic subtypes of SCLC have long been recognized, including the lymphocyte-like (oat cell variety) and the intermediate cell type, in several studies no significant differences have been observed between these types in clinical presentation, response to therapy, or survival (Carney et al. 1980; Ihde et al. 1981).

More recently, Radice et al. (1982) and Hirsch et al. (1983) have reported on the clinical presentations, response to therapy, and survival of patients with SCLC, whose histology revealed a mixed small cell/large cell morphology. Patients with this subtype accounted for 14%-19% of all SCLC patients. In both studies, the overall response and survival was significantly worse in patients with the mixed histology than in those with pure classic SCLC. Based upon our in vitro data and the recognition that many variant cell lines are established from patients with a mixed small cell/large cell morphology, it is most likely that this histologic type of SCLC is the in vivo counterpart of variant cell lines. The in vitro data of variant lines, including their more aggressive growth behavior, the c-*myc* amplification, and their radiation resistance, suggest that patients with this phenotype of SCLC have a much worse prognosis. The finding of the amplified c-*myc* gene in variant lines also suggests a role for the c-*myc* gene in the expression of the variant phenotype and the malignant behavior of human lung cancer.

References

Barnes D, Sato G (1980) Serum free cell culture: a unifying approach. Cell 22: 649-655
Baylin SB, Abeloff MD, Goodwin G, Carney DN, Gazdar AF (1980) Activities of L-dopa decarboxylase and diamine oxidase (histaminase) in human lung cancers: The decarboxylase as a marker for small (oat) cell lung cancer in tissue culture. Cancer Res 40: 1990-1996
Bergh J, Larson E, Zech L, Nilson K (1982) Establishment characterization of two neoplastic cell lines (U-1285 and U-1568) derived from small cell carcinoma of the lung. Acta Pathol Microbiol Immunol Scand [A] 90: 149-158
Carney DN, Minna, JD (1982) Small cell lung cancer. Clin Chest Med 3: 389-398
Carney DN, Matthews MJ, Ihde DC, Bunn PA Jr, Cohen MH, Makuch RW, Gazdar AF, Minna JD (1980) Influence of histologic subtype of small cell carcinoma of the lung on clinical presentation, response to therapy and survival. J Natl Cancer Inst 65: 1225-1229
Carney DN, Bunn PA Jr, Gazdar AF, Pagan JA, Minna JD (1981 a) Selective growth in serum-free hormone-supplemented medium of tumor cells obtained by biopsy from patients with small cell carcinoma of the lung. Proc Natl Acad Sci USA 78: 3185-3189
Carney DN, Gazdar AF, Bunn PA Jr, Guccion JG (1981 b) Demonstration of the stem cell nature of clonogenic tumor cells from lung cancer specimens. Stem Cells 1: 149-164
Carney DN, Marangos PJ, Ihde DC, Bunn PA Jr, Cohen MH, Minna JD, Gazdar AF (1982) Serum neuron-specific enolase: A marker for disease extent and response to therapy in patients with small cell lung cancer. Lancet I: 583-585
Carney DN, Broder L, Edelstein M, Gazdar AF, Hansen M, Havemann K, Matthews MJ, Sorenson GD, Vindelov L (1983 a) Experimental studies of the biology of human small cell lung cancer. Cancer Treat Rep 57: 27-36
Carney DN, Cuttitta FC, Gazdar AF, Minna JD (1983 b) Autocrine clonogenic factor(s) are produced by cell lines of small cell lung cancer. Proc Am Soc Clin Oncol 2: 14
Carney DN, Mitchell JB, Kinsella TJ (1983 c) In vitro radiation and chemosensitivity of established cell lines of human small cell lung cancer and its large cell variants. Cancer Res 43: 2806-2811
Carney DN, Brower M, Bertness V, Oie HK (1984 a) The selective growth of human small cell lung cancer cell lines and clinical specimens in serum-free medium. In: Sato G, Barnes D (eds) Methods in molecular and cell biology. Liss, New York, pp 247-262
Carney DN, Nau MM, Minna JD (1984 b) Variability of cell lines from patients with small cell lung cancer. In: Salmon, Trent, (eds) Human tumor cell cloning. Grune and Stratton, New York, pp 67-81
Gazdar AF, Carney DN, Russel EK, Simms HL, Baylin SB, Bunn PA Jr, Guccion JG, Minna JD

(1980) Small cell carcinoma of the lung: Establishment of continuous clonable cell lines having APUD properties. Cancer Res 40: 3502–3507

Gazdar AF, Zweig MH, Carney DN, Van Stierteghen AC, Baylin SB, Minna JD (1981) Levels of creatine kinase and its isozyme in lung cancer tumors and cultures. Cancer Res 41: 2773–2777

Hirsch FR, Osterlind K, Hansen HH (1983) The prognostic significance of histopathologic subtyping of small cell carcinoma of the lung according to the World Health Organization. Cancer 52: 2144–2150

Ihde DC, Makuch RW, Carney DN, Bunn PA Jr, Cohen MH, Matthews MJ, Minna JD (1981) Prognostic implications of stage of disease and sites of metastases in patients with small cell carcinoma of the lung treated with intensive chemotherapy. Am Rev Respir Dis 123: 500–507

Little CD, Nau MM, Carney DN, Gazdar AF, Minna JD (1983) Amplification and expression of the c-myc oncogene in human lung cancer cell lines. Nature 306: 194–196

Marangos PJ, Gazdar AF, Carney DN (1982) Neuron specific enolase in human small cell carcinoma cultures. Cancer Lett 15: 67–71

Minna JD, Higgins GA, Glatstein EJ (1981) Cancer of the Lung. In: DeVita VT, Hellman S, Rosenberg SA (eds) Principles and Practice of oncology. Lippincott, Philadelphia, pp 396–474

Minna JD, Carney DN, Oie H, Bunn PA Jr, Gazdar AF (1982) Growth of small cell lung cancer in defined medium. In: Pardee A, Sato G (eds) Growth of cells in a defined medium, vol 6. Cold Spring Harbor Press, New York, pp 627–639

Moody TW, Pert CB, Gazdar AF, Carney DN, Minna JD (1981) High levels of intracellular bombesin characterize human small cell lung cancer. Science 214: 1246–1248

Moody TW, Bertness V, Carney DN (1983) Bombesin-like peptides and receptors in human tumors. Peptides 4: 683–686

Morstyn G, Russo A, Carney DN, Karawya E, Wilson SH, Mitchell I (1984) Heterogeneity in the radiation survival curves and biochemical properties of human lung cancer cell lines. J Natl Cancer Inst 50: 801–807

Nau MM, Carney DN, Battey J, Johnson B, Little C, Gazdar AF, Minna JD (1984) Amplification, expression and rearrangement of c-myc and n-myc oncogenes in human lung cancer. Curr Top Microbiol Immunol 113: 172–177

Oboshi S, Tsugawa S, Seido T, Shimosato, Koide T, Ishikawa S (1971) A new floating cell line derived from human pulmonary carcinoma of the oat cell type. Gann 62: 505–514

Ohara H, Okamoto T (1977) A new in vitro cell line established from human oat cell carcinoma of the lung. Cancer Res 37: 3088–3095

Oie HK, Brower M, Carney DN (1984) Growth factor requirements for in vitro growth of small cell and other lung cancers in serum-free defined medium. In: Becker K, Gazdar AF (eds) Endocrine lung in health and disease. Saunders, pp 469–475

Pettengill OS, Sorenson GD, Wurster-Hill DH, et al (1980) Isolation and growth characteristics of continuous cell lines from small cell carcinoma of the lung. Cancer 45: 906–918

Radice PA, Matthews MJ, Ihde DC, Gazdar AF, Carney DN, Bunn PA Jr, Cohen MH, Fossieck BE, Makuch RW, Minna JD (1982) The clinical behavior of mixed small cell/large cell bronchogenic carcinoma compared to pure small cell subtypes. Cancer 50: 2894–2902

Simms E, Gazdar AF, Abrams PAPA, Minna JD (1980) Growth of human small cell (oat cell) carcinoma of the lung in serum-free growth factor-supplemented medium. Cancer Res 40: 4356–4361

Expression of Peptide and Other Markers in Lung Cancer Cell Lines

A. F. Gazdar, D. N. Carney, K. L. Becker, L. J. Deftos, V. Liang, W. Go,
P. J. Marangos, T. W. Moody, A. R. Wolfsen, and M. H. Zweig

NCI-Navy Medical Oncology Branch, National Cancer Institute and Naval Hospital,
Bethesda, MD 20814, USA

Introduction

Lung cancer is occurring in near-epidemic proportions in many parts of the world. In the
United States of America it is the leading cause of cancer deaths, being diagnosed in
about 135000 patients every year, nearly 90% of whom will die of their disease (Minna et
al. 1982).

For multiple reasons, some of which will be discussed below, lung cancers are frequently divided into two major groups: small cell lung cancer (SCLC), having a characteristic
morphology, early metastatic spread and responsiveness to cytotoxic therapy; and non-
SCLC lung cancers (adeno-, squamous cell and large cell carcinomas), which have diverse
histologies but are relatively resistant to cytotoxic therapy. Thus, the management and
therapy of SCLC and non-SCLC are markedly different, and depend on accurate diagnosis (Minna et al. 1982).

Numerous studies have documented the presence of markers in lung cancer. One reason for these studies is that SCLC is the human tumor that is most often associated with
paraneoplastic syndromes (Minna and Bunn 1982). In addition to overt production and
secretion of markers by SCLC, there is an even higher frequency of occult secretion of
peptides and other markers (Hansen 1981; Silva 1984). Humoral markers may play an important role in early diagnosis, staging, and monitoring of response to therapy (Hansen
1981; Silva 1984). However, humoral markers in cancer patients may have origins other
than from tumor tissue (Becker et al. 1984).

We have studied the expression of intracellular markers in vitro using cell lines established from lung cancers. There are several reasons for these studies: (a) any markers
found originate in the cultured tumor cells; (b) some markers may not be actively secreted; (c) cell lines offer uniform, viable, rapidly dividing tumor cell populations free of con-
taminating stromal cells; and (d) physiological studies such as clonal expression, induction by secretogogues, and gene expression are easier to perform. The major disadvantage
of in vitro studies is that selective retention or loss of markers may occur.

In vitro studies conducted by ourselves and others have resulted in major advances in
our understanding of the biology of lung cancer, especially SCLC (Gazdar et al. 1983;
Gazdar 1984). We now know that SCLC is an endocrine tumor and expresses all the properties of APUD cells (Gazdar and Carney 1984). These properties include the presence of
the key APUD cell enzymes L-dopa decarboxylase (DDC) (Baylin et al. 1980; Gazdar et
al. 1980) and neuron-specific enolase (NSE) (Marangos et al. 1982) and the frequent se-

cretion of one or more peptides (Sorenson et al. 1981). A peptide hormone present in almost all SCLC tumors and cultures is the amphibian neuropeptide bombesin, or its larger molecular weight mammalian homologue gastrin-releasing peptide (Moody et al. 1981; Wood et al. 1981; Tsutsumi et al. 1983; Abe et al. 1984). Until the peptide present in SCLC is sequenced we will refer to it as bombesin-like immunoreactivity (BLI).

In addition to APUD cell properties and peptides, another near-constant nonendocrine marker present in exceedingly high concentrations in virtually all SCLC tumors and lines is the brain (BB) isoenzyme of creatine kinase (CK-BB) (Gazdar et al. 1981).

Most SCLC cell lines express characteristic morphological and biochemical properties, and are referred to as classic SCLC (SCLC-C) lines. However, some have altered (non-SCLC) morphologies, and lack some of the biochemical properties (Gazdar 1984). These lines are referred to as variant SCLC (SCLC-V) lines (Carney et al., this volume).

We have used an extensive panel of well-characterized SCLC-C, SCLC-V, and non-SCLC cell lines to study marker expression in vitro and to define the characteristic biochemical profile of these cell types (Carney et al., this volume).

Materials and Methods

Cell Lines

Methods for establishing, maintaining, characterizing, and cloning cell lines from human lung cancers in serum-containing and serum-free fully defined media have been described elsewhere (Gazdar et al. 1980; Carney et al. 1981; Pettengill et al. 1984). A panel of 50 cell lines (33 SCLC-C, 10 SCLC-V, and 7 non-SCLC) were used for most of the marker studies; in all cases at least 40 lines were studied. Most SCLC lines were initiated from metastatic tumors obtained from both newly diagnosed and previously treated patients. The non-SCLC lines included representatives of all of the major cell types, and most were started from primary tumors obtained from newly diagnosed cases.

Assays

Cell pellets were harvested from actively growing cultures 48 h after a medium change, washed three times, and frozen at −70 °C prior to assay. In some instances, supernatant fluids were also harvested, clarified, and assayed. Cell extractions and assays were performed in the laboratories of the various collaborating authors using previously published radioimmunoassay methods with polyclonal antibodies. DDC, for which no suitable antibody is available, was assayed radiometrically (Gazdar et al. 1980).

Results

Definition and Characterization of Classic and Variant SCLC Lines

Of the four biochemical markers previously associated with SCLC, high concentrations of CK-BB (400 ng/mg soluble protein) and NSE (100 ng/mg) were present in all SCLC lines. In contrast, DDC and BLI were expressed in 70% of the SCLC lines (these two markers were expressed together with a concordance of 98%). All four markers were ab-

Table 1. Biochemical phenotypes of lung cancer cell lines

Cell line type	Marker			
	NSE	CK-BB	DDC	BLI
SCLC-C	+	+	+	+
SCLC-V	+	+	−	−
Non-SCLC	−	−	−	−

Table 2. Further characterization of SCLC-C and SCLC-V cell lines

Characteristic	SCLC-C	SCLC-V
Cytological appearance	SCLC	Frequently non-SCLC
Histology of xenografts	SCLC, intermediate subtype	Frequently non-SCLC
Substrate adherence	Absent	Usually absent
Culture appearance	Tight aggregates	Loose aggregates
Doubling time	Long	Short
Colony-forming efficiency	Low	Relatively high
Amplification of c-*myc* oncogene	Absent	Frequently present

sent from non-SCLC lines. On the basis of these studies, we can divide our panel of lung cancer lines into three phenotypes (Table 1): Classic SCLC (SCLC-C) lines expressing all four markers; variant SCLC (SCLC-V) lines, which express NSE and CK-BB but lack DDC and BLI; and non-SCLC lines, which lack all four markers

Further characterization of the SCLC cell lines (Table 2) indicated that SCLC-C lines had characteristic SCLC cytology; that xenografts induced by inoculation of cultured cells into athymic nude mice histologically resembled SCLC, intermediate subtype; they lacked substrate adherence and grew as tightly packed floating cell aggregates; they had long doubling times and cloning efficiencies; and they lacked amplification of the c-*myc* oncogene. SCLC-V lines with altered morphology and/or biochemistry could be further divided into two subclasses: (a) Biochemical variants which had the caracteristic variant biochemical phenotype but which otherwise were similar to SCLC-C lines in morphology and growth characteristics; and (b) morphological variants, which cytologically and histologically resembled large cell undifferentiated carcinoma; which grew as loose floating aggregates or attached to the substrate; which had relatively short doubling times and high cloning efficiencies; and which had amplification and increased expression of the c-*myc* oncogene (Little et al. 1983). While there are major biological (and perhaps clinical) differences between these two variant subclasses, there were no significant differences in marker expression, and their data are presented together in this communication. In a larger series of 50 SCLC lines the incidence of these subclasses was as follows: SCLC-C 70%, SCLC biochemical variants 10%, SCLC morphological variants 20%.

Table 3. Peptide expression in lung cancer cell lines[a]

Peptide	Cell line type (%positive)		
	SCLC-C	SCLC-V	Non-SCLC
BLI	100	0	0
Calcitonin	67	20	0
AVP	67	20	29
Lipotropin	58	10	14
Neurotensin	55	0	0
ACTH	48	40	29

[a] Up to 50 cell lines (SCLC-C=33, SCLC-V=10, non-SCLC=7) were tested for the presence of intracellular peptides. In addition to the peptides presented above, the following peptides were not detected in any line of any cell type: met-enkephalin, CCK-gastrin, GIP, VIP, and substance P

Table 4. Relative concentrations of peptides expressed in SCLC-C cell lines

Peptide	Relative concentration[a]
BLI	1.0
Calcitonin	0.06
ACTH	0.04
Neurotensin	0.03
Lipotropin	0.01
Arginine vasopressin	0.004

[a] The relative molar concentrations of peptides frequently present in SCLC-C cell lines are expressed as fractions of the BLI concentration (1840 fmol/mg soluble protein)

Table 5. Expression of non-APUD cell markers in lung cancer cell lines

Marker	Cell line type (% positive)		
	SCLC-C	SCLC-V	Non-SCLC
CEA	60	0	0
Alpha HCG	14	0	0
Beta HCG	7	0	0
PTH	0	0	0

The same cell lines used to generate the data presented in Table 3 were used in this study. Homogenates of cell pellets were examined for the presence of carcinoembryonic antigen and of alpha and beta chains of HCG and parathormone

Peptide Expression in Lung Cancer Cell Lines

We studied expression of 11 peptides associated with APUD cells in homogenates of lung cancer cell lines (Table 3). Five of these (met-enkephalin, cholecystokinin-gastrin, gastrin-inhibitory peptide, vasoactive intestinal peptide, and substance P) were not detected in any cell type. The other six peptides (BLI, calcitonin, arginine vasopressin, lipotropin, neurotensin, ACTH) were frequently expressed in SCLC-C lines (48%–100%), but less often and in lower concentrations in SCLC-V and non-SCLC lines (0–40%). It is of interest that ACTH and lipotropin were expressed concordantly in 82%. BLI and neurotensin were only present in SCLC-C lines.

While six peptides were frequently expressed in SCLC-C lines, their relative mean molar concentrations varied widely (Table 4). BLI was present in the highest concentration (1840 fmol/mg). The mean concentrations of the other peptides were 17- to 250-fold lower.

Expression of Non-APUD Cell Markers

In addition to CK-BB, the following non-APUD cell markers were measured (Table 5). Carcinogenic antigen (CEA), alpha and beta chains of human chorionic gonadotropin (HCG), and parathormone (PTH). None of the markers were present in any SCLC-V or non-SCLC line. Low concentrations of alpha and beta HCG were present (nonconcordantly) in a few SCLC-C cell lines. However, CEA was present in 60% of SCLC-C lines.

Clonal Expression of Markers

We studied over 20 clones derived from 3 SCLC-C lines and 8 clones derived from an SCLC-V line for expression of the 4 basic SCLC markers (NSE, CK-BB, DDC and BLI). All clones retained their respective parental phenotypes.

In addition, an SCLC-C line expressing multiple peptides (BLI, calcitonin, ACTH, and arginine vasopressin) was cloned. All clones expressed BLI, but there was clonal heterogeneity of expression of the other peptides.

Other Studies with BLI and Calcitonin

We performed further studies with BLI and calcitonin, the two peptides expressed most frequently and in the highest concentrations in SCLC-C lines. Pellets and supernatant fluids of 24 SCLC-C lines in culture for periods of 4–60 months were studied. There was excellent concordance between peptide expression and secretion (80% for BLI, 96% for calcitonin).

BLI was expressed in all SCLC-C lines irrespective of culture time. However, calcitonin expression decreased with culture time, from 90% in newly established cultures (mean culture time 6 months to 50% in long-term cultures (mean culture time 54 months).

Discussion

Lung cancers can be divided into SCLC and non-SCLC types for a number of pathological, clinical, therapeutic and biologic reasons. However, a significant number of SCLC tumors contain non-SCLC subpopulations on diagnosis, and the percentage is much higher at autopsy (Gazdar 1984). These mixed small cell-large cell carcinomas have significantly inferior responses to therapy and shorter survival times (Radice et al. 1982) than SCLC-C tumors. We have identified what we believe to be the in vitro equivalent of these mixed tumors, and refer to them as morphological variants of SCLC. The latter have altered morphologies, growth characteristics, and biochemical profiles. Most are associated with amplification and increased expression of the c-*myc* oncogene (Little et al. 1983). In addition, we have identified a second variant subclass, biochemical variants, which have the characteristic variant biochemical profile but which otherwise resemble SCLC-C lines. While we do not know the clinical and therapeutic significance (if any) of biochemical variants, morphological variants are radioresistant in vitro (Carney et al. 1983). However, because we did not detect significant differences in marker expression between the two variant subclasses, we have combined their data in this study.

Use of the four basic SCLC markers, NSE, CK-BB, DDC, and BLI, reveals that each lung cancer cell type (SCLC-C, SCLC-V, and non-SCLC) has its own distinctive biochemical profile. These and other studies have convincingly demonstrated that SCLC is an endocrine tumor and belongs in the APUD cell category of Pearse (1969). In fact, SCLC is by far the commonest APUD cell tumor occurring in humans (Minna et al. 1982). We studied the expression of 11 APUD cell peptides. Five of these were not detected in any type of lung cancer cell line. However, the other six were present frequently in SCLC-C lines (48%–100%), and occasionally (0–40%) in SCLC-V and non-SCLC lines. Two peptides, BLI and neurotensin, were detected only in SCLC-C lines. It is of interest that three of the six peptiedes detected in SCLC-C lines are also present in pulmonary endocrine cells found in normal or dysplastic bronchi (Cutz et al. 1981). Thus, secretion of these peptides by SCLC-C should be recognized as eutopic rather than ectopic secretion. There was a high degree of concordance between DDC and BLI expression in SCLC-C and SCLC-V lines, suggesting that peptide and amine production may be coregulated in APUD cells.

While many peptides may be present in SCLC-C tumors and cell lines, several data suggest that BLI expression may be in a class by itself. First, it is present in all, or virtually all, SCLC-C tumors in concentrations that are 1–2 logs higher than the other peptides. Second, it is clonally and temporally retained, along with three other basic SCLC markers. BLI appears to be an example of autocrine secretion, that is to say, SCLC cells produce it (Moody et al. 1981), secrete it (Moody et al. 1982), and have receptors for it (Moody et al. 1983), and it serves as a growth-stimulatory factor (Oie et al. 1984).

In addition to CK-BB, another non-APUD marker, CEA, was present in 60% of SCLC-C lines, but was not found in other lung cancer types in vitro. This is a suprising finding, and it does not correlate with clinical studies. It may represent selective retention of CEA by SCLC-C cultures for reasons as yet unknown.

In vitro marker studies using continuous lung cancer cell lines are frequently (but not always) representative of fresh tumor samples. They have greatly expanded our knowledge of the biology of lung cancer, and have provided important new diagnostic tools for cell typing and therapy selection.

References

Abe K, Kamey T, Yamaguchi K, Kikuchi K, Adachi I, Tanaka M, Kimura S, Kodama T, Shimosato Y, Ishikawa S (1984) Hormone producing lung cancers. In Becker KL, Gazdar AF (eds) The endocrine lung in health and disease. Saunders, Philadelphia, pp 549–595

Baylin SB, Abeloff MD, Goodwin G, Carney DN, Gazdar AF (1980) Activities of L-dopa decarboxylase and diamine oxidase (histaminase) in human lung cancers: The decarboxylase as a marker for small (oat) cell lung cancer in tissue culture. Cander Res 40: 1990–1996

Becker KL, Silva OL, Gazdar AF, Snider RH, Moore CF (1984) Calcitonin and small cell carcinoma of the lung. In: Becker KL, Gazdar AF (eds) The endocrine lung in health and disease. Saunders, Philadelphia, pp 528–548

Carney DN, Bunn PA, Gazdar AF, Pagan JA, Minna JD (1981) Selective growth in serum-free hormone supplemented medium of tumor cells obtained by biopsy from patients with small cell carcinoma of the lung. Proc Natl Acad Sci USA 78: 3185–3189

Carney DN, Mitchell JB, Kinsella TJ (1983) In vitro radiation and chemosensitivity of established human small cell lung cancer and its large cell variants. Cancer Res 43: 2806–2811

Cutz E, Chan W, Track NS (1981) Bombesin, calcitonin and leu-enkephalin immunoreactivity in endocrine cells of human lung. Experientia 37: 765–772

Gazdar AF (1984) The biology of endocrine tumors of the lung. In: Becker KL, Gazdar AF (eds) The endocrine lung in health and disease. Saunders, Philadelphia, pp 448–459

Gazdar AF, Carney DN (1984) Endocrine properties of small cell carcinoma of the lung. In: Becker KL, Gazdar AF (eds) The endocrine lung in health and disease. Saunders, Philadelphia, pp 501–508

Gazdar AF, Carney DN, Russell EK, Simms HL, Baylin SB, Bunn PA, Guccion JG, Minna JD (1980) Small cell carcinoma of the lung: Establishment of continuous clonable cell lines having APUD properties. Cancer Res 40: 3502–3507

Gazdar AF, Zweig MH, Carney DN, Van Stierteghen AC, Baylin SB, Minna JD (1981) Levels of creatine kinase and its brain isoenzyme in lung cancer tumor and cultures. Cancer Res 41: 2773–2777

Hansen H (1981) Clinical implications of ectopic hormone production in small cell carcinoma of the lung. Dan Med Bull 28: 221–236

Little CD, Nau MM, Carney DN, Gazdar AF, Minna JD (1983) Amplification and expression of the c-myc oncogene in human lung cancer cell lines. Nature 306: 194–196

Marangos PJ, Gazdar AF, Carney DN (1982) Neuron specific enolase in human small cell carcinoma cultures. Cancer Lett 15: 67–71

Minna JD, Bunn PA (1982) Paraneoplastic syndromes. In: De Vita VT, Hellman S, Rosenberg SA (eds) Principles and practice of oncology. Lippincott, Philadelphia, pp 1476–1517

Minna JD, Higgins GA, Glatstein EJ (1982) Cancer of the lung. In: De Vita VT, Hellman S, Rosenberg SA (eds) Principles and Practice of oncology. Lippincott, Philadelphia, pp 96–474

Moody TW, Pert CB, Gazdar AF, Carney DN, Minna JD (1981) High levels of intracellular bombesin characterize human small cell lung cancer. Science 214: 1246–1248

Moody TW, Russell EK, O'Donohue TL, Linden CD, Gazdar AF (1982) Bombesinlike peptides in small cell lung cancer: Biochemical characterization and secretion from a cell line. Life Sci 32: 487–493

Moody TW, Bertness V, Carney DN (1983) Bombesin-like peptides and receptors in human tumors. Peptides 4: 683–686

Oie HK, Brower M, Carney DN (1984) Growth factor requirements for in vitro growth of endocrine and nonendocrine lung cancers in serum-free defined media. In: Becker KL, Gazdar AF (eds) The endocrine lung in health and disease. Saunders, Philadelphia, pp 469–475

Pearse AGE (1969) The cytochemistry and ultrastructure of polypeptide hormone producing cells of the APUD series and the embryologic, physiologic and pathologic implications of the concept. J Histochem Cytochem 17: 303–313

Pettengill OS, Carney DN, Sorenson GD, Gazdar AF (1984) Establishment and characterization of

cell lines from human small cell carcinoma of the lung. In: Becker KL, Gazdar AF (eds) The endocrine lung in health and disease. Saunders, Philadelphia, pp 460–468

Radice PA, Matthews MJ, Ihde DC, Gazdar AF, Carney DN, Bunn PA, Cohen MH, Fossieck BE,
Makuch RW, Minna JD (1982) The clinical behavior of mixed small cell/large cell bronchogenic
carcinoma compared to pure small cell subtypes. Cancer 50: 2894–2902

Silva OL (1984) Humoral correlates of small cell carcinoma of the lung in blood and urine. In:
Becker KL, Gazdar AF (eds) The endocrine lung in health and disease. Saunders, Philadelphia,
pp 516–527

Sorenson GD, Pettengill OS, Brinck-Johnson, Cate CC, Maurer LH (1981) Hormone production by
cultures of small cell carcinoma of the lung. Cancer 47: 1289–1296

Tsutsumi Y, Osamura Y, Watanabe K, Yanaihara N (1983) Immunohistochemical studies on
gastrin-releasing peptide and adrenocorticotropic hormone containing cells in human lung. Lab
Invest 48: 623–632

Wood SM, Wood JR, Ghatei MA, Lee YC, O'Shaughnessy D, Bloos SR (1981) Bombesin, somatostatin and neurotensin-like immunoreactivity in bronchial carcinoma. J Clin Endocrinol Metab
53: 1310–1312

Neurotensin in Human Small Cell Lung Carcinoma

J. G. Reeve, M. Goedert, P. C. Emson, and N. M. Bleehen

Clinical Oncology and Radiotherapeutics Unit and Neurochemical Pharmacology Unit, MRC Centre, Hills Road, Cambridge, CB2 2QH, Great Britain

A question of fundamental importance in the biology of human lung cancer is whether the hormones produced by lung tumours exert a positive influence on the growth of the tumour through an autocrine function or whether hormone production simply reflects a derepression of genetic material consequent to dedifferentiation. The former possibility would require the presence of hormone receptors on tumour cell membranes.

In the present communication we have characterized and quantified neurotensin-like immunoreactivity (NTLI) in tissue culture material derived from small cell lung carcinomas (SCLC) and non-small cell lung carcinomas (NSCLC) and in post-mortem human lung tissue.

In the light of recent data (Yamaguchi et al. 1983) suggesting that the bombesin-like immunoreactivity exhibited by SCLC tumours (e. g., Moody et al. 1981; Sorenson et al. 1982) is attributable to the presence of the structurally related mammalian peptide hormone, gastrin-releasing peptide (GRP), we have also measured the levels in lung tumour material of the amphibian peptide, xenopsin, which is structurally related to mammalian neurotensin.

Extracts of lung tumour cell pellets and post-mortem lung tissue were assayed for neurotensin and xenopsin using antisera directed against the amino- and carboxy-terminus end of mammalian neurotensin and of amphibian xenopsin, as described previously (Emson et al. 1982; Greco and Oldham 1979).

For characterization of the immunoreactive material, samples were applied to a Sephadex G25 column and fractions were assayed using antisera directed against the amino- and the carboxy-terminus or neurotensin. The material corresponding to the NTLI peak was also subjected to high-performance liquid chromatography (HPLC). Fractions were again assayed for neurotensin by using amino- and carboxy-terminus-directed antisera.

According to these methods, high levels (> 4000 fmol/mg protein) of NTLI were present in extracts of all the SCLC cell lines investigated. With both gel filtration and reverse-phase HPLC the immunoreactivity emerged as a single peak in the same position as synthetic neurotensin. The amphibian peptide xenopsin could not be detected in any of the SCLC lines. In contrast to the SCLC cell cultures, only very low levels (< 25 fmol/mg protein) of NTLI were found in post-mortem human lung tissue, and no NTLI was detectable in NSCLC cultures.

To determine whether or not neurotensin in SCLC has an autocrine mode of action on the tumour cells, some of the SCLC cultures were examined for the presence of neurotensin receptors by using a test-tube binding assay (Goedert et al. 1984). Washed SCLC membrane preparations were incubated in $2\,nM\,^3$H-neurotensin. Non-specific binding was defined as binding in the presence of $1\,\mu M$ neurotensin. Membranes were then filtered and washed, and the radioactivity was determined by liquid scintillation spectrometry. This assay yielded no evidence for the presence of neurotensin receptors in any of the small cell carcinoma lines examined.

On the basis of these findings it seems unlikely that neurotensin has an autocrine mode of action in the SCLC cultures examined thus far. As for GRP, it is the mammalian peptide, neurotensin, and not the structurally related amphibian peptide xenopsin that is present in the SCLC cultures. The possible function of NTLI in human SCLC is unknown at present. However, it is conceivable that the immunoreactive material is released into the blood stream where it may be implicated in the pathogenesis of paraneoplastic syndrome.

References

Emson PC, Goedert M, Horsfield P, Rioux F, St Pierre S (1982) The regional distribution and chromatographic characterization of neurotensin-like immunoreactivity in the rat central nervous system. J Neurochem 38: 992–999

Goedert M, Sturmey N, Williams BJ, Emson PC (1984) The comparative distribution of xenopsin- and neurotensin-like immunoreactivity in *Xenopus laevis* and rat tissues. Brain Res 308: 273–280

Greco FA, Oldham RK (1979) Small cell lung cancer. N Engl J Med 301: 355–358

Moody TW, Pert CB, Gazdar RF, Carney DN, Minna JD (1981) High levels of intracellular bombesin characterize human small cell lung carcinoma. Science 214: 1246–1248

Sorenson GD, Bloom SR, Ghatei MA, Del Prete SA, Cate CC, Pettingill OS (1982) Bombesin production by human small cell carcinoma of the lung. Regul Pept 4: 59–66

Yamaguchi K, Abe K, Kameya T, Adachi I, Taguchi S, Otsubo K, Yanaihara N (1983) Production and molecular size heterogeneity of immunoreactive gastrin releasing peptide in fetal and adult lungs and primary lung tumours. Cancer Res 43: 3932–3939

Clinical Implications of Peptide Hormone Production in Lung Cancer

Peptide Hormones as Tumor Markers in Lung Cancer Patients

S. Krauss

University of Tennessee, Memorial Research Centre, Knoxville, TN, USA

Clinicians have long sought simple, reproducible, and specific laboratory tests to detect malignant neoplasms early and monitor their course. In this regard, a variety of tumor products have been termed biomarkers, since their estimation in blood or other body fluids may indicate the existence of a malignancy and the body burden of tumor. In addition, the study of these substances may shed light on the origin, histogenesis, biological behavior, and interrelationships of the tumor cells producing them. Such biomarkers include several polypeptide hormones, enzymes, biogenic amines, and tumor-associated antigens. To be useful to the clinician, an ideal tumor marker should have the following characteristics

1. It should be produced only by tumor cells
2. Its concentration in blood or urine should reflect the body burden of tumor
3. Measurement should be sensitive enough to detect microscopic or subclinical disease
4. The assay should be convenient, inexpensive, and reproducible
5. The marker should occur with sufficient frequency to make measurement worthwhile
6. The practical value of the marker will depend on the availability of effective treatment for the cancer.

Clinicians treating lung cancer are faced with a disease of epidemic proportions, which is the most frequent cause of cancer deaths in men and has now become the second most frequent cause of cancer deaths in women. With the use of cigarettes being heavily promoted in Third World countries, we can expect the epidemic to reach global proportions.

The potential areas in which biomarkers could be useful in lung cancers include the following:

1. Screening and early diagnosis
2. Clinical staging
3. Identification of tumor cell type, especially where mixed cell types may be present in the same patient
4. Monitoring response to therapy and detecting subclinical disease in relapse
5. As a component of in vivo or in vitro drug sensitivity assays for detection of antitumor activity against established tumor lines or fresh tumors from patients.

The candidates for consideration as potentially useful biomarkers in lung cancer are many (Table 1). These fall into several categories, including polypeptide hormones, enzymes, biogenic amines, tumor-associated antigens, and morphological markers.

Recent Results in Cancer Research. Vol. 99
© Springer-Verlag Berlin · Heidelberg 1985

Table 1. Candidates for tumor biomarkers in lung cancer

Substance	Frequency of occurrence in (cell type)	
	Small cell	Non-small cell
Polypeptide hormones		
ACTH[a]	+ +	+
ADH (Arginine vasopressin)[a,b] neurophysins	+ +	±
Calcitonin[b]	+ + +	+
Bombesin[c] (gastrin-releasing peptides	+ + +	+
Human placental lactogen		
Human chorionic gonadotropin	+	+
Parathyroid hormone	−	+
Growth hormone		
Enzymes		
Dopa decarboxylase[c]	+ +	−
Neuron-specific enolase[b]	+ +	−
Histaminase[c]	+ +	−
Creatine kinase[c]	+ +	−
Biogenic amines[c]		
Serotonin	+	−
5-Hydroxytryptophan	+	−
Histamine	+	−
Tumor-associated antigens		
Carcinoembryonic antigen[b]	+ +	+ +
Cell surface proteins[c]		
Morphological markers[c]		
Electron-dense granules[b]	+ +	±
Specific chromosome abnormality	+	−

[a] Clinical syndrome associated with elevated marker level
[b] Clinically useful for monitoring therapy
[c] Primarily identified in tumor tissue

As indicated in Table 1, a large number of polypeptide hormones have been identified in the plasma and tumor tissue of lung cancer patients. Although most are found in the greatest frequency and concentration in small cell lung cancer (SCC), no one polypeptide marker appears to be unique to a given cell type. In the individual case, any combination of markers may be identified. The current clinical status of biomarkers in the management of a patient with lung cancer may be appreciated by comparing immunoreactive calcitonin as a prototype, with the hypothetical "ideal" biomarker for which all the following questions would be answered in the affirmative:

Has calcitonin been identified as a tumor product? Yes, both by immunohistochemical methods and by its release by cultured small cell lines.

Is it found in the plasma in a high proportion of patients with lung cancer? Yes, in 57%–83% of patients with SCC, although it is associated with the other forms of lung cancer as well.

Do plasma levels reflect the body burden of tumor? There is a rough quantitative correlation between elevated levels and the clinical stage in patients with SCC; the highest levels and the greatest frequency of elevation are seen in patients with extensive disease.

Can subclinical disease be detected? Rising levels of hormone may precede clinical relapse by 2–3 months.

Is a clinical response to cytotoxic therapy accompanied by a decrease in the plasma level of the hormone? Yes, frequently.

Is the species of immunoreactive calcitonin detected unique to the lung cancer? High-molecular-weight forms of calcitonin have been described, and the development of antibodies to these moieties may make the assay more specific as a tumor marker.

Can calcitonin be used as a screening test for lung cancer in high-risk populations? None of the markers so far described has the specificity or sensitivity to detect cancer patients in susceptible groups.

Concerning the other biomarkers listed in Table 1, ACTH, usually in the form of the biologically inactive "big" ACTH molecule, is not present frequently enough to qualify as a clinically useful marker.

Among the more recently described markers, the polypeptide bombesin, one of the gastrin-releasing peptides, and the enzyme neuron-specific enolase, hold promise as specific markers for SCC patients. The circulating markers described, together with morphological markers such as electron-dense granules, specific cytogenetic abnormalities, and tumor-associated cell surface constituents may help in identifying heterogeneous tumor cell populations, which carry a more grave prognosis in patients with SCC.

The challenge to the clinician treating lung cancer is exemplified by the patient with SCC who ist clinically in complete remission but has a more than 80% likelihood of relapsing and dying of the cancer in 18 months to 2 years. Early detection of relapse in a preclinical phase through the use of sensitive biomarkers might enable the physician to reinstitute therapy sooner and regain control of the tumor despite its more aggressive behavior.

Peptide Hormones in Patients with Lung Cancer

M. Hansen and E. Bork

Department of Internal Medicine C, Bispebjerg Hospital, 2400-Copenhagen, Denmark

Introduction

In the early 1970s, the so-called tumor markers indicating the number of tumor cells were introduced in to the management of some types of cancer. Human chorionic gonadotropin (hCG) and calcitonin were found to be valuable in chorioncarcinoma and medullary carcinoma of the thyroid, respectively, in that they could be used to identify a tumour before it was detectable by other means. Similarly, hCG was found to be useful for monitoring the effect of chemotherapy in women with chorioncarcinoma, and hCG and α-fetoprotein in staging and monitoring responses to therapy in men with testicular germ cell cancer. Similarly, gastrointestinal peptide hormones may indicate the effect of chemotherapy in patients with peptide-producing islet-cell tumours.

During the 1970s, the treatment of small cell bronchogenic carcinoma (SCC) was expanded by the use of chemotherapy. Also, SCC was the tumor type in which ectopic hormone production had been described most frequently. It thus seemed interesting to elucidate the clinical usefulness of hormonal polypeptides as tumor markers in SCC. Some investigators concentrated on SCC, while others included all types of lung cancer in their studies. A number of studies have not related their results in detail to the histological types. In these studies it is therefore difficult to make valid comparisons, since elevated concentrations of peptide hormones are mostly found in patients with SCC.

The present paper will focus on the peptide hormone levels in patients with SCC, which is the main subject of our own experience with peptide hormones in lung cancer. Our primary interest has been to evaluate these peptides with regard to their possible therapeutic value as tumor markers. The initial studies included hormonal peptides and amines, while later investigations have included "new" hormones and enzymes, such as bombesin or GRP, neuron-specific enolase, and creatin kinase (brain fraction). The later studies are still in progress but results obtained by other workers will be briefly included.

Our Own Studies

From May 1975 to February 1977, a number of peptides and amines were measured in 75 unselected patients with SCC (Hansen et al. 1980 a, b). The majority of the patients were staged and included in clinical trials of combination chemotherapy. The analysis of calcitonin was omitted in 1 patient, but a pentagastrin stimulation test was performed in 14 of these patients and in 5 others (Hansen et al. 1978). In separate investigations repeated blood sampling was performed in the first few days after the initiation of treatment in 6 patients, while 21 patients were followed up after response and during clinical progression (Hansen et al. 1980c).

Endocrine Syndromes in SCC

Association between an endocrine syndrome and SCC was suggested in 1928 by Brown, but more than 30 years elapsed before ACTH was demonstrated in SCC tumor tissue (Meador et al. 1962). During the same period it also became known that inappropiate ADH secretion might be caused by SCC. These two syndromes are most often caused by SCC but the syndromes are nevertheless uncommon among patients with SCC (Table 1). The frequency of the syndromes in consecutive groups of patients may vary with the criteria used for diagnosing them, i.e., whether hypokalemia and hyponatremia or clinical signs alone are considered. In none of the studies listed in Table 1 were measurements of ACTH and ADH included for definition of the syndromes. Elevation of these peptides is found in a much higher proportion of patients than clinically detectable disease, however.

Table 1. Incidence of endocrine syndromes in small cell lung cancer

	No. of pts	ACTH syndrome %	SIADH %	Carcinoid %
Azzopardi et al. (1970)	39	3	5	0
Rassam and Anderson (1975)	42	5	2	0
Hansen et al. (1980a)	75	1	1	0
Bondy and Gilby (1982)	106	2	11	–
Lokich (1982)	84	5	10	1
Hainsworth et al. (1983)	250	–	7	–

Table 2. Incidence of elevated serum concentrations of calcitonin in untreated patients with small cell carcinoma[a]

	All patients		Limited		Extensive	
	No.	Rate	No. of pts	Rate	No. of pts	Rate
McKenzie et al. (1977)	28	0.75				
Hansen et al. (1978, 1980b)	74	0.64	29	0.59	33	0.73
Greco et al. (1981)	54	0.51				
Sappino et al. (1981)	40	0.25				
Wallach et al. (1981)	24	0.76	5	0.00	19	0.84
Bierbaum et al. (1982)	70	0.70	41	0.60	29	0.84
Luster et al. (1982)	135	0.56				
All studies	425	0.59	75	0.56	81	0.80

[a] Studies including previously treated patients and studies with ≤ 20 patients are excluded

Incidence of Peptide Hormones

The pretreatment frequency of elevated serum *calcitonin* in patients with SCC has been determined in several studies, seven of which are summarized in Table 2. Taking the studies together, serum calcitonin has been measured in 425 untreated patients, and the concentration was found to be elevated in 59% of these cases. In individual studies the incidence varies from 25% to 76%. This variation is not only random. Differences in patient selection and assays may, in fact, explain most of the variation. It would be of considerable interest to characterize a tumor-specific calcitonin precursor molecule and develop an assay that could be performed early in the course. Luster et al. (1982) suggest that this peptide might be valuable in the differential diagnosis of SCC. It is therefore worth noting that elevated serum concentrations of calcitonin are rare in other types of lung cancer than SCC (McKenzie et al. 1977; Roos et al. 1980; Krauss et al. 1981; Luster et al. 1982).

A similar variation in the frequency of elevated concentrations is not found for plasma *ACTH*. In the five studies included in Table 3 the rate er of elevated concentrations varies from 24% to 30%, with a calculated average of 27%. Two studies including previously treated patients found a lower rate of 11% (Gilby et al. 1975) and a higher rate of 37% (Krauss et al. 1981). In most cases the increased concentrations are only marginal, but in

Table 3. Incidence of elevated plasma concentrations of ACTH in untreated patients with small cell carcinoma[a]

	All patients		Limited		Extensive	
	No.	Rate	No. of pts	Rate	No. of pts	Rate
Gropp et al. (1980)	50	0.30				
Hansen et al. (1980b)	75	0.29	29	0.34	34	0.29
Ratcliffe et al. (1982)	63	0.24	25	0.12	38	0.32
Spaulding et al. (1983)	32	0.25				
Bork et al. (in preparation)	32	0.28				
All studies	252	0.27	54	0.24	72	0.31

[a] Studies including previously treated patients and studies with ≤ 20 patients are excluded

Table 4. Incidence of at least one elevated marker in marker panels in small cell carcinoma

	Markers	No. of pts	Rate
Gropp et al. (1980)	ACTH	50	0.78
	Calcitonin		
	PTH		
	hCG-β		
Hansen et al. (1980a)	ACTH	75	0.84
	ADH		
	Calcitonin		
Sappino et al. (1981)	Calcitonin	40	0.50
	CEA		
	hCG-β		

other histological types of lung cancer than SCC even marginal elevations of plasma ACTH have been found to be rare (Ratcliffe et al. 1982; Davis and May 1982). In contrast, immunoreactive "big" ACTH was frequently found to be elevated in all types of lung cancer (Gewirtz and Yalow 1974; Wolfsen and Odell 1979), and the incidence was even higher than that of plasma ACTH in SCC. Examinations of tumor tissue revealed a higher content of big ACTH in SCC than in other types (Yesner 1978), but further evaluation of big ACTH in plasma related to the histological types has not appeared. It might be that the incidence of big ACTH was much higher in SCC than the incidence of plasma 1–39 ACTH. The difference between these assays has been suggested to be a different pattern of enzymes for degradation of macromolecules in the different histological types of lung cancer (Hansen 1981), or detection of molecular forms other than 1–39 ACTH by the assays for big ACTH (Ratcliffe et al. 1982).

Antidiuretic hormone (ADH) has also been investigated in a number of studies (for review, see Hansen 1981). The concentration of ADH depends on the concomitant osmolality in plasma. Plasma ADH is inappropriately increased in about one-third of patients with SCC, but a water loading test reveals that a much larger number of patients have impaired water balance. It is, however, of considerable interest that elevated plasma concentrations of ADH-neurophysin have been detected in 67 of 103 patients (65%) with SCC (Maurer et al. 1983). This peptide might be more useful as a tumor marker than ADH, and it has been discussed by North et al. (this volume).

A number of *other peptides* and amines have been investigated during the last decade, but in most cases these were found not to be significantly elevated in patients with SCC (Gropp et al. 1980; Hansen et al. 1980b; Bondy and Gilby 1982).

At present, clinical studies on *bombesin* are of interest. This peptide has been demonstrated in pulmonary endocrine cells in human SCC grown in nude mice and in most SCC cell lines. Bombesin was isolated from frog skin in 1971 and it has been found to have a marked stimulatory effect on gastrointestinal hormones. A larger peptide with almost identical biological actions has been isolated from porcine intestinal tissue. This peptide is called gastrin-releasing peptide (GRP) and it is considered to be the mammalian counterpart of bombesin. Preliminary results from our ongoing study have not yet revealed significantly elevated plasma concentrations of GRP.

The most promising marker for SCC found within recent years is not a hormonal peptide but *neuron-specific enolase* (NSE). This isoenzyme has also been demonstrated in SCC cell lines, and in clinical studies the serum concentrations of NSE have been found to be elevated in 65%–80% of patients with SCC (Carney et al. 1982; Ariyoshi et al. 1983). It is also noted in these studies that elevated serum NSE was often found in patients with advanced disease, and that the level of NSE corresponded to the clinical course of the disease during treatment. Further investigation of NSE is obviously of interest.

Peptide Hormones in Cerebrospinal Fluid

A specific use of peptides as tumor markers might be in the diagnosis of the spread of SCC to the CNS. Preliminary data have apparently excluded the use of calcitonin (Hansen et al. 1980d), while a limited role of ACTH and ADH might be defined (Pedersen et al. 1982). Preliminary results with NSE determined in the cerebrospinal fluid appear to be more promising than those recorded with the hormonal peptides (Pedersen et al. 1984). These studies are to be published in detail in the near future.

Therapeutic Implications of Peptide Hormones

An essential area of interest in evaluating the occurrence of peptides in patients with SCC is the determination of markers which can be used in the clinical management of patients with SCC. No single marker for which elevated levels can be measured in all patients with SCC has yet been discovered, and it may be that none exists. It might therefore be worthwhile to look at marker panels so that at least one marker might be detected in every patient. This was considered in the three investigations listed in Table 4. In the present authors' experience, PTH and CEA are not realistic tumor markers in SCC. Measurement of calcitonin was included in all the panels, while hCG-β and ACTH were included in two of the three studies. In all the studies, the incidence of (at least one) elevated marker was, of course, higher than for calcitonin alone, but was still far from 100%. It might be worthwile to consider marker panels including calcitonin, NSE, and ADH-neurophysin.

It appears from Table 2 and, to some extent also from Table 3, that elevated markers are more frequent in patients with extensive disease than in patients with limited disease. Similar results have been obtained with ADH (Comis et al. 1980), neurophysin (Maurer et al. 1983), and NSE (Carney et al. 1982; Ariyoshi et al. 1983). These findings support the view that a relation exists between the tumor burden and the levels of the markers. Accordingly, in the individual patient the marker level may increase with progressing tumor burden, but the absolute level of the marker tells nothing about the tumor burden in the individual patient, since the presence of markers varies from patient to patient and may also vary within the same patient with time.

At the same time, elevated levels of the peptides occur independently of each other (Hansen et al. 1982a). Accordingly, the markers cannot be used to define the stage of disease or the metastatic sites in the individual patient. The main purpose of staging is to obtain an impression of the prognosis, for example for stratification in clinical studies. In some studies, a definite trend against a poorer prognosis with elevated peptide levels has been found (Hansen et al. 1980a; Sappino et al. 1981, Lokich 1982). However, none of the studies indicate that measurements of the hormonal peptide markers might improve on the prognostic evaluation as normally performed with staging and the performance status.

If the levels of markers are higher in patients with extensive disease and response to treatment is less frequent in such patients, responses might be related to the peptide levels. In a recent investigation including 40 patients with SCC (Sappino et al. 1981) it appeared that the response rate achieved was related to the markers, i.e., lower response rates with more than one elevated marker and in particular with elevated serum calcitonin. In 61 patients with SCC this trend was not found with respect to either the number of markers or serum calcitonin (Hansen et al. 1980a).

Serial measurements during chemotherapy have been performed in a number of studies. It is generally accepted that high pretreatment levels are decreased in responding patients.

In the long-term follow-up and particularly in the detection of subclinical relapses, monitoring of ACTH appears to be of little value (Hansen et al. 1980c; Bondy and Gilby 1982; Afrikian and Abramov 1982; Spaulding et al. 1983), but several authors find monitoring of calcitonin to be promising (Wallach et al. 1981; Luster et al. 1982; Bierbaum et al. 1982). Despite some small differences in the interpretation of results obtained in the different studies, a well-defined role of the markers in decision-making relating to therapy has not yet been presented. One of the main problems appears to be that initially elevated peptides do not necessarily increase again when the patient relapses, and it has not been convincingly demonstrated that the increase of initially normal levels in all cases indicates a relapse. These problems may be related to the problem of tumor cell heterogeneity in SCC, and further investigation of this particular problem is necessary before markers can be included in the clinical management of patients with SCC.

Monitoring of therapy by serial measurements of serum calcitonin may be worth evaluating further in combination with other markers. Among such other markers, the literature indicates that the following peptides might be included: neuron-specific enolase (Carney et al. 1982; Ariyoshi et al. 1983), and maybe also γ-MSH (Spaulding et al. 1983) and β-endorphins (Gropp et al. 1980). The most interesting aspect of the use of these peptide markers in monitoring therapy in patients with SCC is whether they can help in defining complete remissions (= cure) and subclinical tumor relapse a considerable time before these become clinically detectable.

References

Afrikian HN, Abramov VF (1982) Functional state of adenohypophyseal-adrenal cortex system in patients with lung cancer. Klin Med (Mosk) 60: 82-85

Ariyoshi Y, Kato K, Ishiguro Y, Ota K, Sato T, Suchi T (1983) Evaluation of serum neuron-specific enolase as a tumor marker for carcinoma of the lung. Gann 74: 219-225

Azzopardi JG, Freeman E, Poole G (1970) Endocrine and metabolic disorders in bronchial carcinoma. Br Med J IV: 528-529

Bierbaum W, Hutoma T, Niederle N, et al (1982) Calcitonin in monitoring patients with small cell carcinoma of the lung during chemotherapy. Cancer Detect Prev 5: 258

Bondy PK, Gilby ED (1982) Endocrine function in small cell undifferentiated carcinoma of the lung. Cancer 50: 2147-2153

Brown WH (1928) A case of pluriglandular syndrome: "diabetes of bearded women". Lancet II: 1022-1023

Carney DN, Ihde DC, Cohen MH, Marangos PJ, Bunn PA Jr, Minna JD, Gazdar AF (1982) Serum neuron-specific enolase: A marker for disease extent and response to therapy of small-cell lung cancer. Lancet I: 583-585

Comis RL, Miller M, Ginsberg SJ (1980) Abnormalities in water homeostasis in small cell anaplastic lung cancer. Cancer 45: 2414-2421

Davis S, May PB (1982) Infrequent elevation of plasma ACTH in patients with bronchogenic carcinoma. Am J Med Sci 283: 2-7

Gewirtz G, Yalow RS (1974) Ectopic ACTH production in carcinoma of the lung. J Clin Invest 53: 1022-32

Gilby ED, Rees LH, Bondy PK (1975) Ectopic hormones as markers of response to therapy in cancer. Excerpta Medica, Amsterdam

Greco FA? Sismani A, Hande K, et al (1981) Plasma calcitonin and antidiuretic hormone (ADH) as markers in small cell lung cancer. Proc Am Assoc Cancer Res 22: 342

Gropp C, Havemann K, Schener A (1980) Extopic hormones at diagnosis and during therapy. Cancer 46: 347-54

Hainsworth JD, Workann R, Greco FA (1983) Management of the syndrome of inappropiate antidiuretic hormone secretion in small cell lung cancer. Cancer 51: 161-165

Hansen M (1981) Clinical implications of ectopic hormone production in small cell carcinoma of the lung. Dan Med Bull 28: 221-236

Hansen M, Hansen HH, Tryding N (1978) Small cell carcinoma of the lung. Serum calcitonin and serum histaminase (diamine oxidase) at basal levels and stimulated by pentagastrin. Acta Med Scand 204: 257-261

Hansen M, Hammer M, Hummer L (1980a) Diagnostic and therapeutic implications of ectopic hormone production in small cell carcinoma of the lung. Thorax 35: 101-106

Hansen M, Hansen HH, Hirsch FR, et al. (1980b) Hormonal polpeptides and amine metabolites in small cell carcinoma of the lung, with special reference to stage and subtypes. Cancer 45: 1432–1437

Hansen M, Hammer M, Hummer L (1980c) ACTH, ADH and calcitonin concentrations as markers of response and relapse in small-cell carcinoma of the lung. Cancer 46: 2062–2067

Hansen M, Hansen HH, Almqvist S, Hummer L (1980d) Cerebrospinal fluid ACTH and calcitonin in patients with CNS metastases from small cell bronchogenic carcinoma. Eur J Cancer 16: 855–857

Krauss S, Macy S, Ichiki AT (1981) A study of immunoreactive calcitonin, adrenocorticotropic hormone and carcinoembryonic antigen in lung cancer and other malignancies. Cancer 47: 2485–2492

Lokich JJ (1982) The frequency and clinical biology of the ectopic hormone syndromes of small cell carcinoma. Cancer 50: 2111–2114

Luster W, Gropp C, Sostmann H, et al (1982) Demonstration of immunoreactive calcitonin in sera and tissues of lung cancer patients. Eur J Cancer Clin Oncol 18: 1275–1283

Maurer LH, O'Donnell JF, Kennedy S, Faulkner CS, Rist K, North WG (1983) Human neuronphysins in carcinoma of the lung: Relation to histology, disease stase, response rate, survival, and syndrome of inappropiate antidiuretic hormone secretion. Cancer Treat Rep 67: 971–976

McKenzie CG, Evans IMA, Hillyard CJ, et al (1977) Biochemical markers in bronchial carcinoma. Br J Cancer 36: 700–707

Meador CF, Liddle GW, Island DP, Nocholson WE, Lucas CP, Nuckton JG, Luetscher JA (1962) Cause of Cushing's syndrome in patients with tumours arising from "non-endocrine" tissue. J Clin Endocrinol Metab 22: 693–703

Pedersen AG, Hansen M, Dombernowsky P, Hammer M, Hummer L (1982) ACTH and ADH as markers of CNS metastases from small cell bronchogenic carcinoma. Proc Am Soc Clin Oncol 1: C-4

Pedersen AG, Marangos P, Gazdar A, Bach F, Hansen HH, Bunn P Jr (1984) Neuron-specific enolase in the cerebrospinal fluid as a marker of CNS metastases in small cell lung cancer. Proc Am Soc Clin Oncol: C-25

Ratcliffe JG, Podmore J, Stack BH, et al (1982) Circulating ACTH and related peptides in lung cancer. Br J Cancer 45: 230–236

Rassam JW, Anderson G (1975) Incidence of paramalignant disorders in bronchogenic carcinoma. Thorax 30: 86–90

Roos BA, Lindall AW, Baylin SB, O'Neil JA, Frelinger AL, Birnbaum RS, Lambert PW (1980) Plasma immunoreactive calcitonin in lung cancer. J Clin Endocrinol Metab 50: 659–666

Sappino AP, Ellison ML, Carter SC, et al (1981) Correlation of 3 tumour markers (calcitonin, CEA and beta-hCG) with response to therapy and extent of disease in small-cell lung cancer. Br J Cancer 44: 300

Spaulding MB, Pederson R, Folts J, et al (1983) A comparison of immunoreactive gamma MSH and ACTH in monitoring carcinoma of the lung. Proc Am Soc Clin Oncol 2: C-25

Wallach SR, Royston I, Taetle R, et al (1981) Plasma calcitonin as a marker of disease activity in patients with small cell carcinoma of the lung. J Clin Endocrinol Metab 53: 602–606

Wolfsen AR, Odell WA (1979) Pro-ACTH: Use for early detection of lung cancer. Am J Med 66: 765–772

Yesner R (1978) Spectrum of lung cancer and ectopic hormones. In: Sommers, Rosen (eds) Pathology annual.

Clinical Evaluation of the Neurophysins as Tumor Markers in Small Cell Lung Cancer*

W. G. North, J. Ware, A. P. Chahinian, M. Perry, J. O'Donnell, and L. H. Maurer

Department of Physiology, Dartmouth Medical School, Hanover, NH 03756, USA

Introduction

Small cell carcinoma of the lung (SCCL) is frequently associated with the ectopic production of peptide hormones, some of which produce significant clinical and laboratory abnormalities in the patient. The syndromes of inappropriate production of antidiuretic hormone (SIADH) and the ectopic secretion of adrenocorticotropic hormone (ACTH) have been well described and occur in 5%–10% of patients suffering from this disease (Kato et al. 1969; Eagan et al. 1974; Lokich 1982). The development of sensitive radioimmunoassays for peptide hormones such as calcitonin, vasopressin, pro-opiomelanocortin and its metabolites (ACTH, lipotropin, B-endorphin), for the neurophysins, and for other proteins, such as carcinoembryonic antigen, neuron-specific enolase, and the BB isoenzyme of creatine kinase, have led to the detection of these substances in SCCL. These peptides and proteins have also been detected in other types of lung cancer, but both the frequency of detection and the quantitative levels are much higher in SCCL. Studies of cell lines of SCCL have contributed to an expansion of the list of hormones detectable in culture. Bombesin is detected in a large percentage of small cell culture lines and may be functioning in an autocrine role (Moody et al. 1981; Sorenson et al. 1982). However, elevated serum levels of bombesin are infrequent (Wood et al. 1982).

The biological significance of hormone production by tumors remains elusive, but the detection of these peptides offers the clinician an opportunity to evaluate them as possible tumor markers. As tumor markers they may provide information about staging, serve as prognostic factors, and prove to be useful in monitoring the effects of therapy. Immunohistochemical techniques when combined with measurement of plasma levels may help define new pathological classifications based on the capability of tumors to produce peptide hormones.

---

* WGN was the recipient of USPHS Research Career Development Award and the investigation was supported by research grants from the National Cancer Institute: grants CA 19613, AM 08469, and CA00552 to WGN; grant CA33601 to JW; grant CA04457 to APC; grant CA12046 to MP; and grant CA04326 to JO'D and LHM.
Some of the data for this paper are derived from CALGB studies 8083 and 8084

Neurophysins

The neurophysins are single-chain peptides with molecular weight of about 10000. They are normally produced in separate neurons in the supraoptic and paraventricular nuclei of the hypothalamus. In the human there are two neurophysins (NPs), one associated with vasopressin (VP) and one with oxytocin (OT). Each neurophysin is qualified by its companion hormone: vasopressin-associated neurophysin (VP-NP) and oxytocin-associated neurophysin (OT-NP). Vasopressin, oxytocin, and their respective neurophysins are initially translated as larger precursors with a molecular weight of about 20000 for VP and VP-HNP (Land et al. 1982) and 15000 for OT and OT-HNP (Land et al. 1983). The prohormone is packaged in neurosecretory granules which migrate down the axon to the neural lobe of the hypophysis. Processing of the prohormones to hormones includes cleavage of the precursor portion of the molecule (North et al. 1983a, b). After appropriate physiologic stimuli the neurophysins are released from the neurosecretory granules along with vasopressin or oxytocin. For vasopressin and VP-NP an increase in plasma osmolality or a decrease in plasma volume will cause release. Nicotine also is a potent stimulus of release in the nonsmoker, but not in habitual smokers. An increased plasma estrogen level is a potent stimulus for oxytocin and OT-HNP secretion (Robinson et al. 1977; North et al. 1980).

North et al. (1980a) have developed radioimmunoassays for VP-HNP and OT-HNP. There are advantages to using these assays in preference to the assays reported for the hormones vasopressin and oxytocin. The assays for the neurophysins can be used with both tissue extracts and in unextracted plasma without interference. The current assays for the hormones do not have this advantage. The cross-reaction for VP-HNP in the assay for OT-HNP is <5% and that for OT-HNP in the assay for VP-HNP is <2%. The combination of the two assays may increase the sensitivity of these peptides as tumor markers.

The Neurophysins as Tumor Markers

Evidence that the neurophysins were present in lung tumors from patients manifesting SIADH was first reported by Hamilton et al. (1972) and was subsequently confirmed by Legros (1975). Sorenson et al. (1981) have reported the neurophysins to be present in many of their small cell tissue cultured cell lines. Yamaji et al. (1981) also studied neurophysins from tumors of five patients with SCCL and SIADH, and found VP-HNP in all five cases and small amounts of OT-HNP in three cases.

The neurophysins were first studied as tumor markers in SCCL by North et al. (1980b). Plasma levels of VP-HNP or OT-HNP or both neurophysins were found to be elevated to more than three times the normal value (>4 SDs for VP-HNP and >3 SDs for OT-HNP) in 65% of patients prior to the onset of therapy; 24% were found to have HNP plasma levels in the normal range (nonsecretors), and an additional 12% had plasma levels which were in an indeterminate range (between 2 and 3 times the normal plasma value). These were classified as possible secretors. In subsequent studies the incidence rates have been similar (North et al. 1983a; Maurer et al. 1983).

As a result of these studies, several conclusions have been reached regarding the neurophysins as tumor markers. Plasma VP-HNP was elevated more frequently than plasma OT-HNP (34% and 13%, respectively), although an additional 20% of patients were found to have an increased level of both neurophysins. Patients with extensive disease had a significantly higher incidence of elevated neurophysins (82%) than patients with limited disease (40%). After correcting for stage, no differences were seen between secretors and nonsecretors with respect to sex, age, response rates to therapy, survival, or histologic subtype of SCCL. When elevated, sequential levels of the plasma neurophysins reflected response to therapy. The observation that both secretors and nonsecretors maintained their secretory status throughout the course of the disease suggests that there may be biochemical differences between tumors presenting as limited disease and those presenting as extensive disease.

Although VP-HNP does not itself cause SIADH, in every instance of this syndrome VP-HNP was elevated. There was a strong correlation with plasma concentration of VP-HNP and SIADH, with values being five to six times greater in the patients with SIADH (Maurer et al. 1983). SIADH was seen in 33% of patients with extensive disease, but only in 7% of those with limited disease.

In patients with non-small cell carcinoma approximately 20% of patients had elevated plasma levels of one or both neurophysins, but quantitatively the levels were much lower than in SCCL (Maurer et al. 1983; North et al. 1983a). The likelihood of making a correct diagnosis in patients with lung cancer on the basis of the presence of an elevated neurophysin (>3 times normal) prior to therapy was 83%, but it could be increased to 95% if cutoff levels for elevated plasma neurophysins were increased to four times the normal value.

Because of these encouraging results, the Cancer and Leukemia Group B (CALGB) began a prospective evaluation of the neurophysins as tumor markers in 1980. The objectives of their studies were to establish (a) the frequency of production of human neurophysins in patients with SCCL entered on cooperative group trials for limited and extensive disease; (b) the correlation between these markers and disease extent; (c) the value in predicting response to therapy; and (d) the differences between subpopulations of secretors and nonsecretors. The data presented here represent a preliminary analysis of these studies.

Methods and Materials

All patients who entered on CALGB trials 8083 and 8084 for limited and extensive SCCL were eligible for this study. At the time of randomization and prior to the onset of therapy, 15 ml blood was drawn into heparinized tubes. After centrifugation the plasma was separated and frozen. The specimens were shipped frozen to the laboratory of Dr. W. G. North for analysis with a double-blind approach for the protocol chairman, the referring physician, and the testing laboratory. The assay procedures were performed according to the methods developed by North et al. (1980a, b). The results were forwarded to the Central Statistical Office and entered on the Chairman's evaluation sheet. At 6–12 weeks after therapy commenced repeat samples were obtained and the assay repeated.

Each patient was evaluated for sites of metastatic disease using physical examination, chest x-ray, liver function tests, bone marrow aspirates or biopsies, and radionuclide scans of liver, bone, and brain, or computed axial tomography of the brain. Staging was assessed in accordance with CALGB criteria, which defines patients with limited disease as those with tumor confined to the lung, mediastinum, and ipsilateral or contralateral supraclavicular lymph nodes; all other patients had extensive disease (Maurer et al. 1980).

All patients received combination chemotherapy. In the group with limited disease all patients received whole-brain radiation therapy, and were randomized to two schedules of chemotherapy plus radiation therapy to the primary tumor site or to chemotherapy alone

(Perry et al. 1984). In patients with extensive disease, chemotherapy alone was given (Cha-hinian et al. 1984). A complete remission (CR) was defined as disappearance of all mea-surable tumor and all clinical laboratory or radiological evidence of disease. A partial re-sponse (PR) was defined as a 50% or greater reduction in the sum of the products of two diameters of all measurable tumors. Nonresponders had less than a 50% regression in tu-mor measurements or no objective disease progression.

Patients were divided into two categories of presumed tumor secretory status: (a) nonsecretors had plasma neurophysins that were less than three times the mean for the control population; and (b) secretors had plasma neurophysins that were three or more times the mean of the normal population (>4 SDs for VP-HNP and >3 SDs for OT-

Table 1. Baseline neurophysin secretory status

		OT secretor		
		Yes	No	Total
VP-HNP secretor	Yes	29 (11%)	67 (26%)	96
	No	10 (3%)	148 (60%)	158
	Total	39	215	254

Table 2. Neurophysin secretory status by stage of disease

		Stage of disease	
		Limited	Extensive
VP-HNP secretor	Yes	46 (32%)	50 (46%)
	No	100 (68%)	58 (54%)
	Total	146	108
OT-HNP secretor	Yes	16 (11%)	23 (21%)
	No	130 (89%)	85 (79%)
	Total	146	108

Table 3. Secretory status by number of metastatic sites

	Stage of disease		
	Limited	Extensive	
		1 Site	2 Sites or more
Secretory status			
VP-HNP secretors	46 (32%)	20 (34%)	30 (60%)
VP-HNP secretors	100 (68%)	38 (66%)	20 (40%)
OT-HNP secretors	16 (11%)	11 (19%)	12 (24%)
OT-HNP nonsecretors	130 (89%)	47 (81%)	38 (70%)
Total	146	58	50

HNP). Normal plasma values used for VP-HNP and OT-HNP were 73 ± 30 pg/ml and 283 ± 30 pg/ml, respectively (North et al. 1980). VP-HNP secretors were defined as having plasma concentrations of VP-HNP greater than 220 pg/ml and OT-HNP secretors as having plasma concentrations of OT-HNP greater than 850 pg/ml.

Associations between secretory status and other categorical variables were based on the Chi-square test for association (Sendecor and Cochran 1967). Measured outcomes between groups were compared using the Wilcoxon rank-sum test (Steel and Torrie 1980). Survival distributions were compared by the log-rank test (Teto et al. 1977).

Results

At the time of this preliminary analysis 685 evaluable patients have been entered on the treatment protocols. In all, patients have had initial neurophysin assays performed prior to the onset of therapy, and these form the basis for this report. Of the 254 patients 93 (37%) were female; 77% were 55 or older. After staging, limited disease was recorded in 146 patients and 108 patients had extensive disease.

At the time of randomization 106 (41%) of the 254 patients had elevated neurophysins and were classed as secretors: 67 (26%) had elevated VP-HNP and 10 (3%) had elevated OT-HNP; in 29 (11%) both neurophysins were elevated. There was a correlation between the levels of the two peptides in the same patient ($P < 0.0001$), as seen in Table 1.

The extent of disease was the best indicator of whether the patient was a secretor, as seen in Table 2. Secretors for both VP-HNP ($P < 0.02$) and OT-HNP ($P < 0.02$) were more prevalent in extensive disease. There was an increased frequency of elevated neurophysin levels for both peptides with increasing numbers of metastatic sites. These results are displayed in Table 3.

Sex was not correlated with secretory status. The prevalence of VP-HNP secretors was 41% for men and 32% for women, and for OT-HNP the prevalence was 16% for men and 15% for women. Similarly, age was not a factor: 36% of patients younger than 55 and 38% of patients older than 55 were VP-HNP secretors. For OT-HNP 17% of patients younger than 55 were secretors, compared with 15% for the older age group.

The analysis is not mature enough to determine the value of these markers in predicting response outcome and duration of remission in responders. The analysis is also incomplete for survival trends, although secretory status appears to have a negative effect in that VP-HNP secretors have a reduced survival after controlling for the number of metastatic sites ($P < 0.03$).

Discussion

These preliminary analyses confirm previous findings that elevated plasma concentrations of neurophysins are present in a significant percentage of patients with SCCL. While the rate is somewhat lower than that originally reported by North et al. (1980b) and Maurer et al. (1983), this can be explained in part by the selection bias that occurs for entry into cooperative group trials. In our original studies all patients with SCCL were studied.

The present results also confirm that the presence of elevated neurophysins best correlates with the extent of disease and with the number of metastatic sites. If patients with two or more metastatic sites are compared with the patients with extensive disease reported by Maurer et al. (1983) the percentage of patients secreting VP-HNP is essentially identical (60% and 67%). These findings are similar to those recorded with calcitonin (Wallach et al. 1981), ACTH (Ratcliffe et al. 1982), and vasopressin (Greco et al. 1981) in that in all cases the peptide hormones were more frequently present in extensive disease.

While sex appears to have some prognostic significance in the CALGB studies (Maurer and Pajak 1981), no differences have yet been seen in this study between the sexes in the frequency rates of increased secretion of the neurophysins. Age also does not appear to be a factor.

Further analyses of this large, carefully evaluated population of patients will determine the usefulness of these markers in predicting response outcome and duration of remission in responders. It will be important to define whether there are any significant differences in the biology and natural history of the disease between the two populations, secretor and nonsecretor.

The potential of the peptide hormones as tumor markers should continue to be evaluated in studies with carefully defined populations by well-characterized assay procedures, and preferably prior to therapeutic intervention. Panels of assays for several peptide hormones or other tumor products may provide better information about the overall frequency of tumor markers in SCCL. It is hoped that the present study will continue to provide specific information to increase our understanding of the biology of SCCL.

References

Eagan RT, Maurer LH, Forcier RJ, Tulloh M (1974) Small cell carcinoma of the lung: Staging, paraneoplastic syndromes, treatment and survival. Cancer 33: 527–532

Chahinian AP, Ware JH, Zimmer B, Comis RL, Perry MC, Hirsch V, Skarin AT, Raich PC, Weiss RB, Carey RW (1984) Evaluation of anticoagulation with warfarin and of alternating chemotherapy in extensive small cell cancer of the lung (SCCL). Proc Am Soc Clin Oncol 3: 225

Greco FA, Hainsworth J, Sismani A, Richardson RL, Hande KR, Oldham RK (1981) Hormone production and paraneoplastic syndromes. In: Greco FA, Oldham RA, Bunn PA Jr (eds) Small cell lung cancer Grune and Stratton, New York, pp 177–223

Hamilton BPM, Upton GV, Amatruda TT (1972) Evidence for the presence of neurophysin in tumors producing the syndrome of inappropriate antidiuresis. J Clin Endocrinol Metab 35: 764–767

Kato Y, Ferguson TB, Bennett DE, Burford TH (1969) Oat cell carcinoma of the lung: a review of 138 cases. Cancer 23: 517–524

Land H, Schutz G, Schmale H, Richter D (1982) Nucleotide sequence of cloned cDNA encoding bovine arginine vasopressin-neurophysin II precursor. Nature 295: 299–303

Land H, Grez M, Ruppert S, Schmale H, Rehbein M, Richter D, Schutz G (1983) Deduced amino acid sequence from the bovine oxytocin-neurophysin I precursor cDNA. Nature 302: 342–344

Legros JJ (1975) The radioimmunassay of human neurophysins: contributions to the understanding of the physiopathology of neurohypophyseal function. Ann N Y Acad Sci 248: 281–303

Lokich JJ (1982) Frequency and clinical biology of ectopic hormone syndromes of small cell carcinoma. Cancer 50: 2111–2114

Maurer LH, Pajak TF (1981) Prognostic factors in small cell carcinoma of the lung: a CALGB study. Cancer Treat Rep 65: 767–774

Maurer LH, Tulloh M, Weiss RB, Blom J, Leone L, Glidewell O (1980) A randomized combined modality trial in small cell carcinoma of the lung. Comparison of combination chemotherapy – radiation therapy versus cyclophosphamide – radiation therapy, effects of maintainence chemotherapy and prophylactic whole brain irradiation. Cancer 45: 30–39

Maurer LH, O'Donnell JF, Kennedy S, Faulkner CS, Rist K, North WG (1983) Human neurophysins in carcinoma of the lung: Relation to histology, disease stage, response rate, survival, and the syndrome of inappropriate antidiuretic hormone secretion. Cancer Treat Rep 67: 971–976

Moody TW, Pert CB, Gazdar AF, Carney DN, Minna JP (1981) High levels of intracellular bombesin characterize human small cell lung carcinoma. Science 214: 246-247

North WG, LaRochelle FT, Jr, Melton J, Mills RC (1980a) Isolation and partial characterization of two human neurophysins: their use in the development of specific radioimmunoassays. J Clin Endocrinol Metab 51: 884-889

North WG, Maurer LH, Valtin H, O'Donnell J (1980b) Human neurophysins as potential tumor markers for small cell carcinoma of the lung: Application of specific radioimmunassays for vasopressin-associated and oxytocin associated neurophysins. J Clin Endocrinol Metab 51: 892-896

North WG, Maurer LH, O'Donnell JF (1983a) The neurophysins and small cell lung cancer. In: Greco FA (ed) Biology and management of lung cancer. Nijhoff, Boston, pp 143-169

North WG, Valtin H, Cheng S, Hardy GR (1983b) The neurophysins: Production and Turnover. Prog Brain 60: 217-225

Perry MC, Eaton WL, Ware J, Zimmer B, Comis R, Chahinian AP, Skarin A, Carey R, Hirsch V (1984) Chemotherapy (CT) with or without radiation therapy (RT) in limited small cell carcinoma of the lung (SCCL). Proc Am Soc Clin Oncol 3: 230

Ratcliffe JG, Podmore J, Stack BHR, Spilg WGS, Gropp C (1982) Circulating ACTH and related peptides in lung cancer. Br J Cancer 45: 230-236

Robinson AG, Haluszczak C, Wilkins JA, Heullmantel AB, Watson CG (1977) Physiological control of two neurophysins in humans. J Clin Endocrinol Metab 44: 330-339

Sendecor GW, Cochran WG (1967) Statistical methods. Iowa State University Press, Ames

Sorenson GD, Bloom SR, Ghatei MA, Del Prete SA, Cate CC, Pettengill OS (1982) Bombesin production by human small cell carcinoma of the lung. Regul Pept 4: 59-66

Sorenson GD, Pettengill OS, Brinck-Johnson T, Cate CC, Maurer LH (1981) Hormone production by cultures of small cell carcinoma of the lung. Cancer 47: 289-296

Steel RGD, Torrie JH (1980) Principles and procedures of statistics, McGraw Hill, New York

Teto R, Pike MC, Armatage P, et al (1977) Design and analysis of randomized clinical trials requiring prolonged observations of each patient. Br J Cancer 35: 1-39

Wallach SR, Royston I, Taetle R, Wohl H, Deftos LJ (1981) Plasma calcitonin as a marker of disease activity in patients with small cell carcinoma of the lung. J Clin Endocrinol Metabol 53: 602-606

Wood SM, Wood JR, Ghatei MA, Sorenson GD, Bloom SR (1982) Is bombesin a tumor marker for small cell carcinoma? Lancet I: 690

Yamaji T, Ishibashi M, Katayama S (1981) Nature of the immunoreactive neurophysins in ectopic vasopressin-producing oat cell carcinomas of the lung. Demonstration of a putative common precursor to vasopressin and neurophysin. J Clin Invest 68: 388-98

Prospective Multicenter Study of Hormone Markers in Small Cell Lung Cancer

K. Havemann, R. Holle, and C. Gropp

Klinikum der Philipps-Universität, Zentrum für Innere Medizin, Medizinische Klinik und Poliklinik, Schwerpunkt Hämatologie, Onkologie, Immunologie, Baldingerstrasse, 3550 Marburg, FRG

Ectopic production of several peptide hormones in patients with lung cancer has been described repeatedly in recent years. Hormones which are frequently elevated in the sera of patients with small cell lung cancer (SCLC) at diagnosis are summarized in Table 1. Only

Table 1. Peptide hormones and neuron-specific enolase (NSE) in serum or plasma of untreated patients with small cell lung cancer

	No. of patients	Incidence %	Reference
ACTH	75	29	Hansen et al. 1980
	50	30	Gropp et al. 1981
	68	38	Krauss et al. 1981
	63	24	Ratcliffe et al 1982
α-MSH	43	19	Gropp et al. 1981
β-Endorphin	58	45	Gropp et al. 1981
LPH	24	54	Odell et al. 1979
ADH	41	39	Hansen et al. 1980
	61	48	North et al. 1980
	54	17	Greco et al. 1981
	66	30	Gropp et al. 1981
Oxytocin	61	30	North et al. 1980
Calcitonin	75	64	Hansen et al. 1980
	54	48	Gropp et al. 1981
	49	73	Krauss et al. 1981
	54	40	Greco et al. 1981
	135	56	Luster et al. 1982
PTH	43	27	Gropp et al. 1981
β-HCG	39	33	Gropp et al. 1981
Gastrin	69	20	Hansen et al. 1980
Glucagon	46	11	Hansen et al. 1980
Secretin, insulin VIP	46–65	5	Hansen et al. 1980
NSE	94	69	Carney et al. 1982

results based on sufficiently large numbers of patients are included. The incidence of elevated levels of a given peptide hormone varies depending on the assay system, the antibodies used, and the upper limits of the normal range used in the different studies. For many of the hormones listed in Table 1 increased levels can be detected at diagnosis in up to 70% of all patients, and in many cases several hormones are elevated in parallel. Often, however, blood levels are only marginally increased, and evidence of excessive production may be present in only up to 25% (Carney et al. 1982).

Serum levels of these hormones were evaluated as markers for disease extent and response to therapy. ACTH, calcitonin, neurophysins, and other tumor markers, such as CEA, were measured in patients at diagnosis and sequentially during therapy (Hansen et al. 1980; Krauss et al. 1981; North et al. 1980; Ratcliffe et al. 1982). The serum levels were correlated with the disease extent and the observed clinical response. While in some studies the presence of increased levels showed a good correlation with disease extent, no such correlation was observed in others.

A close correlation between tumor response to cytotoxic therapy and a decrease of hormone levels was shown in some studies (Carney et al. 1982). This suggests the possibility of monitoring with a subsequent change of therapy according to the serum peptide hormone levels. These results, however, were obtained in rather small groups of patients and in retrospective investigations.

In this paper we report interim results of a multicenter trial on the treatment of SCLC by chemotherapy and radiation (Havemann et al. 1984). Patients from 14 participating institutions were randomized between the therapeutic protocols A and B. Under protocol A patients received a sequential chemotherapy, while patients in group B were treated with an alternating chemotherapy of different combinations as described in Fig. 1. Responding patients in each protocol group received prophylactic cranial irradiation after three cycles and chest irradiation after eight cycles. No maintenance therapy was administered to patients in complete remission.

Before each cycle and at monthly intervals during the follow-up period deep-frozen serum samples were sent to the central marker laboratory in Marburg for determination of calcitonin, ACTH, and CEA (Harms et al. 1983). Due to organizational problems only 60% of the scheduled samples could be tested for these three markers. However, the samples not tested belonged mainly to the time period after the end of therapy. About 5% of the samples were thawed during transportation. Up to now about 5400 marker analyses of 1816 sera have been performed, and the results together with the clinical data are stored in the computer.

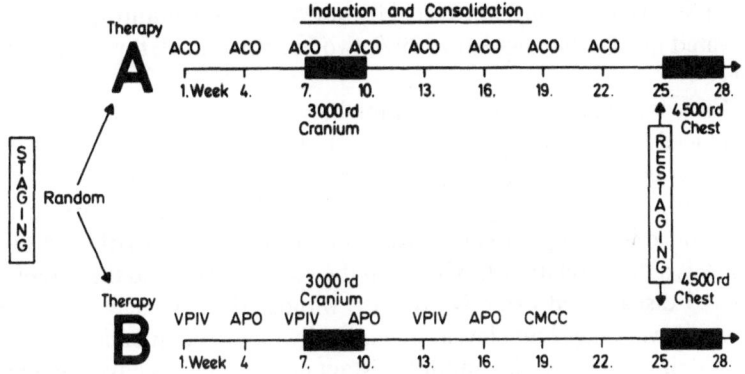

Fig. 1. Treatment plan for the multicenter trial

Table 2. Classification of tumor marker levels

	Normal	Elevated	Pathologic
CEA (ng/ml)	≤ 5	5– 20	> 20
Calcitonin (pg/ml)	≤ 100	100–200	> 200
ACTH	≤ 80	80–150	> 150

Table 3. Tumor markers at diagnosis ($n = 172$)

	Elevated or pathologic (%)	Pathologic (%)
Calcitonin	48	19
ACTH	16	9
CEA	41	20
No marker	30	65
One marker	39	23
Two markers	26	9
All three markers	5	3
	100	100

For practical reasons the marker levels were classified as normal, elevated, or clearly pathologic, according to Table 2. At the time of this interim analysis 250 patients have been included in the multicenter trial.

Tumor Markers Before Therapy

At diagnosis we observed an incidence of elevated or clearly pathologic levels for calcitonin in 48%, for ACTH in 16%, and for CEA in 41% of patients (Table 3). As expected, the clearly pathologic levels were less frequent (viz. 19% calcitonin, 9% ACTH, 20% CEA). Thus, the incidence is lower than in retrospective studies, which could be due to the prospective design of our trial or to the transportation of the samples to the central laboratory (ACTH, especially, is affected by increased temperatures). None of the markers were elevated in 30%, one marker in 39%, two in 26%, and all three markers in only 5% of patients.

There is a slight positive correlation between any two of the markers, an effect possibly produced by the correlation of the marker levels with the extent of disease. The association between markers and prognostic variables at diagnosis is given in Fig. 2. The comparisons are based on the percentages of pathologic values, because differences between prognostic subgroups are more distinct in the upper ranges of CEA and calcitonin values. There is a direct relationship between the marker levels and the extent of disease. Whereas limited disease and extensive disease without distant metastases show nearly identical marker levels, the serum levels are markedly increased in patients with distant metastases. In addition, patients with multiple distant metastatic sites have higher CEA and calcitonin levels than patients with a single distant metastasis. The same is true for patients with liver,

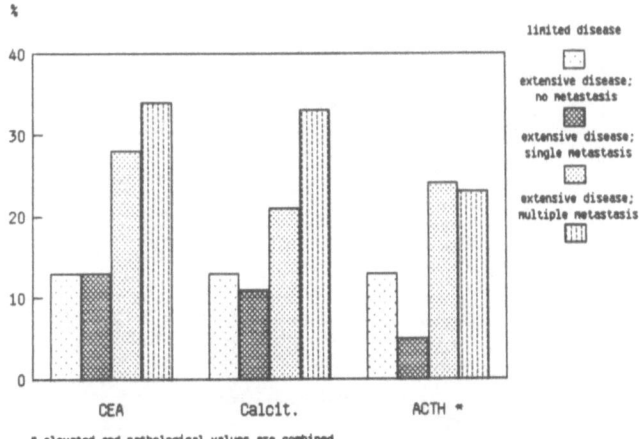

Fig. 2. Percentage of patients with pathologic marker levels

* elevated and pathological values are combined

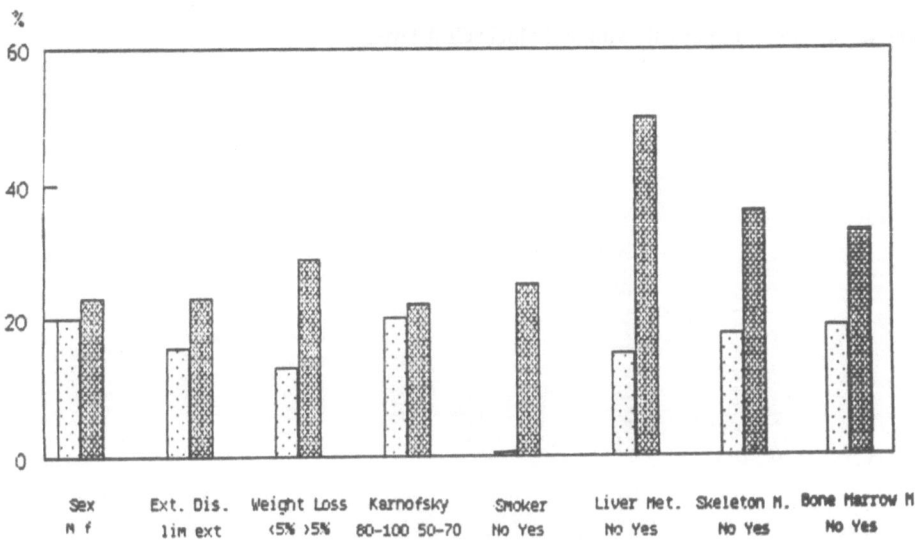

Fig. 3. Percentage of patients with pathologic calcitonin levels

bone, and bone marrow metastases in comparison with patients without metastases in the respective sites (Figs. 3 and 4). Patients with brain metastases exhibit no significant increase (not shown).

As is already known, smoking is strongly correlated with increased levels of CEA and calcitonin. Further positive correlations exist between pathologic CEA levels and weight loss and between pathologic calcitonin levels and reduced Karnofsky index, whereas the other variables show a less prominent correlation in this analysis.

A relationship between marker levels at diagnosis and survival seems to be indirectly induced by the influence of distant metastases on survival. The results of the stratified analysis show that in patients with limited disease or extensive disease without distant metastases no influence of the marker levels on survival can be observed (Fig. 5). Only in the

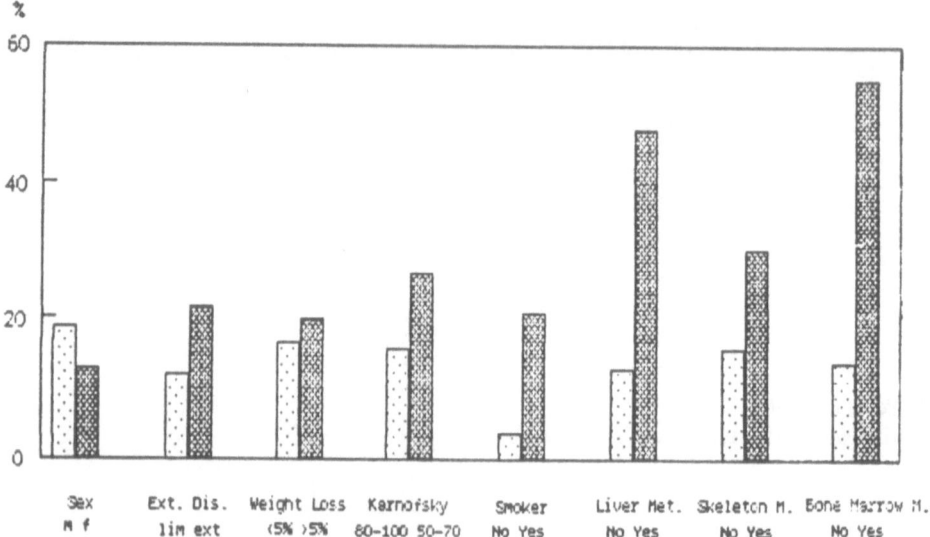

Fig. 4. Percentage of patients with pathologic CEA levels

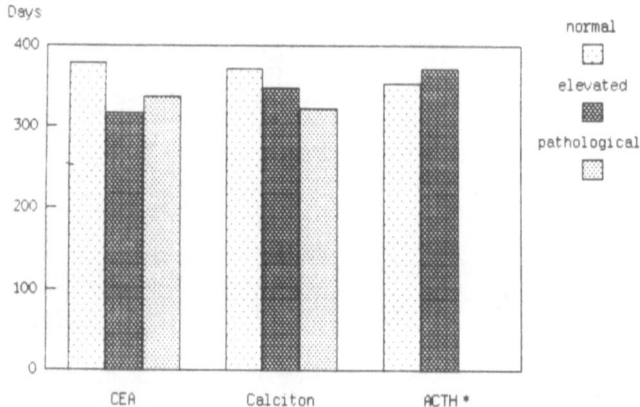

Fig. 5. Median survival time of patients without distant metastases, referred to levels of tumor markers. *Elevated and pathological levels are combined

group of patients with distant metastases do the calcitonin levels seem to be related to survival time (Fig. 6). However, further analysis, including analysis of the number of metastatic sites, may explain this correlation.

Tumor Markers in Relation to Tumor Response

In patients with elevated or pathologic values at diagnosis the change in levels of the markers calcitonin, ACTH, and CEA was compared with the degree of tumor response. In the first analysis the outcome of treatment was classified as CR (complete remission),

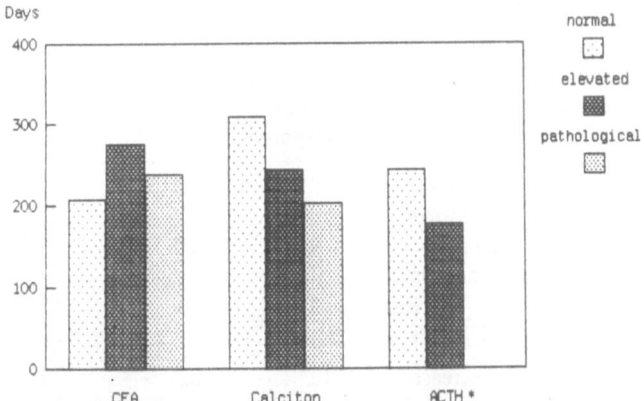

Fig. 6. Median survival time of patients with distant metastases, referred to levels of tumor markers. *Elevated and pathological levels are combined

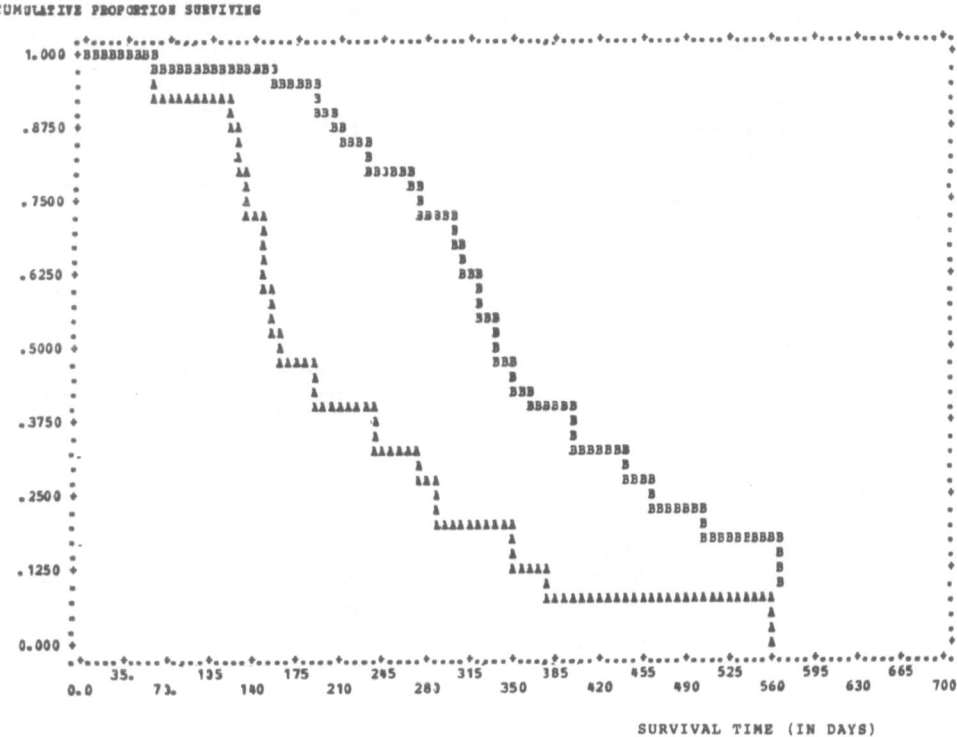

Fig. 7. Decrease in CEA level after second cycle of chemotherapy. *A*, no decrease; *B*, decrease

PR + MR (partial remission, including minimal response), or NR (no response) and compared with the behavior of the three markers within the first 2 months of chemotherapy (Table 4). A strong correspondence becomes apparent, which is most evident in the cases of complete remission and of no response. Among the CR patients only two exceptions

Table 4. Analyses of patients with initially elevated or pathologic marker levels

		Complete remission	Partial or minimal response	No response
CEA	Decrease to normal	7	9	0
	Decrease (not normal)	8	18	2
	No decrease of ≥10%	1	6	5
Calcitonin	Decrease to normal	12	32	1
	Decrease (not normal)	2	13	5
	No decrease of ≥10%	0	5	4
ACTH	Decrease to normal	4	10	0
	Decrease (not normal)	3	0	1
	No decrease of ≥10%	1	4	1

Table 5. Comparison of marker decrease (at least 10%) and chest x-ray findings after the first cycle of chemotherapy

Decrease of marker (≥10%)		Tumor response in chest x-ray	
		Response	No response
Calcitonin	Response	49	10
	No response	5	5
ACTH	Response	9	2
	No response	5	2
Calcitonin or CEA	Response	60	12
	No response	5	7

Table 6. Attainment of normal values for tumor markers

	n	After 1 cycle	After 2 cycles	After 3 cycles or more	Never
CEA	23	8 (35%)	3 (13%)	5 (23%)	7 (30%)
Calcitonin	36	24 (67%)	9 (25%)	0	3 (8%)

Table 7. Median survival (days) and the response reflected in tumor markers and chest x-ray during the first two cycles of chemotherapy

		Decrease median (n)	No decrease median (n)	
CEA	1st cycle	319 (35)	230 (19)	$P > 0.1$
	2nd cycle	336 (33)	161 (15)	$P < 0.001$
Calcitonin	1st cycle	314 (54)	212 (11)	$P > 0.05$
	2nd cycle	319 (50)	154 (8)	$P < 0.001$
All markers	1st cycle	324 (65)	215 (30)	$P < 0.05$
	2nd cycle	344 (60)	177 (23)	$P < 0.001$
Chest x-ray	1st 2nd cycle	347 (120)	219 (45)	$P < 0.1$

with pathologic values of CEA or ACTH at relatively constant levels are found. Only one patient had a marker decrease into the normal range without showing any clinical response, but in this case the initial calcitonin level had only just exceeded the normal limit. When marker determinations from the time period after the 2 months are included in the analysis, patients who achieved a complete remission show a drop of increased values into the normal range in 75% of cases for CEA, in 93% for calcitonin, and in 63% for ACTH.

To investigate whether a response of the tumor during the first weeks of treatment is accompanied by a change in the marker concentration, a decrease of the marker by at least 10% was compared with the chest x-ray after the first cycle of therapy. With the exception of ACTH a significant correlation between the decrease in the marker and the reduction of tumor size in the chest x-ray can be demonstrated, as shown in Table 5. Corresponding results occur for calcitonin in 78%, for CEA in 73%, and for CEA and calcitonin combined in 80%. The lack of agreement in the remaining 20% could be due to a delayed marker decrease or could suggest that in some cases a decrease in hormone and CEA levels is a better indicator of tumor reduction than the chest x-ray.

To determine the speed of marker decrease, markers which had dropped during the first cycle were studied further during the following cycles. It is evident, as shown in Table 6, that calcitonin decreases much faster than CEA, which does not usually reach normal levels until the end of the second or third cycle, or even remains above normal values together, especially in the case of initially pathologic levels. Although the number of patients with elevated ACTH was rather small, the tendency for a response similar to the calcitonin response could be observed (not shown).

In addition, we tested whether the marker levels or the chest x-ray allowed a more reliable prognosis of survival. Table 7 indicates the median survival time of different groups (because of the small sample numbers ACTH is not included). The decrease of markers was determined 3 and 6 weeks after the start of treatment, i.e., after one or two cycles of chemotherapy.

To extend the validity of the results to a larger group, all patients with at least one elevated marker were included in a further analysis. They were only classified as showing a marker decrease if the levels of all elevated markers dropped by 10% or more. The median survival time for patients showing no decrease of CEA, calcitonin, or all elevated markers during the second cycle is significantly shorter than that for patients with a decrease of these markers. The corresponding survival curves, calculated by the Kaplan–Meier method, are shown in Figs. 7–9; the P values were computed by using the log rank test. A re-

CUMULATIVE PROPORTION SURVIVING

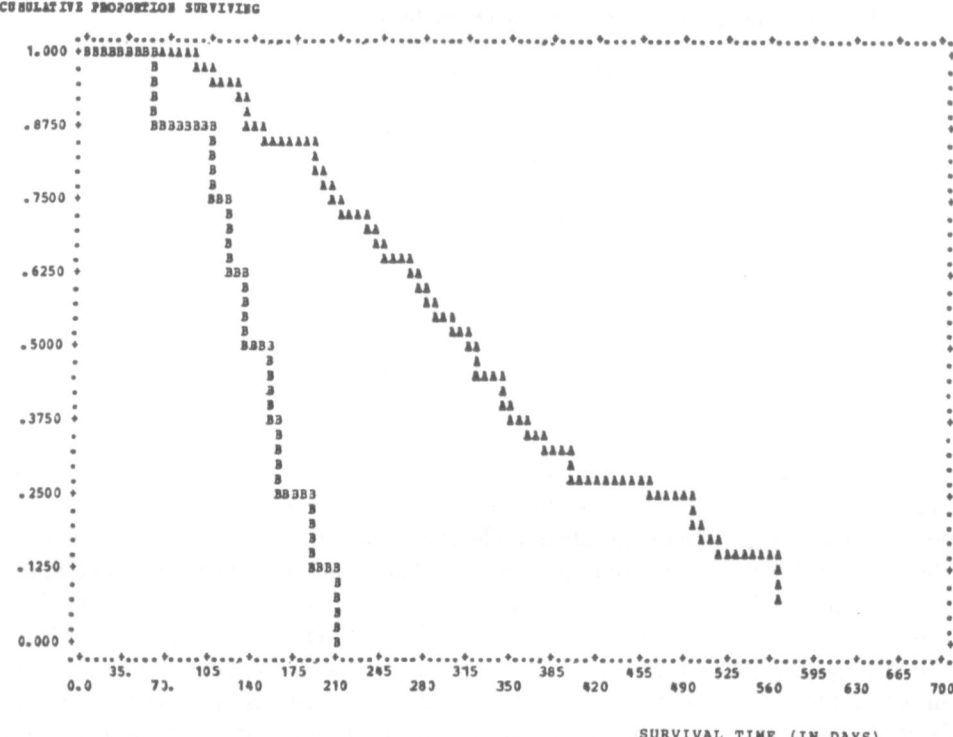

Fig. 8. Decrease in calcitonin level after second cycle of chemotherapy. *A*, decrease; *B*, no decrease

duction of the tumor size in the chest x-ray during the first cycle shows nearly the same influence on survival, while the tumor markers appear to have slightly better prognostic value than the chest x-ray according to a comparison of the respective findings after two cycles of chemotherapy (Fig. 10). Nevertheless, the chest x-ray cannot be replaced as a monitoring aid by the tumor markers, since both methods seem to reflect different aspects of tumor response and since about one-third of the patient population shows normal values of all three markers in question.

It has to be mentioned that the results described in this section are dependent on the arbitrary choice of a 10% decrease for the definition of marker response. Further statistical analyses will probably allow definition of a more suitable value.

Tumor Markers During the Further Course of Therapy

To classify the courses of markers during therapy only those courses could be taken into consideration for which complete series of serum samples were available covering the time from diagnosis up to shortly before the death of the patient. Only 63 patients fulfilled this criterion when the interval between the last serum sample and the patient's death was restricted to less than 6 weeks. Several characteristical types of marker courses can be observed, which can be defined as follows:

Type 1: Initially elevated levels decrease to normal and stay normal
Type 2: Initially elevated levels decrease and increase again during relapse

CUMULATIVE PROPORTION SURVIVING

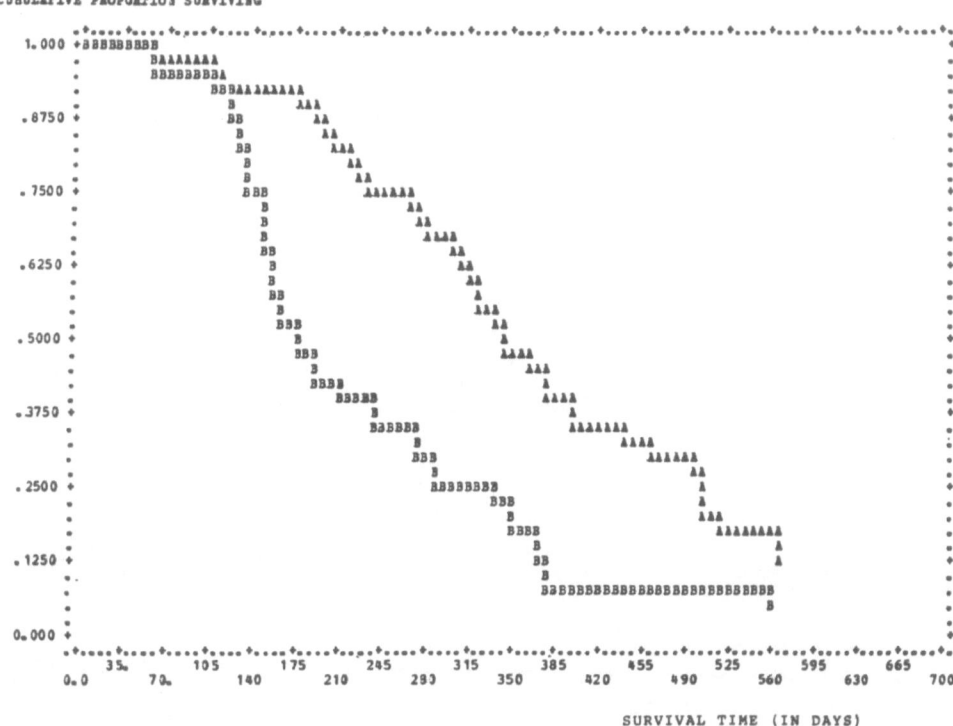

SURVIVAL TIME (IN DAYS)

Fig. 9. Decrease in levels of all markers after second cycle of chemotherapy. *A*, decrease; *B*, no decrease

Table 8. Types of marker response during therapy

Type:	1	2	3	4	5	n
	(Initially elevated)			(Initially normal)		
CEA	3	7	17	10	24	61
Calcitonin	9	10	13	16	15	63
ACTH	1	2	4	4	52	63

Type 3: Increased marker levels remain elevated
Type 4: Initially normal markers increase to elevated levels
Type 5: Markers remain normal throughout.

The numbers of patients with marker courses classified according to this definition are contained in Table 8. It has to be noted that the frequencies of types 2 and 4 have supposedly been underestimated in favor of types 1 and 5, respectively, because a final rise in markers could have been missed when the last serum sample of a patient was collected 4–6 weeks before death. This speculation is supported by the fact that in some of these cases a rise of marker levels to the upper normal range can be observed.

Among patients with increased pretreatment levels type 3 is most common. However, this group is heterogeneous, because it includes nonresponding patients as well as patients

CUMULATIVE PROPORTION SURVIVING

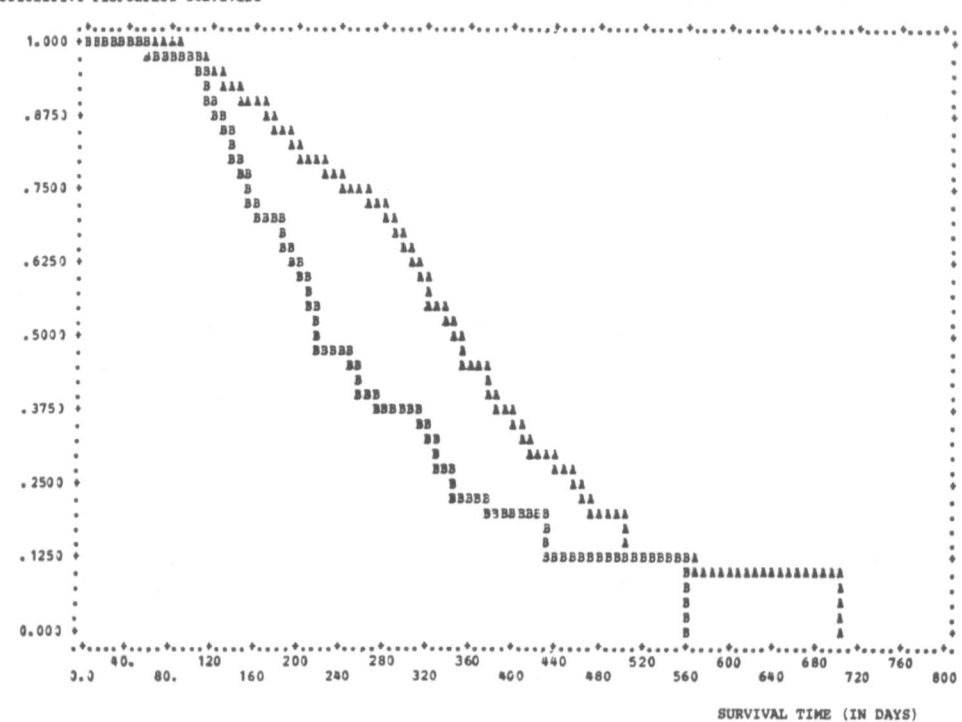

Fig. 10. Response seen in chest x-ray after first two cycles of chemotherapy. *A*, size reduction. *B*, no size reduction

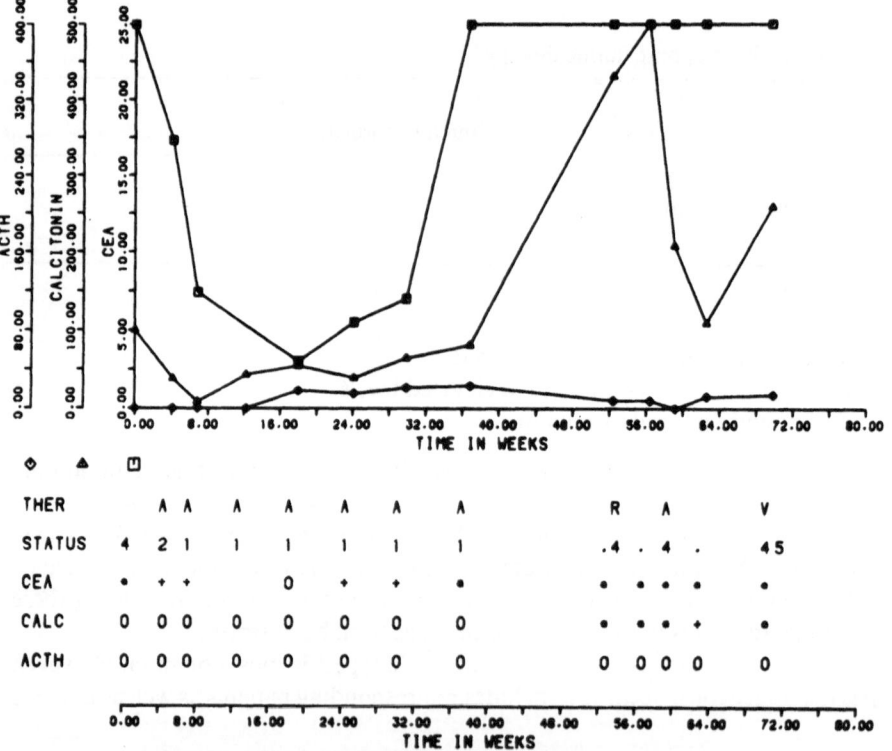

Fig. 11. Serial marker levels recorded in patient 98

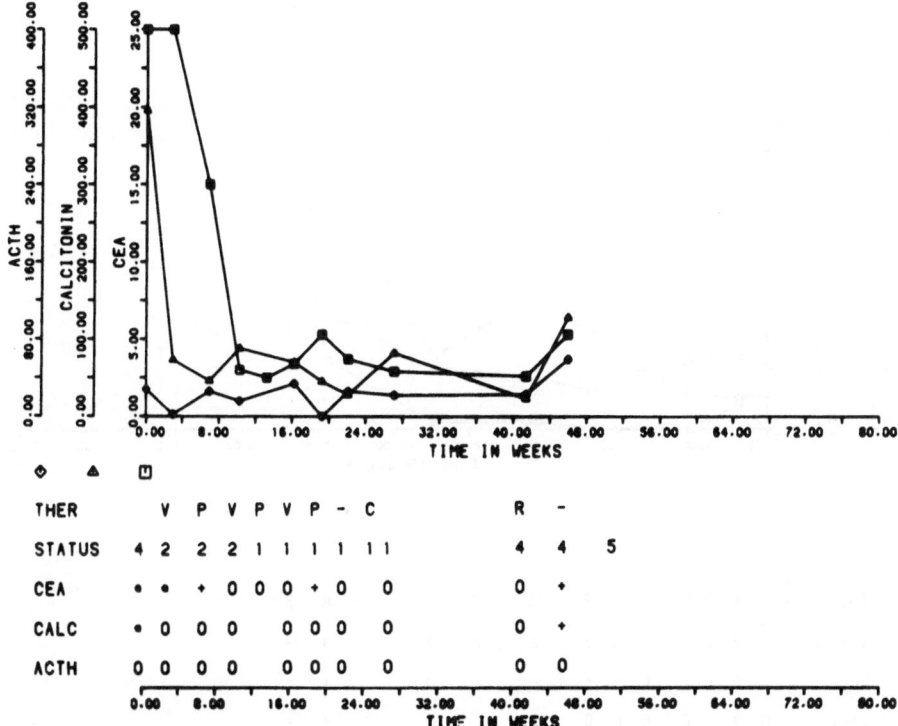

Fig. 12. Serial marker levels recorded in patient 102

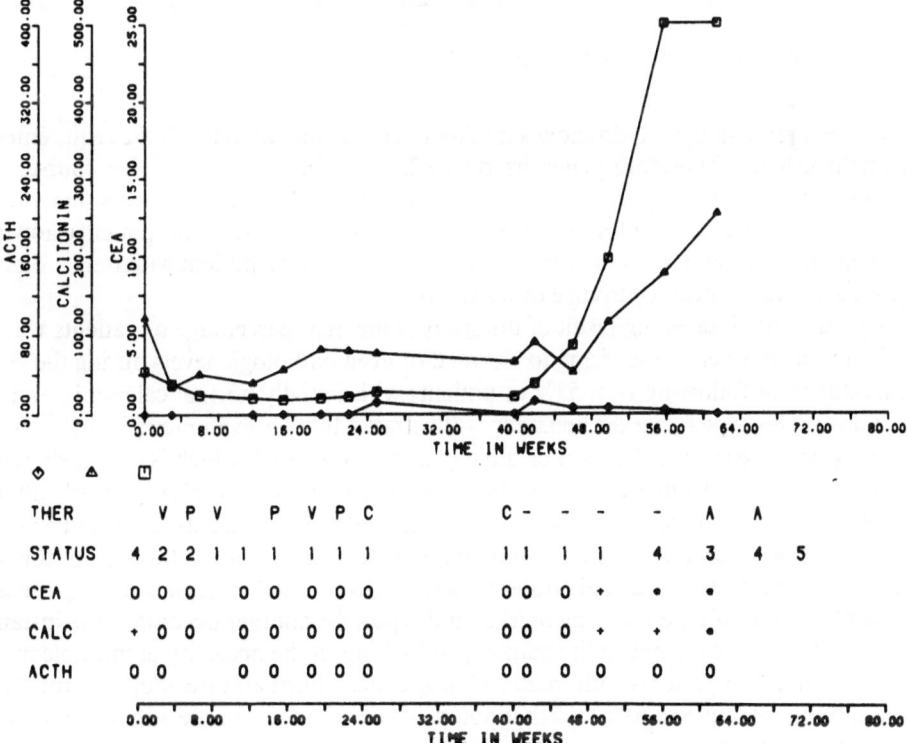

Fig. 13. Serial marker levels recorded in patient 74

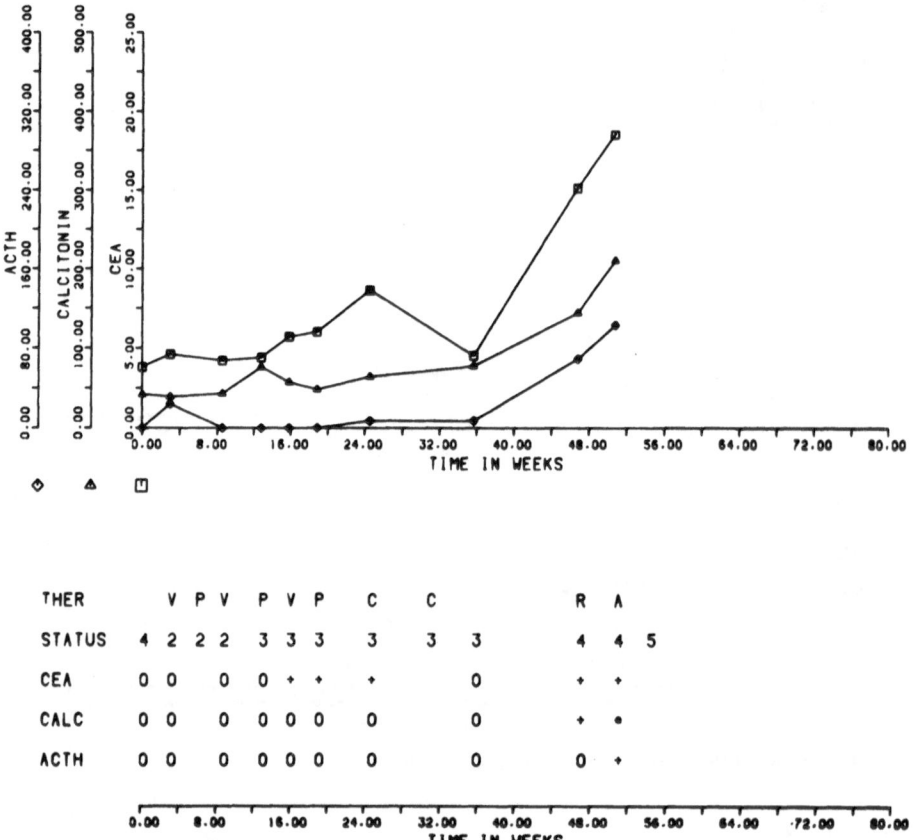

THER		V	P	V	P	V	P		C		C		R	A	
STATUS	4	2	2	2		3	3	3		3		3		4	4	5
CEA	0	0		0	0	+	+		+		0		+	+	
CALC	0	0		0	0	0	0		0		0		+	•	
ACTH	0	0		0	0	0	0		0		0		0	+	

Fig. 14. Serial marker levels recorded in patient 85

with tumor remission, who do show a marker decrease but fail to fulfill the requirement of normalized levels. Therefore, generalized type 2, i. e. bathtub-shaped marker courses without reference to normal values, represents an even larger number of patients than indicated by Table 8. Two examples are presented in Figs. 11 and 12: One patient showing a marked increase many weeks before his relapse and a second patient with only a slight increase after the clinical recurrence of his tumor.

A particularly interesting result of the study is the high percentage of patients with initially normal marker values rising to elevated or even pathologic levels during the course of treatment or follow-up (viz. 52% of patients with initially normal calcitonin and 29% for CEA). This type 4 marker behavior is illustrated by two examples (Figs. 13 and 14), showing an increase of either two or three markers in parallel, which is obviously related to tumor growth. A further analysis of the type 4 CEA courses reveals that nearly all these patients have pretreatment levels in the upper normal range (i. e., between 3 and 5 ng/ml). In the case of type 4 (calcitonin courses) no similar result could be found, but for some patients the marker rise was only transient and not connected with clinical recurrence.

Combinations of type 1 for one marker and type 4 for another occur in some instances. This possibility of a change in the marker profile hints at the necessity of multiple marker determinations for patients with small cell lung cancer. Furthermore, multivariant statistical analyses are required to establish a potential use of tumor markers in the early detection of tumor relapse.

So far, all efforts to use tumor markers as a reliable and clinically useful monitoring aid in patients staying in tumor remission have failed. Although in many cases increased levels of tumor markers can be observed at the time of recurrence, there are still more exceptions, where markers rise too late, indistinctly, or not at all. As in the case of pretreatment marker levels, high levels of one or more markers during relapse can mainly be seen if the recurrence includes distant metastases and not only the primary tumor. Again, the most pronounced increase occurs in patients with liver and bone metastases, while patients with brain metastases and no others show no marker elevation.

Conclusion

In our prospective study there is a lower incidence of elevated tumor markers than has been found in retrospective analyses. Even so, the markers are of some prognostic use and may be helpful in the treatment of patients with small cell lung cancer.

A marked increase of ACTH, calcitonin, and CEA indicates the presence of distant metastases. This may allow a better classification of the extent of disease. This finding suggests that there is a direct correlation between tumor burden and the marker levels. Since the tumor burden is negatively correlated to survival there is also a negative correlation between increased marker levels and survival.

There is a positive correlation between the marker values and smoking. The multivariant analyses also showed a correlation with weight loss and performance status, whereas correlations with other variables could not be found.

The most important finding of this study is that during the first two or three cycles of chemotherapy immediate monitoring of treatment success by calcitonin and CEA determinations as well as the chest x-ray may allow early detection of tumor resistance in the individual patient and may suggest a change to another treatment program.

So far, monitoring tumor markers during the further course of treatment to detect an early relapse is rather disappointing, because of the multiple response types observed and the number of cases not showing an increase of the marker until during or after clinical relapse. Nevertheless, further research might help to define a more sensitive criterion of marker response.

In a considerable number of patients initially normal values increased during the later course parallel with tumor growth, particularly for calcitonin. This finding may indicate an overgrowth of a resistant hormone-producing tumor cell clone.

References

Carney DN, Marangos PJ, Ihde DC (1982) Serum neuron-specific enolase: a marker of disease extent and response to therapy in patients with small cell lung cancer. Lancet I: 583-585

Greco AF, Hainsworth J, Sismann A (1981) Hormone production and paraneoplastic syndromes. In: Greco AF, Oldham RK, Bunn PA (eds) Small cell lung cancer. Grune and Stratton, New York, pp 177-224

Gropp C, Luster W, Havemann K, Lehmann FG (1981) ACTH, calcitonin, a-MSH, β-endorphin, parathormone and β-HCG in sera of patients with lung cancer. In: Uhlenbruck G, Wintzer G (eds) CEA und andere Tumormarker. Tumor-Diagnostik, Leonberg, pp 358-363

Hansen M, Hammer M, Hummer L (1980) ACTH, ADH, and calcitonin concentrations as marker of response and relapse in small cell carcinoma of the lung. Cancer 46: 2062-2067

Havemann K, Harms V, Gropp C, Heinemann B, Holle R, Victor N, Thomas C, Drings P, Manke HG, Bayer G, Dirks P, Georgii A, Mende S, Mitrou PS, Graubner M, Hans K, Heim M, Pfann-schmidt G, Schroeder M, Wehr E, Weiss J (1984) A randomized multicenter trial comparing sequential chemotherapy (ACO) with cyclic alternating polychemotherapy (APO, VPIV, CMCC) in small cell lung cancer. Verh Dtsch Krebs Ges 5: 357-360

Krauss S, Macy S, Ichiki AT (1981) A study of immunoreactive calcitonin, ACTH and CEA in lung cancer and other malignancies. Cancer 47: 2485-2492

Luster W, Gropp C, Sostmann H, Kalbfleisch H, Havemann K (1982) Demonstration of immunoreactive calcitonin in sera and tissues of lung cancer patients. Eur J Cancer Clin Oncol 18: 1275-1283

North WG, Maurer H, Valtin H, O'Donell JF (1980) Human neurophysins as potential tumor markers for small cell carcinoma of the lung: application of specific radioimmunoassays. J Clin Endocrinol Metab 51: 892-897

Odell WD, Wolfsen AR, Bachelot I, Hirose FM (1979) Ectopic production of lipotropin by cancer. Am J Med 66: 631-638

Ratcliffe JG, Podmore J, Stack BHR, Spilg WGS, Gropp C (1982) Circulating ACTH and related peptides in lung cancer. Br J Cancer 45: 230-238

Syndromes Associated with Inappropriate Hormone Synthesis by Tumors: An Evolutionary Interpretation

D. LeRoith and J. Roth

Diabetes Branch, NIASSK, Bethesda, MD, USA

Introduction

A wide range of tumors derived from nonendocrine cells synthesize and secrete inappropriately large amounts of bioactive peptides, producing in their hosts syndromes typical of hormone excess (Odell and Wolfsen 1982; Fmura 1980; Baylin and Mendelsohn 1980). Table 1 lists some of the theories that have been proposed to explain "ectopic hormone production," the mysterious ability of these tumors to produce hormonal peptides.

In an attempt to explain these phenomena we have used a phyletic approach and searched for the evolutionary origins of classic vertebrate-type peptide hormones. In this chapter, we summarize the evidence for the presence in unicellular organisms of materials closely resembling these peptides. On the basis of these findings, we propose a unifying hypothesis to rationalize a variety of previously unrelated vertebrate phenomena, including syndromes of inappropriate hormone production caused by tumors.

Nonvertebrate Tissues Produce Vertebrate-Type Hormonal Peptides

Traditionally, vertebrate-type hormonal peptides were considered unique products of typical endocrine glands. More recently, however, nervous and neoplastic tissues and also non-neural, nonmalignant mammalian tissues have been shown to make hormonal peptides. For example, multiple studies indicate that hCG, ACTH, and vasopressin are made by many types of normal cells (Braunstein et al. 1975; Chen et al. 1976; Liotta et al. 1977; Larsson 1978; Lin et al. 1984; Nussey et al. 1984). Recent studies suggest the possibility that insulin is produced by mammalian nervous tissue, salivary gland, putuitary, and pla-

Table 1. Proposed mechanisms underlying ectopic hormone production

1. "Sponge" theory: Selective uptake and accumulation of circulating hormones by tumors with release of these hormones upon death of tumor cells
2. Random mutational events in the genome: occurring following neoplastic transformation
3. Tissue dedifferentiation theory: neoplastic transformation causes tissues to revert to states of more primitive gene expression
4. Gene derepression theory: tumor transformation allows previously· repressed genes to become transcribed

Recent Results in Cancer Research. Vol. 99
© Springer-Verlag Berlin · Heidelberg 1985

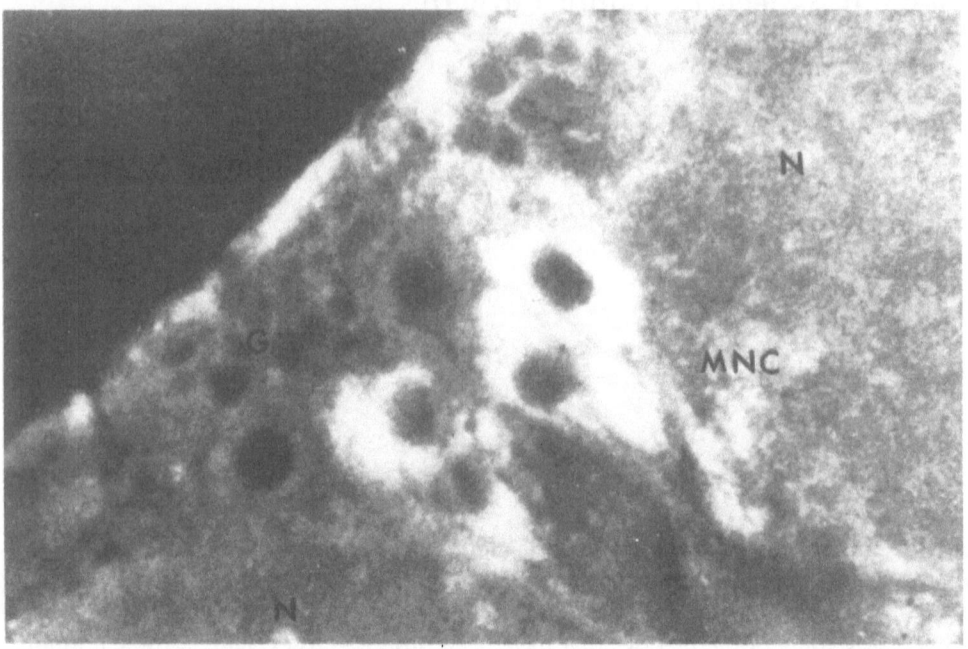

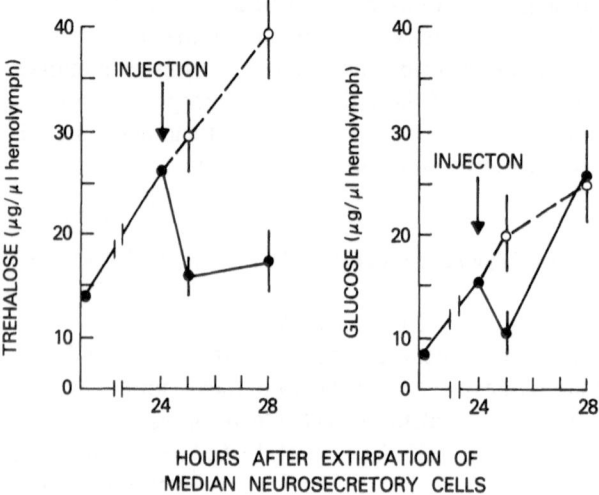

HOURS AFTER EXTIRPATION OF
MEDIAN NEUROSECRETORY CELLS

Fig. 1. Insulin-related material in insect brain. *Above:* Median neurosecretory cells (MNC) in the pars intercerebralis of the blowfly, containing specific reactivity with antibovine insulin antiserum. *N,* neuropil; *G,* glial cells. *Below:* After extirpation of the MNC cells, both glucose and trehalose levels in hemolymph rise (O-----O). Injection of partially purified extracts of MNC cells normalizes the levels of these carbohydrate intermediates (●-----●). (Adapted from Duve and Thorpe 1979 and Duve et al. 1979, with permission)

centa (LeRoith et al. 1983a; Murakami et al. 1982; Budd et al. 1983; Younes et al. 1984; Birch et al. 1984).

Because typical mammalian endocrine glands are absent from nonvertebrates, and because the assay methods available were insensitive, vertebrate-type hormonal peptides in

Table 2. Unicellular organisms contain materials closely resembling vertebrate-type peptide hormones

Prokaryotes		
TSH	*Clostridium*	(Macchia et al. 1967)
hCG	Many	(Acevedo et al. 1978; Maruo et al. 1979; Livingstone and Livingstone 1974; Backus and Affronti 1981)
Insulin	*E. coli*	(LeRoith et al. 1981)
Somatostatin	*E. coli*	(LeRoith et al. 1983b)
Neurotensin	Many	(Bhatnagar and Carraway 1981)
Calcitonin	*E. coli*	(Perez-Cano et al. 1982)
Eukaryotes		
Insulin	*Tetrahymena*, fungi	(LeRoith et al. 1980)
ACTH, β-endorphin	*Tetrahymena*	(LeRoith et al. 1982a)
Somatostatin	*Tetrahymena*	(Berelowitz et al. 1982)
Relaxin	*Tetrahymena*	(Schwabe et al. 1983)
Calcitonin	*Tetrahymena, Candida*	(Deftos et al. 1984; Perez-Cano et al. 1982)

multicellular nonvertebrates were largely overlooked earlier. However, following the discovery of hormonal peptides in nonendocrine tissues and improved assay methods, investigators intensified their search for these peptides in nonvertebrates, especially in nervous tisues. Numerous reports have demonstrated the presence of vertebrate-type hormonal peptides and neuropeptides in multicellular nonvertebrate tissues (Duve and Thorpe 1979; Duve et al. 1979; Plisetskaya et al. 1978; El-Salhy et al. 1980; Falkmer et al. 1973, 1981; Fritsch et al. 1976). Interestingly, insulin-related material is present in the blowfly brain (Fig. 1). In a freshwater mollusk, on the other hand, insulin-related material is present in cells that form part of the lining of the gut (Plisetskaya et al. 1978). In both organisms, studies suggest the insulin-related material is a regulator of carbohydrate metabolism.

Hormone-Related Material in Unicellular Organisms

In extending the phyletic search for vertebrate-type peptide hormones and neuropeptides, we, as well as other workers, examined unicellular organisms. Macchia et al. (1967) demonstrated the presence of material closely resembling TSH in cultures of *Clostridium perfringens,* whereas other investigators have identified hCG-like material in numerous strains of bacteria isolated from patients suffering from cancer (Acevedo et al. 1978; Maruo et al. 1979; Livingstone and Livingstone 1974; Backus and Affronti 1981). We have identified material closely resembling insulin, ACTH, β-endorphin, somatostatin, relaxin, and salmon-type calcitonin in extracts of unicellular eukaryotes and bacteria grown in defined, synthetic medium (Table 2) (LeRoith et al. 1980, 1981, 1982a; Berelowitz et al. 1982; Schwabe et al. 1983; Deftos et al. 1984).

The insulin-related materials from *Tetrahymena pyriformis* (a ciliated protozoan) and *E. coli* were purified using numerous chromatographic techniques, including Sephadex G-50, DEAE ion exchange, and reverse-phase hydrophobic chromatography with HPLC. The material reacted specifically in a mammalian insulin immunoassay and demonstrated activity in a bioassay for insulin using isolated rat adipocytes. Furthermore, the bioactivity

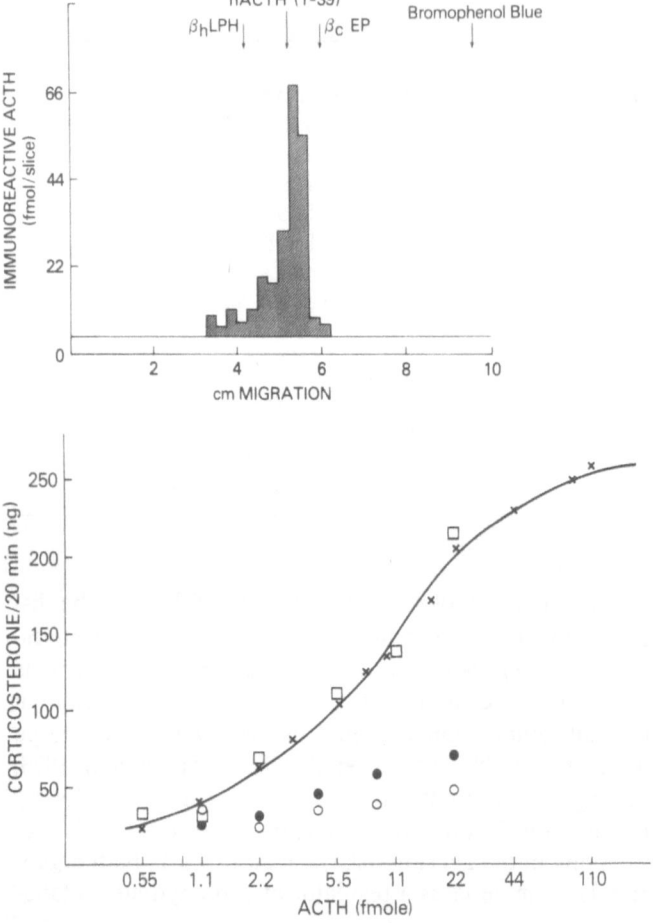

Fig. 2. ACTH-related material in *Tetrahymena pyriformis*. Following growth in defined synthetic medium, *T. pyriformis* cells were separated from the medium by continuous centrifugation. The cells were homogenized in 0.1 *M* hydrochloric acid/0.22 *M* formic acid and mixed overnight at 4 °C. Following centrifugation, the supernatant was separated from the precipitated material, defatted, and concentrated on Sep-Pak C_{18} cartridges. The material that eluted from the Sep-Pak with 1% formic acid: 60% acetonitrile was applied to a column of Sephadex G-50 and eluted with 0.22 *M* formic acid; those fractions containing ACTH-related immunoactivity were electrophoresed on SDS-polyacrylamide gel. Elution was performed in 0.2-cm cuts and the material was tested for ACTH-related immunoactivity *(upper panel)*. *βhLPH*, human lipotropin; *βcEP*, camel β endorphin.

The peak fractions containing ACTH immunoactivity were also tested for bioactivity *(lower panel)*. The bioassay measures corticosterone release from dispersed adrenal cells of the rat. Aliquots of the purified extract were tested over a 40-fold dilution range (□, ×) in comparison with standard ACTH (1–39). Immunodepletion using anti-ACTH antiserum removed almost all biological activity in the extracts (○,●). (Reprinted from LeRoith et al. 1982a)

was neutralized using an anti-insulin antibody as well as by an antibody which blocks insulin's effect on the insulin receptor (LeRoith et al. 1980; 1981). ACTH- and β-endorphin-like materials were recovered from extracts of *T. pyriformis*. The immunoactive ACTH- and β-endorphin-like materials behaved chromatographically in a similar way to authentic ACTH (1-39) and to β-endorphin standards on Sep-Pak C_{18}, Sephadex G-50, and HPLC columns and SDS-polyacrylamide gel electrophoresis (Fig. 2). The ACTH-like material was biologically active, i.e., stimulated corticosterone release from dispersed adrenal cells of the rat, and the bioactivity was removed by exposure to antibodies produced against synthetic ACTH (1-39). The β-endorphin-like material reacted in a radioreceptor assay for opiate peptides. In addition, the extracts contained materials of high molecular weight, i.e., less retarded on gel filtration that had both ACTH and β-endorphin-related moieties on the same molecule. This larger molecular weight material was approximately equal in size (on SDS-polyacrylamide gels) to pro-opiomelanocortin (POMC), the common biosynthetic precursor of ACTH and β-endorphin in vertebrates (LeRoith et al. 1982a).

Possible Role of Hormonal Peptides in Unicellular Organisms

We performed numerous experiments to demonstrate that the vertebrate-type peptide hormones present in the extracts of unicellular organisms were not artifacts of the system introduced by exogenous contamination, but were indeed native to unicellular organisms. Final confirmation of these findings will require the purification and sequence analysis of each peptide or of the gene coding for the peptide.

In addition, we have not yet demonstrated a function for these hormone-like peptides in the unicellular organisms. We suggest, however, that their function is possibly related to cellular communication, for the following reasons.

1. There is conservation of the surface structures of the peptides isolated from microbes, since they are recognized by antibodies against their vertebrate counterparts as well as by receptors on mammalian tissues (LeRoith et al. 1980, 1981, 1982a; Berelowitz et al. 1982). Evolutionary conservation of a biological region of the molecule usually suggests some important biological function is acting as a constraint against random mutations.
2. Multiple studies have suggested that vertebrate hormones affect unicellular organisms. Josefsson and Johansson (1979) showed that the feeding behavior (pinocytosis) of amoebae is affected by both opioid peptides and opiate alkaloids. This effect is reversed in the presence of naloxone, a specific opioid receptor inhibitor, suggesting the presence in amoebae of an opioid-like receptor and effector mechanism similar to that found in vertebrates (Josefsson and Johansson 1979).
3. Systems of intercellular communication serve important functions in unicellular organisms (Table 3) (Stephens et al. 1982; Sarker et al. 1979; Dunny et al. 1979). Peptides may serve as messenger molecules in some of these systems. Interestingly, α-mating factor of the yeast *Saccharomyces cerevisae* has an N-terminal amino acid sequence homology with GnRH of mammals (Fig. 3). In addition, yeast α-mating factor binds to GnRH receptors on rat pituitary cells and stimulates LH release from dispersed rat pituitary cells (Loumaye et al. 1982). Also, it has been reported that GnRH and α-mating factor both inhibit adenylate cyclase in yeast.

A Unifying Hypothesis in an Attempt to Explain Ectopic Hormone Production

Extraglandular production of hormones by vertebrate tissues has only become appreciated during the past 2 decades. Initially, production of hormones by nonendocrine tissues was only recognized because certain tumors overproduced these hormones, causing par-

Table 3. Systems of intercellular communication in unicellular organisms

Food-related	Sex-related
Myxobacteria (Stephens et al. 1982)	*Streptococcus fecalis* (Dunny et al. 1979)
Slime mold (Bonner 1971)	*Saccharomyces cervisiae* (Ciejek et al. 1977)
Bacillus brevis (Sarker et al. 1979)	

SACCHAROMYCES

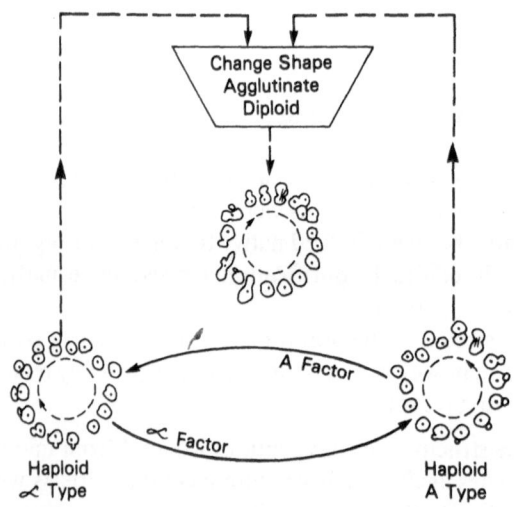

MAMMALIAN LHRH:

<GLU – HIS – TRP – SER – TYR – GLY – LEU – ARG – PRO – GLY – NH$_2$

TRP – HIS – TRP – LEU – ---- – GLN – LEU – LYS – PRO – GLY

YEAST α FACTOR:

Fig. 3. Sex pheromone in *Saccharomyces cerevisiae*. The life cycle of *S. cerevisiae* involves two mating-type haploids (*α* and A) as well as diploids. Sexual conjugation is facilitated by mating factors that induce specific changes in the other haploid cell *(upper panel)*. Amino acid sequences of mammalian luteinizing hormone releasing hormone (LHRH) and the N-terminus of *α*-mating factor of *S. cerevisiae*

ticularly well-characterized syndromes. The evidence presented in this paper strongly suggests that vertebrate-type messenger molecules are produced widely by nonvertebrate tissues and even in ancestral undifferentiated unicellular organisms. We propose that synthesis of hormonal peptides is not a unique property of glandular tissues, but that many types of cells are capable of making these peptides, albeit in much smaller amounts. Inter-

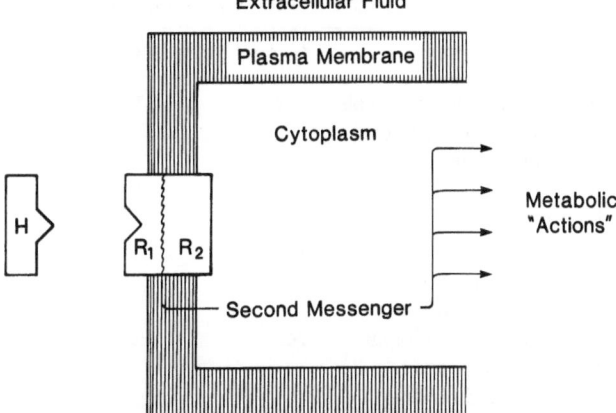

Fig. 4. Peptide hormone action on target cells. Soluble peptide hormones *(H)* bind to a specific binding site *(R 1)* on the plasma membrane receptor. *(R 2)* represents regions of the receptor not directly involved in ligand binding. Following this hormone–receptor interaction, activation of multiple early and late postreceptor events occurred, with in metabolic consequences (action)

estingly, hCG, originally thought to be produced only by placental tissue and tumors, has recently been found in testes, colon, and livers of normal males (Braunstein et al. 1975; Chen et al. 1976). Similarly, ACTH is made not only by the pituitary and brain but also by normal placenta, lung and adrenal glands (Liotta et al. 1977; Larsson 1978). Apparently, tumors not associated with ectopic hormone syndromes often contain similar quantities of ACTH or hCG to those found in the tumors that are associated with the syndromes. The difference is that the latter secrete excess amounts of these hormonal peptides, whereas the former do not secrete the hormone they contain (Ratcliff et al. 1972; Odell and Wolfsen 1978). Thus, ectopic production (and release) of hormones by tumors may represent part of a continuum, and the major difference between normal tissue, cancer cells without syndromes, tumors with syndromes, and normal endocrine glands is the amount of peptide each of the cells can synthesize, store, and release into the bloodstream and in response to what stimuli this takes place.

Syndromes Associated with Tumor Overproduction of Hormones

Upon presentation of the clinical syndrome associated with tumor overproduction of hormones (Fig. 4) an attempt can be made to identify the causative agent using an approach outlined in Table 4. A few illustrative samples are discussed below and summarized in Table 4. A speculative extension of this approach is presented by Rosenzweig et al. (1983).

Fasting Hypoglycemia

Tumor-induced hypoglycemia may be caused by excess insulin and or proinsulin released by a neoplasm of β cells of the pancreatic islets (group 1 syndrome in Table 4). When caused by non-islet tumors, hypoglycemia is probably also mediated via the metabolic events associated with insulin-like action at the target cell (Kahn 1980). In these cases in-

Table 4. Tumor-associated syndromes of hormone excess: investigation of causative agent

Type of syndrome	Group I	Group 2[a]	Group 3[a]	Group 4
1. Radioimmunoassay reveals inappropriately high level of circulating hormone	Yes	No	No	No
2. Reactive in (in vivo) bioassay	Yes	Yes	Yes	Yes
3. An excess of specific antibody to hormone neutralizes biologic effect	Yes	Yes	No	No
4. Neutralization of biologic effect by antagonist to specific ligand binding site on receptor	Yes	Yes	Yes	No
Hypoglycemia	Insulin	–	Insulin-like growth factors	Anti-insulin receptor antibody
Examples of tumor- Hyponatremia	Arginine vasopressin	Arginine vasotocin	–	–
induced: Hypercalcemia	Parathyroid hormone	–	"PTH-like"	Prostaglandin (PGE)

[a] Difference between group 2 and group 3 is the degree of reactivity in the RIA; excess antibody will cross-react with molecule causing group 2 syndrome, but not with agent causing group 3 syndrome

sulin immunoactivity is characteristically not elevated, but in approximately one-third of patients affected the blood has an elevation of an insulin-like growth factor (IGF, formerly referred to as nonsuppressible insulin-like activity or NSILA). The IGF is totally unreactive with anti-insulin antibodies, but binds to the receptor for insulin (in addition to binding to specific IGF receptors). Furthermore, the IGF-related material is probably closer to IGF-II than to IGF-I, and in excess amounts it may cause insulin-like metabolic events acting via the insulin receptors (Megyesi et al. 1974; Gorden et al. 1981). These patients correspond to group 3 in Table 4. In patients with non-islet cell tumors associated with hypoglycemia and in whom neither insulin nor IGF molecules have been identified the hypoglycemia may be caused by other insulin-related molecules that bind to and activate the insulin receptor-effector pathways.

In rare patients with autoimmunity polyclonal antibodies directed against any of multiple sites on the insulin receptor are capable of stimulating insulin-mediated events resulting in hypoglycemia (Taylor et al. 1982). Theoretically, a monovalent antibody directed towards the insulin receptor produced by lymphoma or myeloma could cause hypoglycemia, which would fall into group 4 of Table 4.

Syndrome of Inappropriate Antidiuretic Hormone (SIADH)

Abnormal secretion by tumors of antidiuretic hormone causing hyponatremia has been well described, and elevated or inappropriate circulating levels of arginine vasopressin (AVP) may be found on radioimmunoassay in 30%–50% of these cases (Robertson 1978), corresponding to group 1 in Table 4. In the remaining cases plasma AVP levels revealed by RIA are suppressed and the causative agent in these cases is as yet unidentified. Possi-

Table 5. Vertebrate phenomena explained by the proposed unifying hypothesis

1. Endocrine and nervous systems overlap: classic gastroenteropancreatic hormones are produced by nervous tissue, e. g., CCK-PZ, whereas classic neuropeptides are produced by endocrine cells, e. g., TRH in the pancreas
2. Exocrine and endocrine systems overlap: amphibian skin makes many mammalian-type peptides, e. g., bombesin. In addition, hormones may reach their target tissues via exocrine secretions, e. g., Prolactin, GnRH, and ovarian steroids in milk
3. Tissue-specific growth factors and paracrine agents: hormone-like peptides, e. g., EGF and NGF, act as signal messengers for intercellular communication
4. Hormonal peptide is produced by nonglandular tissues

ble mediators of the syndrome include substances working through totally different receptor effector pathways or alternatively, as suggested by Zerbe and Robertson (1981), molecules sufficiently different not to be detected by the AVP-specific radioimmunoassay. Potential candidates include arginine-vasotocin-related material[1] (group 2 syndrome); a substance like oxytocin which cross-reacts negligibly with the AVP antibody but binds to the AVP receptor could in theory produce a group 3 syndrome, whereas a group 4 syndrome could theoretically be caused by tumor production of an antibody directed towards the AVP receptor.

Hypercalcemia

Hypercalcemia is another extremely troublesome complication associated with tumors (Spiegel et al. 1983). Exact mechanisms remain unclear. However, two basic processes are currently held to be responsible: (a) Direct invasion of bone by metastases; and (b) production by the tumor of humoral factors causing bone resorption, e. g., parathyroid hormone (group 1 syndrome). In a number of cases prostaglandins and vitamin D-related steroids have been implicated as the humoral substance (Breslau et al. 1984; Metz et al. 1981). Both PGE and vitamin D cause hypercalcemia via pathways unrelated to parathyroid hormone receptor-effector mechanisms, and serum PTH levels are appropriately suppressed in these cases (group 4 syndrome). Of the remaining cases, many are associated with clinical and laboratory features very similar to those seen in parathyroid-induced hypercalcemia, except for the absence of parathyroid-like immunoactivity. Indeed, Rodam et al. (1982) and Tam et al. (1984) have postulated that the causative humoral factor is a substance with PTH-like biological activity without PTH-like immunoactivity (group 2 or group 3 syndrome).

Acknowledgement. We wish to thank Violet Katz for her expert secretarial assistance.

References

Acevedo HF, Slifkin M, Pouchet GR, Pardo M (1978) Immunocytochemical localization of a choriogonadotropin-like protein in bacteria isolated from cancer patients. Cancer 41: 1217–1219
Backus BT, Affronti LF (1981) Tumor-associated bacteria capable of producing a human choriogonadotropin-like substance. Infect Immun 32: 1211–1215

---

1 Arginine vasotocin-related material has been found in pineal, CSF, and brain of normal mammals

Baylin SB, Mendelsohn G (1980) Ectopic (inappropriate) hormone production by tumors: mechanisms involved and the biological and clinical implications. Endocr Rev 1: 45–77

Berelowitz M, LeRoith D, Von Schenk H, Newgard C, Szabo M, Frohman LA, Shiloach J, Roth J (1982) Somatostatin-like immunoactivity and biological activity is present in *T. pyriformis*, a ciliated Protozoan. Endocrinology 110: 1939–1944

Bhatnagar YM, Carraway R (1981) Bacterial peptides with C-terminal similarities to bovine neurotensin. Peptides 2: 51–59

Birch NP, Christie DL, Renwick AGC (1984) Proinsulin-like material in mouse fetal brain cell cultures. FEBS Lett 168: 299–302

Bonner JT (1971) Aggregation and differentiation in the cellular slime molds. Annu Rev Microbiol 25: 75–92

Braunstein GD, Rasor J, Wade ME (1975) Presence in normal human testes of a chorionic-gonadotropin-like substance distinct from human lutenizing hormone. N Engl J Med 293: 1339–1343

Breslau NA, McGuire JL, Zerwekh JE, Frenkel EP, Pak CYC (1984) Hypercalcemia associated with increased serum calcitriol levels in three patients with lymphoma. Ann Intern Med 100: 1–7

Budd GC, Pansky B, Cordell B (1983) Insulin or insulin-like peptides in the pituitary gland. Cell Biol (abstract) 83

Chen HC, Hodgen GD, Matsuura S, Lin SJ, Gross E, Reichert LE, Birken S, Canfield RE, Ross GT (1976) Evidence for a gonadotropin from nonpregnant subjects that has physical, immunological and biological similarities to human choronic gonadotropin. Proc Natl Acad Sci USA 73: 2885–2888

Ciejek E, Thorner J, Geier M (1977) Solid phase peptide synthesis of α factor, a yeast mating pheromone. Biochem Biophys Res Commun 78: 952–961

Deftos L, LeRoith D, Shiloach J, Roth J (1984) Salmon calcitonin-like immunoactivity in extracts of *Tetrahymena pyriformis*. Horm Metab Res (in press)

Dunny GM, Craig RA, Carron RL, Clewell DB (1979) Plasmid transfer in *Streptococcus fecalis*: production of multiple sex pheromones by recipients. Plasmid 2: 454–465

Duve H, Thorpe A (1979) Immunofluorescent localization of insulin-like material in the median neurosecretory cells of the blowfly *Calliphora vomitoria*. Cell Tissue Res 200: 187–191

Duve H, Thorpe A, Lazarus NR (1979) Isolation of material displaying insulin-like immunological and biological activity from the brain of the blowfly *Calliphora vomitoria*. Biochem J 284: 21–27

El-Salhy M, Abou-El-Ela R, Falkmer S, Grimelius L, Wilander E (1980) Immunohistochemical evidence of gastro-enteropancreatic neurohormonal peptides of vertebrate type in the nervous system of the larva of a dipteron insect, the hoverfly, *Eristalis aeneus*. Reg Pept 1: 187–204

Falkmer S, Emdin S, Havu N, Lundgren G, Marques M, Ostberg Y, Steiner DF, Thomas NW (1973) Insulin in invertebrates and cyclostomes. Am Zool 13: 625–628

Falkmer S, Carraway RE, El-Salhy M, Emden SO, Grimelius L, Rehfeld JF, Reinecke M, Schwartz TW (1981) Phylogeny of the gastroenteropancreatic neuroendocrine system: a review. In: Grossman MI, Brazier AB, Lechago J (eds) Cellular basis of chemical messengers in the digestive system. Academic, New York, pp 21–42

Fritsch HAR, Van Noorden S, Pearse AGE (1976) Cytochemical and immunofluorescence investigations of insulin-like producing cells in the intestine of *Mytelus edulis* (Bivalvia). Cell Tissue Res 165: 365–369

Gorden P, Hendricks CM, Kahn CR, Megyesi K, Roth J (1981) Hypoglycemia associated with non-islet-cell tumor and insulin-like growth factors. N Engl J Med 305: 1452–1455

Imura H (1980) Ectopic hormone syndromes. Clin Endocrinol Metab 9: 235–260

Josefsson JO, Johansson P (1979) Naloxone-reversible effects of opioids on pinocytosis in *Amoeba proteus*. Nature 282: 78–80

Kahn CR (1980) The riddle of tumor hypoglycemia revisited. Clin Endocrinol Metab 9: 335–360

Larsson LI (1978) Distribution of ACTH-like immunoreactivity in rat brain and gastrointestinal tract. Histochemistry 55: 225–233

LeRoith D, Shiloach J, Roth J, Lesniak MA (1980) Evolutionary origins of vertebrate hormones: substances similar to mammalian insulins are native to unicellular organisms. Proc Natl Acad Sci USA 77: 6184–6188

LeRoith D, Shiloach J, Roth J, Lesniak MA (1981) Insulin or a closely related molecule is native to *Escherichia coli*. J Biol Chem 256: 6533–6536

LeRoith D, Liotta AS, Roth J, Shiloach J, Lewis ME, Pert CB, Krieger DT (1982a) Corticotropin and β-endorphin-like materials are native to unicellular organisms. Proc Natl Acad Sci USA 79: 2086–2090

LeRoith D, Shiloach J, Roth J (1982b) Is there an earlier phylogenetic precursor that is common to both the nervous and endocrine systems? Peptides 3: 211–2154

LeRoith D, Hendricks SA, Lesniak MA, Rishi S, Becker KL, Havrankova J, Rosenzweig JL, Brownstein MJ, Roth J (1983a) Insulin in brain and other extrapancreatic tissues of vertebrates and non-vertebrates. Adv Metab Disord 10: 303–340

LeRoith D, Berelowitz M, Pickens W, Crosby LK, Shiloach J (1983b) Somatostatin-related material in *E. coli:* evidence for two molecular forms. Clin Res 31: 739 A (abstract)

Lim ATW, Lolait SJ, Barlow JW, Autelitano DJ, Toh BH, Boublik J, Abraham J, Johnston CI, Funder JW (1984) Immunoreactive ariginine-vasopressin in Brattleboro rat ovary. Nature 310: 61–63

Liotta AS, Osathanondh R, Ryan KJ, Kreiger DT (1977) Presence of corticotropin in human placenta: demonstration of in vitro synthesis. Endocrinology 101: 1552–1558

Livingston V, Livingston AM (1974) Some cultural, immunological and chemical properties of *Progenitor cryptocides*. Trans NY Acad Sci 36: 569–582

Loumaye E, Thorner J, Catt KJ (1982) Yeast mating pheromone activates mammalian gonadotrophs: evolutionary conservation of a reproductive hormone. Science 218: 1324–1325

Macchia V, Bates RW, Pastan I (1967) Purification and properties of thyroid stimulating factor isolated from *Clostridium perfringens*. J Biol Chem 242: 3726–3730

Maruo T, Cohen H, Segal SJ, Koide SS (1979) Production of choriogonadotropin-like factor by a microorganism. Proc Natl Acad Sci USA 76: 6622–6626

Megyesi K, Kahn CR, Roth J, Gorden P (1974) Hypoglycemia in association with extrapancreatic tumors: demonstration of elevated plasma NSILA-S by a new radioreceptor assay. J Clin Endocrinol Metab 38: 931–934

Metz SA, McRae JR, Robertson RP (1981) Prostaglandins as mediators of paraneoplastic syndromes: review and update. Metabolism 30: 299–316

Murakami K, Taniguchi H, Baba S (1982) Presence of insulin-like immunoreactivity and biosynthesis in rat and human parotid gland. Diabetologia 22: 358–362

Nussey SS, Ang VTY, Jenkins JS, Chowdrey HS, Bisset GW (1984) Brattleboro rat adrenal contains vasopressin. Nature 310: 64–66

Odell WD, Wolfsen AR (1978) Hormones from tumors: are they ubiquitous? Am J Med 68: 317–318

Odell WD, Wolfsen AR (1982) Humoral syndromes associated with cancer: ectopic hormone production. Prog Clin Cancer 8: 57–74

Perez-Cano R, Murphy PK, Girgis SI, Arnett TR, Blankharn I, MacIntyre I (1982) Unicellular organisms contain a molecule resembling human calcitonin. Endocrinology 110: 673 (abstract)

Plisetskaya E, Kazakov VK, Solititakaya L, Leibson LG (1978) Insulin producing cells in the gut of freshwater bivalve molluscs *Anodonta cygnea* and *Unio pictorum* and the role of insulin in the regulation of their carbohydrate metabolism. Gen Comp Endocrinol 35: 133–45

Ratcliff JG, Knight RA, Besser GM, Landon J, Stansfeld AG (1972) Tumor and plasma ACTH concentration of patients with and without the ectopic ACTH syndrome. Clin Endocrinol 1: 27–44

Robertson GL (1978) Cancer and inappropriate antidiuresis. In: Ruddon RW (ed) Biological markers of neoplasia: basic and applied aspects. Elsevier North Holland, Amsterdam, pp 277–293

Rodam SB, Insogna KL, Vignery AMC, Stewart AF, Broadus AE, D'Souza SM, Bertolini DR, Mundy GR, Bodam GA (1982) Factors associated with humoral hypercalcemia of malignancy stimulate adenylate cyclase in osteoblastic cells. J Clin Invest 72: 1511–1515

Rosenzweig JL, LeRoith D, Lesniak MA, MacIntyre I, Sawyer WH, Roth J (1983) Two distinct insulins in the guinea pig: the broad relevance of these findings to evolution of peptide hormones. Fed Proc 42: 2608–2614

Roth J, LeRoith D, Shiloach J, Rosenzweig JL, Lesniak MA, Havrankova J (1982) The evolutionary origins of hormones, neurotransmitters. N Engl J Med 306: 523–527

Sarker N, Langley D, Paulus H (1979) Studies on the mechanism and specificity of inhibiton of ribo-
nucleic acid polymerase by linear gramicidin. Biochemistry 18: 4536–4541

Schwabe C, LeRoith D, Thompson RP, Shiloach J, Roth J (1983) Relaxin extracted from protozoa
(Tetrahymena pyriformis). J Biol Chem 258: 2778–2781

Spiegel AM, Saxe AW, Deftos LJ, Brennan MF (1983) Humoral hypercalcenia caused by a rat Ley-
dig-cell tumor is associated with suppressed parathyroid hormone secretion and increased urinary
cAMP excretion. Horm Metab Res 15: 299–304

Stephens K, Hegeman GD, White D (1982) Pheromone produced by the Myxobacterium *Stigmatel-
la aurantiaca*. J Bacteriol 149: 739–747

Tam CS, Heersche JNM, Santoa A, Spiegel AM (1984) Skeletal response in rats following the im-
plantation of hypercalcemia-producing Leydig cell tumors. Metabolism 33: 50–53

Taylor SI, Grunberger G, Marcus-Samuels B, Underhill LH, Dons RF, Ryan J, Rodden RF,
Rupe CE, Gorden P (1982) Hypoglycemia associated with antibodies to the insulin receptor. N
Engl J Med 307: 1422–1426

Younes MA, D'Agnostino JB, Fazier ML, Besch PK (1984) mRNA in human placenta homologous
to insulin mRNA. Diabetes [Suppl] 1: 161 (abstract)

Zerbe RL, Robertson GL (1981) Arginine Vasotocin: identification and biological actions in mammals. In: Fotherby K, Pal SB (eds) Hormones in normal and abnormal tissues, vol 2. de Gruyter,
New York pp 165–186

Genes and Gene Products Involved in Growth Regulation of Tumor Cells*

U. R. Rapp, T. I. Bonner, K. Moelling, H. W. Jansen, K. Bister, and J. Ihle

Laboratory of Viral Carcinogenesis, National Cancer Institute, Frederick Cancer Research Facility, Frederick, MD 21701, USA

Introduction

Protooncogenes, the cellular homologs of retroviral oncogenes, are cellular genes that control the growth and affect the differentiation of eukaryotic cells. Long before the discovery of viral oncogenes and their cellular homologs (Bishop 1983), it was clear that malignantly transformed cells were genetically altered with regard to their growth factor requirements for propogation in culture (Temin 1967, 1970). Moreover, the fact that cells transformed by oncogene transducing mammalian retroviruses were blocked for the binding of epidermal growth factor (DeLarco and Todaro 1978) and released transforming growth factors into the culture medium (Todaro et al. 1979, 1980; Marquardt et al. 1983) suggested early on that some viral oncogenes might be coding for or regulating the expression of such growth factors by infected cells. These observations have led to the proposal that transformed cells maintain autonomous growth by autocrine secretion (Sporn and Todaro 1980). Substance was added to such hypotheses by the recent finding that one oncogene, v-*sis,* was derived from a cellular growth factor gene, that for platelet-derived growth factor (PDGF) (Doolittle et al. 1983; Johnsson et al. 1984). Moreover, another viral oncogene, *erb*-B, was identified as a portion of the receptor gene for epidermal grwoth factor (EGF) (Downward et al. 1984; Ullrich et al. 1984). Evidence is accumulating that other oncogenes also code for portions of growth factor receptors or for proteins that affect the signal transmission from growth factor receptors to the nucleus. Analogous to the observation that certain combinations of peptide growth factors may have synergistic effects on cell growth properties, it was recently demonstrated that combinations of "complementary" oncogenes may act synergistically to transform a particular cell type (Land et al. 1983 a, b).

Two oncogenes which may have been implicated in human lung carcinoma are *raf* and *myc*. The avian counterparts of these two genes naturally occur in the avian carcinoma virus MH$_2$ (Jansen et al. 1983 a, b, 1984; Sutrave et al. 1984, Kan et al. 1984). Experiments to be discussed indicate that this combination increases the transforming activity of either oncogene.

* This project has been funded, at least in part, with Federal funds from the Department of Health and Human Services, under contract number N01-CO-23909 with Litton Bionetics, Inc. The contents of this publication do not necessarily reflect the views or policies of the Department of Health and Human Services, nor does mention of trade names, commercial products, or organizations imply endosement by the U. S. Government

Elements of Growth Control in Normal and Transformed Cells

Growth of normal cells in culture is controlled by exogenous growth factors. Proliferation of such cells ceases once the culture medium is depleted of mitogens. Transformed cells have a decreased requirement for exogenous mitogenic signals. This is due either to the secretion of growth factors that act on their own receptors (autocrine secretion) (Sporn and Tadoro 1980) or by the constitutive expression of proximal effectors in the pathway along which mitogenic signals are translated into cell division. To the extent that continued cell division is incompatible with differentiation, such a transformed cell will be unable to differentiate.

Oncogenes comprise a family of genes which appear to include components from every level in the pathway of growth factor signal transmission. A summary of oncogenes and their location in the cell is shown in Fig. 1. The v-*sis* oncogene (Robbins et al. 1981) is an example of the ligand category and is derived from the gene for the PDGF-B chain (Doolittle et al. 1983; Waterfield et al. 1983). The *src* family genes, on the other hand, appear to belong in the signal transducer category. One of the *src* family genes, the v-*erb* B gene, was derived from the cellular gene for the EGF receptor (Downward et al. 1984; Ullrich et al. 1984). Several of the other *src* family v-*onc* proteins, including *src, fms,* and *abl,* are associated with the plasma membrane (Schultz and Oroszlan 1984; Bishop 1983) in addition to their presence in the cytoplasm. Proteins encoded by the viral oncogenes *raf, rel* (Rice, personal communication), and *mos* (Papkoff et al. 1983) are predominantly cytoplasmic, with a minor fraction being membrane-associated in the case of c-*raf*-transformed fibroblasts (Mölders et al. 1985). Many of the *src* family genes have previously been shown to code for protein kinases, most with specificity for tyrosine (Bishop 1983), while others also (Gilmore personal communication) or exclusively (Moelling et al. 1984; Rice, personal communication) phosphorylate at serine and threonine. The *gag-raf* and *gag-mil* fusion protein and the *raf* protein expressed in cells containing an LTR insertion into the mouse c-*raf* locus (Mölders et al. 1985), while negative for tyrosine-specific kinase activity (Rapp et al. 1983a), have associated kinase activity specific for serine and threonine demonstrated by use of affinity-purified proteins (Moelling et al. 1984; Schultz et al., in prepara-

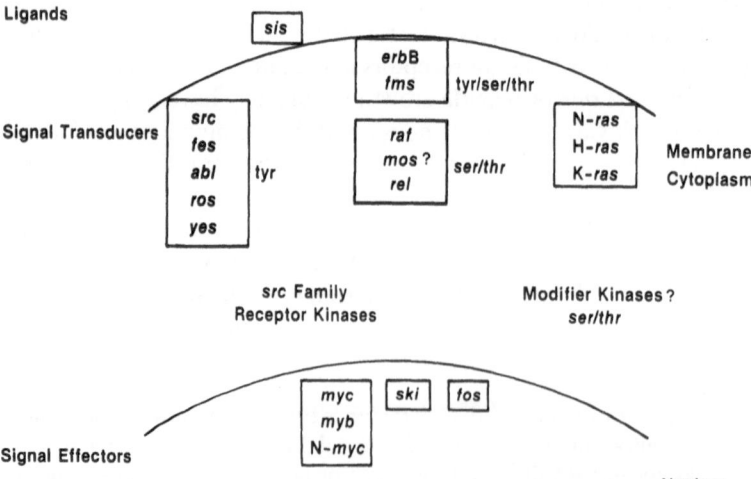

Fig. 1. Oncogenes. Genes were grouped according to their amino acid sequence relatedness and location in the cell

tion). Such an activity may also be associated with the v-*rel* (Rice, personal communication) and v-*mos* proteins (Arlinghaus, personal communication).

Another family of cytoplasmic oncogene proteins, the *ras* family, which includes Kirsten-*ras,* Harvey-*ras,* and N-*ras* (Bishop 1983), is located at the inner side of the plasma membrane, and at least two have associated autophosphorylating activity. No definitive data are available that would put any member of this family into the receptor category. There is some evidence, however, that associates their protein products with receptors. c-*ras* Proteins were found to coprecipitate with the transferrin receptor (Finkel and Cooper 1984). Also, the guanidine triphosphate (GTP)-binding activity of *ras* proteins is reminiscent of the GTP-binding regulatory subunits of adenylase cyclase N_s and N_i as well as the adenylase cyclase-regulating binding protein transducin (Houselay 1984). A third possibility for the function of *ras* proteins, again based on their GTP-binding activity, was raised by Michell (1984), who suggested a possible role in the regulation of phosphatidylinositol 4,5-biphosphate (Ptd Ins 4,5 P_2) breakdown. Hydrolysis of Ptd Ins (4,5 P_2) yields 1,2-diacylglycerol (1,2-DG) and inositol-triphosphate (Ins 3), two cellular messenger molecules which, respectively, activate protein kinase C and mobilize Ca^{2+} ions from intracellular stores (Nishizuka 1984), thereby stimulating cell proliferation. Protein kinase C in turn may increase or decrease, respectively, the tyrosine kinase and ligand binding activity of EGF (Cochet et al. 1984) and insulin (Jacobs et al. 1983) receptors. We have therefore tentatively grouped *ras* family proteins into the category of modifier kinases (Fig. 1).

A third level in the pathway of growth factor signal transmission may use oncogenic products that act in the nucleus. Such oncogenes include a gene family consisting of *myc, myb,* and perhaps N-*myc* genes as well as the unrelated genes *ski* (Stavnezer, personal communication) and *fos* (Bishop 1983). Nothing is known about the function of their proteins except for *myc,* which appears to be a DNA-binding protein (Donner et al. 1982) that can modulate transcription.

There are functional interactions within and between different growth factor categories as there are interactions between different oncogenes. Figure 2 shows a schematic representation of some of these connections as they might exist in normal and transformed cells. Examples of two apparently distinct classes of growth factor receptors are shown, one of which induces competence for cell division represented by the receptor for PDGF (Pledger et al. 1977; Stiles et al. 1979) and perhaps IL-3 (Ihle and Weinstein 1983). Others appear to transmit progression signals, a receptor category which would be represented by EGF and IGF-1 receptors. The protein kinase and ligand binding activities of these receptors may be modified by other protein kinases, which we have called here modifier kinases and which include protein kinase C and perhaps the *ras* proteins as mentioned above. Signal transmission through the PDGF receptor has been shown to induce transcription of *myc* (Kelly et al. 1983), which may in turn enhance the transcription of other genes linked with cell division – and perhaps differentiation. Ligand binding to the EGF receptor, on the other hand, may trigger transcription more directly, since in this case the receptor or an associated protein appears to be able to act on DNA as a topoisomerase (Mroszkowski et al., unpublished work). PDGF and IGF-1 cooperate in the induction of cell division (Clemmons et al. 1981). Similarly, the *myc* oncogene, which may be the proximal effector of PDGF action, has been shown to complement other oncogenes, such as *ras* and *src,* in transformation of normal embryonal fibroblasts (Land et al. 1983a; Ruley 1983), suggesting to us that complementation more generally may occur between genes in the competence and progression inducing pathways. There is another example for a specific combination of peptide factors that jointly brings about sweeping changes in cellular phenotype including both morphological transformation and growth promotion. Such a

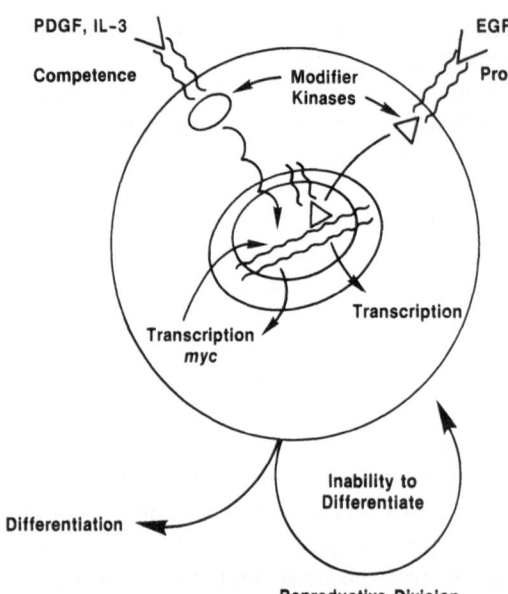

Fig. 2. Schematic representation of some oncogene-related elements controlling cell growth. *PDGF,* platelet-derived growth factor; *IL-3,* interleukin 3; *EGF,* epidermal growth factor; *IGF-1,* insulin-like growth factor 1; ≈, receptors

complementation was observed between the transforming growth factor (TGF) *α,* which binds to the EGF receptor, is mitogenic, and does not induce morphological transformation, and TGF *β,* which has its own receptor (Anzano et al. 1983), appears not to be a good mitogen, and on its own does not phenotypically transform cells. Thus, it appears that factors involved in normal growth control may be subverted to support the altered regulation and malignant growth of cells by being constitutively expressed, perhaps in specific combinations in the same cell.

The *raf* oncogene

A specific example of a complementation is the *raf* oncogene and its interaction with a second oncogene, v-*myc.* Table 1 summarizes the major structural and functional characteristics of the *raf* oncogene. v-*raf* was originally isolated from the genome of 3611 MSV, an oncogenic retrovirus that was generated as part of a systematic effort to obtain new oncogenes by retrovirus transduction (Rapp and Todaro 1978a, b, 1980; Rapp et al. 1984). At the time of its isolation, v-*raf* was found to be unrelated to all previously known oncogenes (Rapp et al. 1983a, b) except for its distant relationship (40%) with *src* family genes, which became apparent only after comparison of their deduced amino acid sequences (Mark and Rapp 1984). The avian homolog of v-*raf,* v-*mil,* was subsequently discovered in the genome of the avian carcinoma virus MH 2 (Jansen 1983a, b; Kan et al. 1983; Cole et al. 1983). v-*raf* and v-*mil* have an 80% nucleic acid sequence homology and share 94% of their amino acid sequence (Jansen et al. 1984; Sutrave et al. 1984; Kan et al. 1984).

Two genes with homology to v-*raf* are present in human DNA, a pseudogene which is located on chromosome 4 and an active gene c-*raf*-1 on the short arm of chromosome 3 at 3p25 (Bonner et al. 1984). The active gene is transcribed into a major mRNA species of 3.5 kb and a minor species of 5.5 kb (Mölders 1985). The 3.5 kb mRNA appears to have a coding capacity of 648 amino acids (Bonner et al. unpublished), which corresponds well with the size of the major cytosolic *raf* protein, 60–65 kilodaltons (Schultz et al. unpublish-

Table 1. Properties of the *raf* oncogene

Transforming gene of 3611 MSV	Isolated from a mouse with lung and peritoneal tumors. Induces fibrosarcoma/histiocytoma in newborn mice
Avian homolog of v-*raf* is part of MH 2	An avian carcinoma virus which also contains v-*myc*
Member of the *src* family	Has associated *ser/thr*-specific protein kinase activity
Two homologous genes in man, an active gene and a pseudogene	The active gene maps on chromosome 3 at p 25. Site is specifically altered in small cell lung carcinoma, ovarian carcinoma, and mixed salivary gland tumors
A third gene, *δ-raf,* is distantly related to c-*raf*-1	*δ-raf* is located on human chromosome 7 near the centromere
Active gene has 16 introns transcribed into ≥ 2 mRNAs	A 3.5 kb and minor 5.5 kb mRNA
Major mRNA codes for 648AA protein	A 74 and a 60- to 65 kilodalton protein are the major *raf* gene products Located predominantly in cytoplasm

ed). Both fluorescence and cell fractionation studies suggest that a minor fraction of cellular *raf* proteins is associated with the plasma membrane (Schultz et al., unpublished work). C-*raf*-1 is part of a subgroup within the *src* family. A distantly related gene, *δ-raf,* was isolated from a mouse cDNA library by virtue of its cross hybridization to v-*raf* (Goldsborough et al., in preparation). The human *δ-raf* gene is located on chromosome 7 near the centromere (Croce et al., in preparation).

Cells Transformed by 3611 MSV Produce Transforming Growth Factor

A characteristic of several acute transforming mammalian retroviruses is their ability to induce transformed fibroblasts to produce and secrete transforming growth factors (Todaro et al. 1979). FRE rat fibroblasts transformed by 3611 MSV share this trait with those transformed by Moloney and Kirsten MSV (Todaro et al. 1979), Snyder Theilen feline sarcoma virus (FeSV) (Marquardt et al. 1983), and Abelson leukemia virus (Twardzik et al. 1982). FRE cells nonproductively transformed by 3611 MSV are blocked for binding of epidermal growth factor (EGF) and release TGF α and TGF β into the culture medium (unpublished data). Moreover, secretion of TGFs can be demonstrated without concentration of culture fluid by seeding a single transformed nonproducer cell together with 10^5 untransformed FRE cells into soft agar, where recruitment to soft agar growth can be observed in the vicinity of the large colony of virus-transformed cells (Fig. 3).

The Active Human v-*raf* homolog, c-*raf*-1, Has Transforming Potential

The human c-*raf*-1 protooncogene has been implicated in the development of several epithelial tissue neoplasias. This was initially suggested by cytogenetic data which demonstrated an alteration of the c-*raf*-1 locus in approximately 90% of small cell lung carcinoma (Bonner et al. 1984, Table 1). Consistent with a role of c-*raf*-1, these are high levels of expression of *raf* mRNA and proteins in cell lines derived from small cell lung carcinoma (Mark, Rapp and Harris, unpublished data). To test the biological activity of c-*raf*-1 DNA directly, we have isolated this gene from a DNA library of human DNA and determined

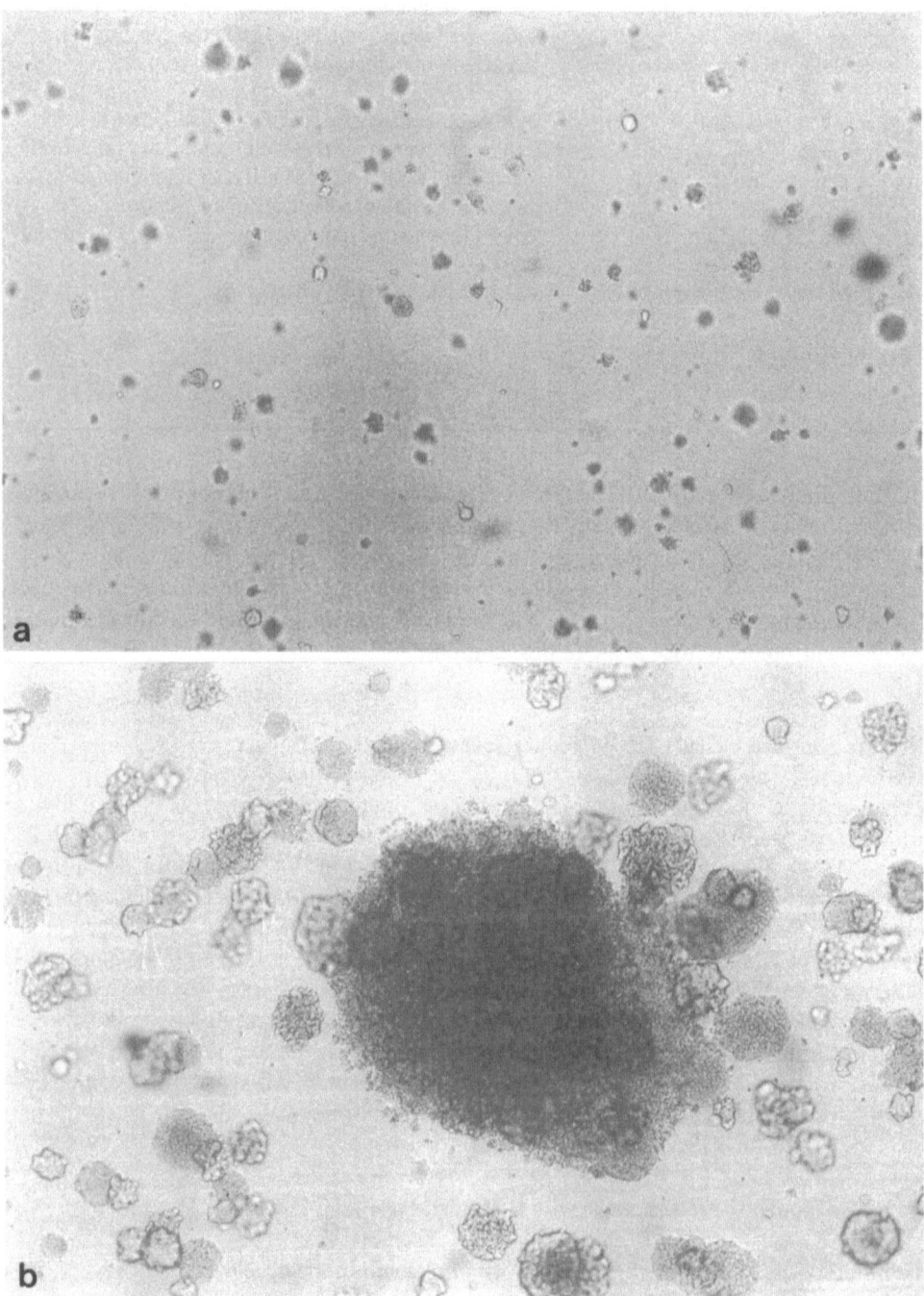

Fig. 3 a, b. Release of transforming growth factors by 3611 MSV FRE nonproducer cells. Control
FRE cells are shown for comparison (**a**). Transformed nonproducer cells were mixed with untransformed FRE cells at a ratio of $1/10^5$ (**b**). Cells were suspended in 0.3% noble agar in DMEM substi-
tuted with 10% FCS and seeded on top of a 5-ml 0.5% agar base in 60-mm petri dishes. Colonies
were scored after 10 days

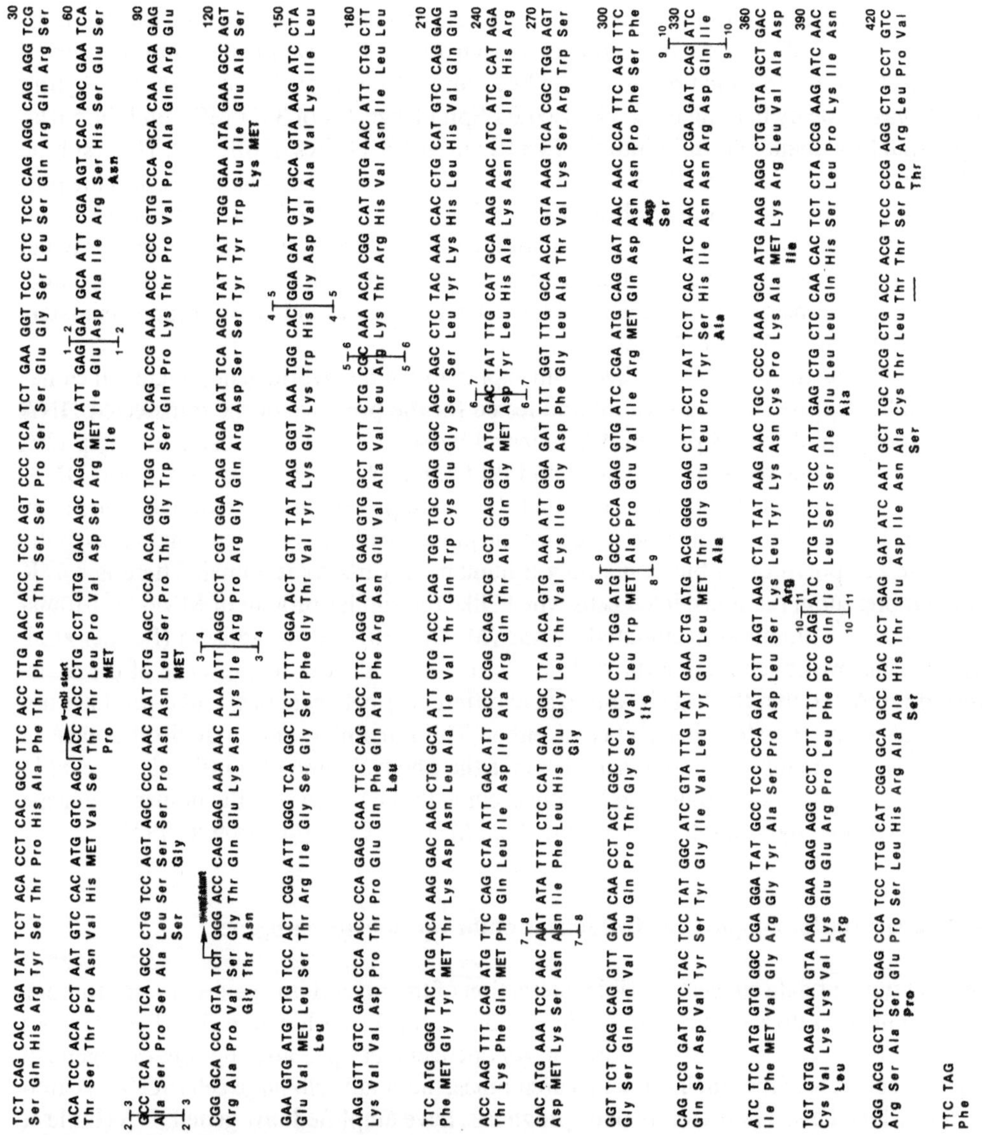

Fig. 4. Amino acid sequence of c-*raf*-1. Positions at which the v-*mil* and v-*raf* sequence differ from c-*raf*-1 are indicated below the c-*raf*-1 sequence. Isolation of v-*raf*-related chromosomal DNA and DNA sequencing are described elsewhere. (Bonner et al. 1985)

its nucleic acid sequence (Bonner et al. 1985). The 5' end of c-*raf*-1 coding sequences and the location of the poly A addition site at the 3' end were determined from the sequence of cDNA clones isolated from placenta and fetal liver cDNA libraries (Bonner et al., unpublished work). The c-*raf*-1 gene contains 11 exons which are homologous to v-*mil*, nine of which are also homologous to v-*raf*. These exons span more than 21 kb and account for 1.26 kb of coding sequence and 0.96 kb of 3' untranslated sequence. The amino acid sequence encoded by exons 1–11 of the c-*raf*-1 gene shows seven amino acid changes relative to v-*raf* and 21 amino acid changes relative to v-*mil* (Fig. 4). Of the changes relative to v-*raf*, one is in exon 5 and the remaining six are clustered at the carboxy terminal and in

exons 9–11 (Fig. 4). To test whether any of the last six changes are essential for the transforming activity of v-*raf*, we have constructed a hybrid DNA making use of the conserved
Sph I restriction site which occurs at the beginning of the exon 7 in c-*raf*-1. This hybrid
DNA was made by ligating the 2.3 kb Eco RI-Sph I fragment of 3611 MSV and the 5.6 kb
Sph I-Sal I fragment of c-*raf*-1 into the Eco RI and the Sal I sites of pBR322. The resulting
plasmid thus contains the 5' LTR of 3611 MSV, the viral *gag* gene sequences, and the portion of v-*raf* which is 5' of the Sph site followed by exons 7–11 of c-*raf*-1, the poly (A) ad-
dition site at the end of exon 11 and an additional 1.5 kb of 3' flanking sequence. When
NIH/3T3 cells were transfected with plasmid DNA, transformed colonies appeared.
However, the transforming efficiency of the hybrid DNA was much lower (approximately
1000-fold) than 3611 MSV DNA.

To verify that the colonies resulted from expression of the hybrid gene, single cell clones
from individual foci were grown up and tested for the presence of the transfected DNA
and expression of hybrid protein. When tested by Southern blotting using a v-*raf* probe,
all foci showed the presence of Eco RI and Pst I bands characteristic of the hybrid DNA.
The intensity of these bands indicated that they were present at approximately one copy
per cell. The cell clones were also tested for the presence of a polyprotein analogous to the
3611 *gag-raf* polyprotein but containing a human *raf* carboxy terminal. There is a 75K
protein present in the transfected cells which, like the similar protein of MSV-3611-trans-
formed cells, is immunoprecipitated by *gag* (MoLV p30) and *raf* antisera (Bonner et al.
1985). Thus, we conclude that the cells are transformed due to the presence of the transfected DNA and that the last six amino acid differences between v-*raf* and c-*raf*-1 are not
essential to the transforming ability of either. The remaining amino acid difference in
exon 5 is presumably also not essential, since the phenyalanine of c-*raf*-1 also occurs in
v-*mil*. This supposition would also be consistent with the report that the mouse c-*raf* gene
can be activated by promoter insertion with an MoLV LTR (Mölders et al. 1985).

Increased Transforming Activity by Linkage of the *raf* and *myc* oncogenes

Several types of data suggest possible interaction of the *raf* and *myc* genes in some tumors.
Human mixed salivary gland tumors have a 3p25; 8q23 translocation (Mark et al. 1980)
which may affect the activity of both the *raf* oncogene at 3p25 and the *myc* oncogene at
8q24. A subclass of the small cell lung carcinomas, the small cell/large cell (SC/LC) carcinomas which have a particularly poor prognosis, have amplified *myc* gene DNA (Little et
al. 1984) and also express the c-*raf*-1 gene. Morever, the avian homolog of v-*raf*, v-*mil*, is
naturally linked to v-*myc* in the carcinoma virus MH2 (Jansen et al. 1983 a, 1984; Sutrave
et al. 1984), where it presumably contributes to carcinoma induction since avian acute leukemia viruses which only contain v-*myc* have a much lower incidence of this tumor type.
It therefore became important to test the effect of v-*myc* on the oncogeneity of 3611 MSV.
We approached this question by making constructs between DNA from 3611 MSV, MH 2,
and the v-*myc* transducing virus MC 29 (Bishop 1983). The strategies used for these constructions are outlined in Fig. 5. Two constructs have been compared, pHWJ-1, which contained a hybrid *raf/mil* oncogene but a defective *myc* gene, and pHWJ-2, in which the defect in the v-*myc* gene was corrected by addition of the 3' half of MC29 v-*myc*. Foci that
appeared after cotransfection of NIH/3T3 cells with construct DNA in the presence of
helper virus DNA were isolated for growth of high-titer virus stocks and the virus was ti-
trated on NIH/3T3 cells. The transformed morphology of foci induced by pHWJ-2 was
dramatically enhanced relative to those induced by pHWJ-1 (Fig. 6). To determine the ex-

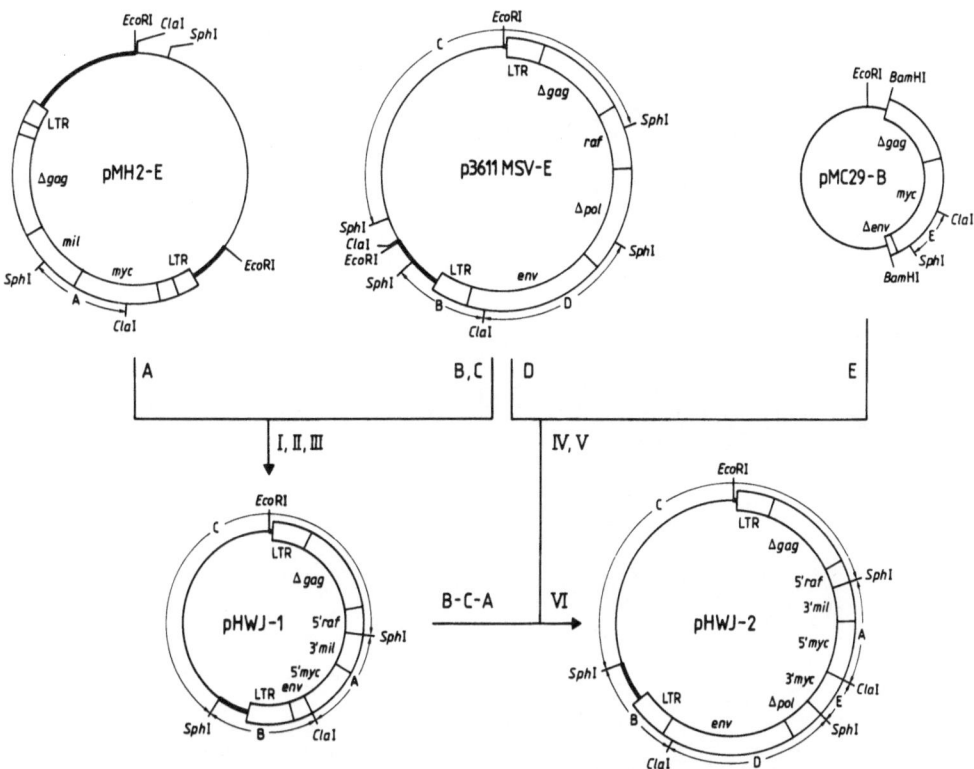

Fig. 5. For the assembly of pHWJ-1, which contains a complete *raf/mil* hybrid as well as oncogene and the 5' half of MH2 v-*myc*, 3611 MSV and MH 2 DNA was cleaved with restriction enzymes C1a I and Sph I and fragments A, B, and C were isolated. Fragments A and B were ligated and the ligated DNA cut with Sph 1. Fragment C was dephosphorylated and ligated to the A-B DNA. The resulting plasmid was selected for size and orientation and used for transfection of NIH/3T3 cells). To complete the v-*myc* gene in pHWJ-1, fragments D and E were prepared by digestion of 3611 MSV and MC 29 DNA with restriction enzymes C1a I and Sph I and ligated followed by cleavage with C1a I. The ligated D-E fragment was isolated and ligated with B-C-A DNA from pHWJ-1 which had been dephosphorylated after cleavage with C1a I. After selection for size and orientation DNA from plasmid pHWJ-2 was prepared for transfection of NIH/3T3 cells

pression of the *raf/mil* and *myc* hybrid oncogenes in these cells, we have performed immunoprecipitation experiments with antisera specific for p30, v-*raf/mil* and v-*myc*, respectively. The gag-*raf/mil* specific p75/p90 fusion proteins were detected in both transformed cells, and a v-*myc*-specific p61/63 doublet band was additionally present in pHWJ-2 transformed cells. The level of gag-*raf/mil* fusion protein expression was comparable in cells transformed by either construct (unpublished data). Consistent with these immunoprecipitation data was the pattern of expression as determined by indirect immunofluorescence with a polyvalent anti-v-*raf* serum (Oppermann and Rapp, unpublished) and a polyvalent anti-*myc* serum (Moelling and Schaller, unpublished). Uninfected NIH/3T3 cells were weakly positive for either protein, *raf* predominantly in the cytoplasm and *myc* predominantly in the cell nucleus (Fig. 7). Cells infected with 3611 MSV had increased cytoplasmic fluorescence with anti-*raf* antibodies (Fig. 7 B). Infection with pHWJ-1 and pHWJ-2 virus also leads to increased cytoplasmic *raf*-specific fluorescence (Fig. 7 C, D). The latter cells also have a bright nuclear and weak cytoplasmic and mem-

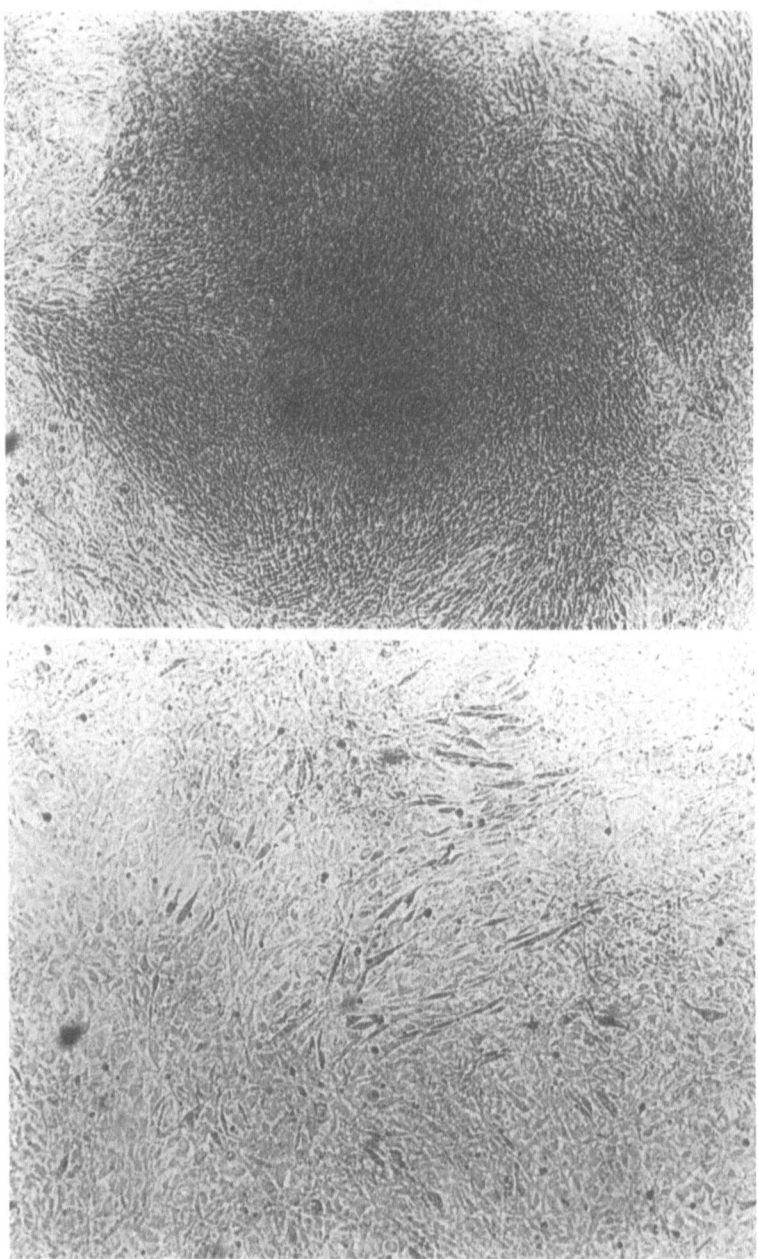

Fig. 6. Morphology of NIH/3T3 cell foci induced by pHWJ-1 *(above panel)* and pHWJ-2 *(below panel)*. Conditions for infection were as previously described (Rapp et al. 1978a). Comparison was between foci from dishes containing a comparable number of foci

brane-associated *myc*-specific fluorescence (Fig. 7 H). Having verified the proper expression of the v-*raf/mil* and v-*myc* oncogenes by both constructs we tested their ability to transform hematopoietic cells in vitro and induce tumors in newborn NSF/N mice. Consistent with their differential ability to transform fibroblasts in culture we observed a strik-

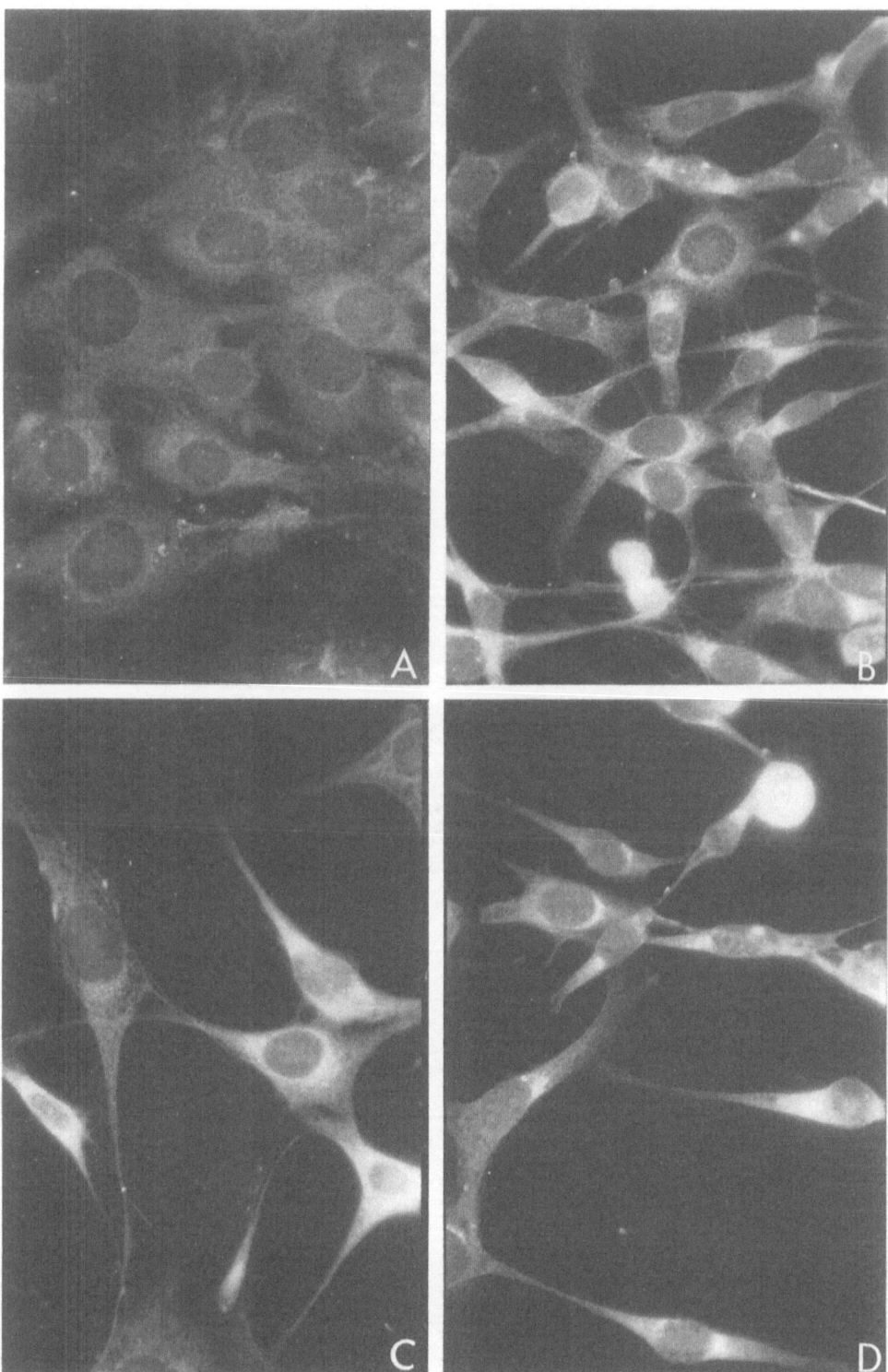

Fig. 7 A–H. Expression of v-*raf* and v-*myc* oncogenes in NIH/3T3 cells transformed by pHWJ-1 and pHWJ-2 virus, respectively. **A–D** Indirect immunofluorescence of acetone fixed cells with poly-valent rabbit α-v-*raf* protein serum (Oppermann and Rapp, unpublished).

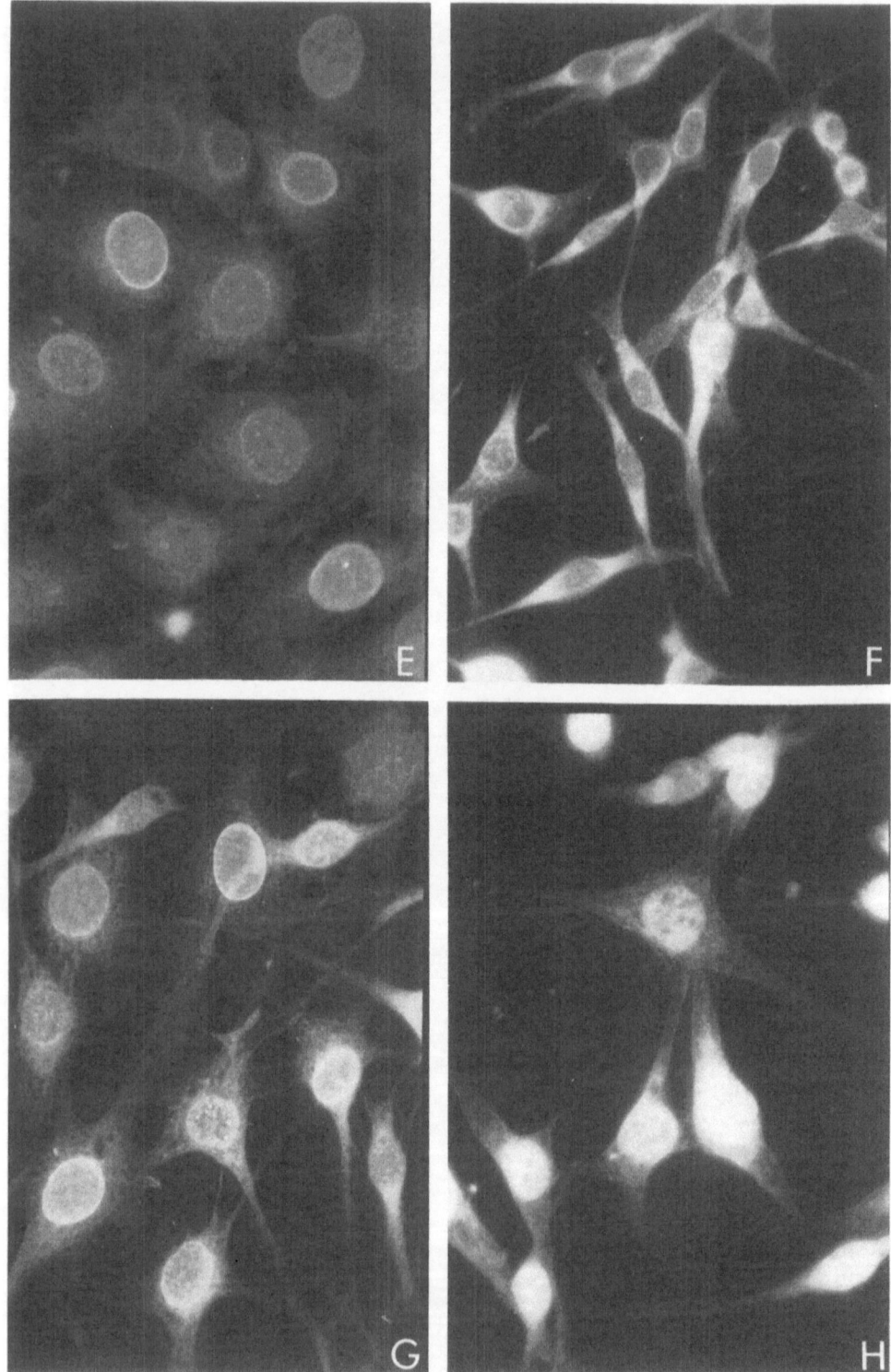

Table 2. Frequency of establishing long-term lines[a]

Virus	Oncogene	In vivo primary tumors		In vitro HPT cultures	
		+IL3	−IL3	+IL3	−IL3
3611	raf	10/10	0/10	3/3	0/3
HaSV	ras	3/3	0/3	10/10	0/10
pHWJ-2	raf + myc	10/10	10/10	2/2	2/2

[a] For tumor induction, new NSF/N mice were inoculated with 3611/MSV Harvey SV and pHWJ-2 virus as described previously (Rapp et al. 1978a). Infection of hematopoietic tissue cells (HPT) in culture included fetal liver (3611 MSV, Harvey SV and pHWJ-2) and bone marrow (3611 MSV, pHWJ-2). Culture medium was supplemented with IL-3 at 20 units/ml (Ihle and Weinstein 1983)

ing difference in the transformation of bone marrow cells. Virus derived from pHWJ-2 rapidly induced the emergence of transformed cells in these cultures in the absence of added CSF-1 or IL-3 in the culture medium, whereas pHW-1 virus did not. Similarly, pHWJ-2 virus induced rapid tumor development (2–3 weeks' latency) in 100% of the inoculated newborn mice, whereas pHWJ-1 virus had a latency > 7 weeks. The pathology of pHWJ-2-induced tumors differed from those induced by 3611 MSV in that malignant lymphomas and adenocarcinomas developed in pancreas liver and lung, in addition to the less frequent histiocytomas, fibrosarcoma, and erythroleukemias which also develop with 3611 MSV (Rapp et al. 1983a, and unpublished data). In parallel with the difference in transforming activity of pHWJ-1 and -2 for fibroblasts and hematopoietic cells in vitro, we found that permanent cell lines could be readily established from tumors induced by pHW-2 virus, whereas 3611 virus-induced tumors required IL-3 for maintenance of lines in culture (Table 2). The latter phenomenon, i.e., growth factor dependence of cells from virus-induced hematopoietic tumors, is not unique to 3611 MSV having also been observed with Harvey sarcoma virus (Table 2) and Moloney sarcoma virus (Ihle and Rein, unpublished data). Typing of tumor cell lines with a panel of differentiation specific markers has indicated that pHWJ-2 induced tumors contain T, B, and erythroid lineage cells. Cells from 3611 MSV- and Ha-MSV-induced tumors as well as from in vitro-infected fetal liver cultures which had been established in the presence of IL-3 initially consist of a mixture of pre-B cells, pre-T cells, macrophages, and mast cells, which develop into homogeneous IL-3-dependent permanent mast cell lines that continue to express virus-specific oncogene products. Bone marrow cells transformed in vitro with pHWJ-2 in the absence of a growth factor supplement resemble the tumor-derived cells and can belong to the myeloid or the B cell lineage. We conclude that combination of the v-raf with the v-myc oncogene increases the transforming activity and alters the range of target cells for virus-induced transformation. In the presence of both genes, cells are transformed in vivo and in vitro to autonomous, factor-independent growth, whereas the v-raf gene alone transforms hematopoietic cells in vivo to immortalized factor (IL-3)-dependent lines.

The mechanism that underlies the cooperative interaction between the v-raf and v-myc oncogenes in fibroblasts and hematopoietic cells is not understood. It appears that either oncogene provides, directly or indirectly, a function that complements the other. The v-raf

◀ **Fig. 7E–H.** See p. 232. Fluorescence with polyvalent rabbit α myc serum (Moelling and Schaller, unpublished work). **A, E** Uninfected NIH/3T3 control; **B, F** NIH/3T3 cells nonproductively transformed by 3611 MSV clone F4; **C, G** NIH/3T3 cells transformed by pHWJ-1 virus; **D, H** NIH/3T3 cells transformed by pHWJ-2 virus. All *panels* represent equal exposures and developing times

oncogene (Fig. 3) and the Ki-*ras* and several *src* family oncogenes, including v-*mos*, v-*abl*, and v-*fes*, are known to induce production of mitogenic signals such as TGFs upon transformation of fibroblast cells. v-*myc*, which does not morphologically transform fibroblast cell lines to a similar degree, is not known to cause secretion of such factors in these cells. Because of the induction of v-*myc* in fibroblastic and lymphoid cells (Kelly et al. 1983) upon exposure to PDGF, it may be speculated that the complementation is between a progression signal coded for or induced by v-*raf* and a competence signal provided by v-*myc* (Fig. 2).

There are currently three oncogenes implicated in the development of small cell lung carcinoma. An activated Ki-*ras* gene was isolated by DNA transfection (Santos et al. 1984) and may be involved in a small fraction of these tumors. The *myc* protooncogene was found to be amplified in small cell/large cell carcinoma (Little et al. 1984). And according to cytogenetic (Bonner et al. 1984) observations, the c-*raf*-1 protooncogene may be involved in 90% of small cell lung carcinoma. It is tempting to speculate that two or more of these genes act jointly in this tumor type in a manner similar to the joint action of v-*raf* and v-*myc* in the virus constructs that we have presented. It is therefore becoming important to study the transformation of primary human lung cells in culture with these constructs in the hope of gaining insights into mechanisms involved in the transformation-associated ectopic hormone production that may be typical for these tumors.

Acknowledgements. We gratefully acknowledge Mindy Goldsborough for her assistance in the preparation of the manuscript.

References

Anzano MA, Roberts AB, Smith JM, Sporn MB, De Larco JE (1983) Sarcoma growth factor from conditioned medium of virally transformed cells is composed of both type α and type β transforming growth factors. Proc Natl Acad Sci USA 80: 6264–6268

Bishop JM (1983) Cellular oncogenes and retroviruses. Annu Rev Biochem 52: 301–354

Bonner T, O'Brien SJ, Nash WG, Rapp UR, Morton CC, Leder P (1984) The human homologs of the *raf (mil)* oncogene are located on human chromosomes 3 and 4. Science 223: 71–74

Bonner TI, Kerby S, Sutrave P, Gunnell M, Mark GE, Rapp UR (1985) The structure and biological activity of the human homologues of the *raf/mil* oncogene. Mol Cell Biol (in press)

Clemmons DR, Underwood LE, Van Wyk JJ (1981) Hormonal control of immunoreactive somatomedin production by cultured human fibroblasts. J Clin Invest 67: 10–19

Cochet C, Gill GN, Mersenholder J, Cooper JA, Hunter T (1984) C-kinase phosphorylates the epidermal growth factor receptor and reduces its epidermal growth factor-stimulated tyrosine protein kinase activity. J Biol Chem 259: 2553–2558

Cole JM, Righi C de Toisne, Dissous C, Gegonne G, Stehelin D (1983) Molecular cloning of the avian acute transforming retrovirus MH₂ reveals a novel cell derived sequence (v-*mil*) in addition to the myc oncogene. EMBO J 2: 2189–2194

De Larco JE, Todaro GJ (1978) Growth factors from murine sarcoma virus-transformed cells. Proc Natl Acad Sci USA 75: 4001–4005

Donner P, Greiser-Wilke I, Moelling K (1982) Nuclear localization and DNA binding of the transforming gene product of avian myelocytomatosis virus. Nature 296: 262–266

Doolittle RF, Hunkapiller MW, Good LE, Devare SG, Robbins KC, Aaronson SA, Antoniades HN (1983) Simian sarcoma virus *onc* gene, v-*sis*, is derived from the gene (or genes) encoding a platelet-derived growth factor. Science 221: 275–277

Downward J, Yarden Y, Mayes E, Scrace G, Totty N, Stockwell P, Ullrich A, Schlessinger J, Waterfield MD (1984) Close similarity of epidermal growth factor receptor and v-*erb*-B oncogene protein sequences. Nature 207: 521–527

Finkel T, Cooper GM (1984) Detection of a molecular complex between *ras* proteins and transferrin receptor. Cell 136: 1115–1121

Houselay MD (1984) A family of guanine nucleotide regulatory proteins. Trends Biochem Sci 9: 39–45

Ihle JN, Weinstein Y (1983) Interleukin 3: Regulation of a lineage of lymphoid cells characterized by the expression of 20 hydroxysteroid dehydrogenase. In: Watson JD, Marbrook J (eds) Recognition and regulation in cell-mediated immunity. Dekker, New York (in press)

Jacobs S, Sahyoun NE, Saltiel AR, Cuatrecasas P (1983) Phorbol esters stimulate the phosphorylation of receptors for insulin and somatomedin C. Proc Natl Acad Sci USA 80: 6211–6213

Jansen HW, Ruckert B, Lurz R, Bister K (1983a) Two unrelated cell-derived sequences in the genome of avian leukemia and carcinoma inducing retrovirus MH$_2$. EMBO J 2: 1969–1975

Jansen HW, Patschinsky T, Bister KJ (1983b) Avian oncovirus MH$_2$: Molecular cloning of proviral DNA and structural analysis of viral RNA and protein. J Virol 48: 61–73

Jansen H, Lurz R, Bister K, Bonner TI, Mark GE, Rapp UR (1984) Homologous cell-derived oncogenes in avian carcinoma virus MH$_2$ and murine sarcoma virus 3611. Nature 307: 281–284

Johnsson A, Heldin CH, Wasteson A, Westermark B, Deuel TF, Huang JS, Seeburg DH, Gray E, Ullrich A, Scrace G, Stroobant P, Waterfield MD (1984) The c-sis gene encodes a precursor of the B chain of platelet-derived growth factor. EMBO J (in press)

Kan NC, Flordellis CS, Garon CF, Duesberg PH, Papas TS (1983) Avian carcinoma virus MH$_2$ contains a transformation specific sequence, *mht,* and shares the *myc* sequence with ML29, CMII and OK10 viruses. Proc Natl Acad Sci USA 80: 6566–6570

Kan NC, Flordellis CS, Mark GE, Duesberg PH, Papas TS (1984) A common onc gene sequence transduced by avian carcinoma virus MH$_2$ and murine sarcoma virus 3611. Science 223: 813–816

Kelly K, Cochran BH, Stiles CD, Leder P (1983) Cell-specific regulation of the c-*myc* gene by lymphocyte mitogens and platelet-derived growth factor. Cell 35: 603–610

Land H, Parada LF, Weinberg RA (1983a) Tumorigenic conversion of primary embryo fibroblasts requires at least two cooperating oncogenes. Nature 304: 596–602

Land H, Parada LF, Weinberg RA (1983b) Cellular oncogenes and multistep carcinogenesis. Science 222: 771–778

Little CD, Nau MM, Carny DN, Gazdan AF, Minna JD (1984) Amplification and expression of the c-*myc* oncogene in human lung cancer cell lines. Nature 306: 194–196

Mark GE, Rapp UR (1984) Primary structure of v-*raf:* Relatedness to the *src* family of oncogenes. Science 224: 285–288

Mark J, Dahlenfors R, Ededahl C, Stenman G (1980) The mixed salivary gland tumor – a normally benign human neoplasm frequently showing specific chromosomal abnormalities. Cancer Genet Cytogenet 2: 231

Marquardt H, Hunkapiller MW, Hood LE, Twardzik DR, De Larco JE, Stephenson JR, Todaro GJ (1983) Transforming growth factors produced by retrovirus-transformed rodent fibroblasts and human melanoma cells: amino acid sequence homology with epidermal growth factor. Proc Natl Acad Sci USA 80: 4684–4688

Michell B (1984) Oncogenes and inosito lipids. Nature 308: 770

Moelders H, Defesche J, Müller D, Bonner TI, Rapp UR, Müller R (1985) Integration of transfected LTR sequences into the c-*raf* proto-oncogene: activation by promoter insertion. EMBO J (in press)

Moelling K, Heiman B, Bunte T, Rapp UR (1984) Association of a *ser/thr*-specific protein kinase with purified *gag-mil* and *gag-raf* proteins in vitro. Nature 312: 558–561

Nishizuka Y (1984) The role of protein kinase C in cell surface signal transduction and tumor promotion. Nature 308: 693–698

Papkoff J, Lai MH, Hunter T (1983) Analysis of v-mos encoded proteins in cells transformed by several related murine sarcoma viruses. In: Scolnick EM, Levine AJ (eds) tumor viruses and differentiation. (UCLA Symposia on molecular and cellular biology, new series, vol 5) Liss, New York, pp 121–134

Pledger WJ, Stiles CD, Antoniades HN, Scher CD (1977) Induction of DNA synthesis in BALB/c 3T3 cells by serum components: reevaluation of the commitment process. Proc Natl Acad Sci USA 74: 4481–4485

Rapp UR, Todaro GJ (1978a) Generation of oncogenic type C viruses derived from C3H mouse cells in vivo and in vitro. Proc. Natl Acad Sci USA 75: 2468–2472

Rapp UR, Todaro GJ (1978b) Generation of new mouse sarcoma viruses in cell culture. Science 201: 821–824

Rapp UR, Todaro GJ (1980) Generation of oncogenic mouse type C viruses: In vitro selection of carcinoma-inducing variants. Proc Natl Acad Sci USA 77: 624–628

Rapp UR, Reynolds FH Jr, Stephenson JR (1983a) New mammalian transforming retrovirus: Demonstration of a polyprotein gene product. J Virol 45: 914–924

Rapp UR, Goldsborough MD, Mark GE, Bonner TI, Groffen J, Reynolds FH Jr, Stephenson JR (1983b) Structure and biological activity of v-raf, a unique oncogene transduced by a retrovirus. Proc Natl Acad Sci USA 80: 4218–4222

Rapp UR, Reynolds FH Jr, Stephenson JR (1984) Isolation or new mammalian type C transforming viruses. In: Pearson ML, Sternberg NL (eds) Gene transfer and cancer. Raven, New York, pp 169–177

Robbins KC, Devare SG, Aaronson SA (1981) Molecular cloning of integrated simian sarcoma virus: genome organization of infectious DNA clones. Proc Natl Acad Sci USA 78: 2918–2922

Ruley HE (1983) Adenovirus early region 1 A enables viral and cellular transforming gnes to transform primary cells in culture. Nature 304: 602–606

Santos E, Martin-Zanca D, Reddy EP, Pierotti MA, Della Porta G, Barbacid M (1984) Malignant activation of a k-ras oncogene in lung carcinoma but not in normal tissue of the same patient. Science 223: 661–664

Schultz A, Oroszlan S (1984) Myristylation of gag-onc fusion proteins in mammalian transforming retroviruses. Virology 133: 431–437

Sporn MB, Todaro GJ (1980) Autocrine secretion and malignant transformation of cells. N Engl J Med 303: 878–880

Stiles CED, Capone GT, Scher CD, Antoniades HN, Van Wyk JJ, Pledger WJ (1979) Dual control of cell growth by somatomedins and platelet-derived growth factor. Proc Natl Acad Sci USA 76: 1279–1283

Sutrave P, Bonner TI, Rapp UR, Jansen HW, Patschinsky T, Bisten K (1984) Nucleotide sequence of avian retroviral oncogene v-mil: homologue of murine retroviral oncogene v-raf. Nature 309: 85–88

Temin HM (1967) Control by factors in serum of multiplication of uninfected cells and cells infected and converted by avian sarcoma viruses. In: Growth regulating substances for animal cells in cultures. Wistar Institute, Philadelphia, pp 103–116 (Wistar symposium monographs, no. 7)

Temin HM (1970) Control of multiplication of uninfected rat cells and rat cells converted by murine sarcoma virus. J Cell Physiol 75: 107–120

Todaro GJ, De Larco JE, Marquardt H, Bryant ML, Sherwin SA, Sliski AH (1979) Polypeptide growth factors produced by tumor cells and virustransformed cells: a possible growth advantage for the producter cells. In: Sato GH, Ross R (eds) Hormones and cell culture, Book A. Cold Spring Harbor Press, New York, pp 113–127 (Cold Spring Harbor conferences on cell proliferation, vol 6)

Todaro GJ, Fryling C, De Larco JE (1980) Transforming growth factors produced by certain human tumor cells: polypeptides that interact with epidermal grwoth factor receptors. Proc Natl Acad Sci USA 77: 5258–5262

Twardzik DR, Todaro GJ, Marquardt H, Reynolds FH Jr, Stephenson JR (1982) Transformation induced by Abelson murine leukemia virus involves Production of a polypeptide growth factor. Science 216: 894–896

Ullrich A, Coussens L, Hayflick JS, Dull TJ, Gray A, Tam AW, Lee J, Yarden Y, Libermann TA, Schlessinger J, Downward J, Mayes ELV, Whittle N, Waterfied MD, Seeburg PH (1984) Human epidermal growth factor receptor cDNA sequence and aberrant expression of the amplified gene in A431 epidermoid carcinoma cells. Nature 309: 418–425

Waterfield MD, Scrace T, Whittle N, Stroobant P, Johnsson A, Wasteson A, Westermark B, Heldin C-H, Huang JS, Deuel TF (1983) Platelet-derived growth factor is structurally related to the putative transforming protein p28 of simian sarcoma virus. Nature 304: 35–39

Oncogene Expression in Human Small Cell Lung Carcinoma*

C. A. Griffin and S. B. Baylin

Johns Hopkins University Oncology Center, Endocrine Oncology, 600 North Wolfe Street, Baltimore, MD 21205, USA

Introduction

We have recently reviewed the potential dynamics of cellular differentiation in the bronchial epithelium of man, which underly the evolution of the four major types of lung cancer (squamous cell, adeno-, large cell undifferentiated, and small cell carcinomas) and which determine how the tumor types express properties characteristic of particular differentiated cell types in the normal mucosa (Baylin 1983, 1984; Goodwin et al. 1983). The accumulating evidence that small cell lung carcinoma (SCLC), which accounts for 25% of human lung cancers and is a distinct, aggressive neoplasm with neuroendocrine properties, is linked by way of a common cell lineage to the non-SCLC lung tumors has also been stressed (Baylin 1983, 1984; Goodwin et al. 1983). Particular importance in this regard attaches to the in vitro data suggesting the potential of SCLC to change towards other lung cancer histologies with time (Baylin 1983, 1984; Goodwin et al. 1983; Gazdar et al. 1981 a). An important transition step in this process may involve a variant cell form (SCLC-V) which has lost the neuroendocrine properties characteristic of SCLC.

It is critically important for our understanding of the biology and clinical behavior of lung cancer to examine further the molecular events involved in the formation and progression of these tumors. In such studies, elucidation of the genes which determine these steps is an essential line of investigation. We review here some of the early work concerning the expression of cellular oncogenes in cultures of human lung cancers (Little et al. 1983; Griffin and Baylin 1984). Recently, a variant form of SCLC has been shown to have amplification of the c-myc oncogene and increased amounts of c-myc RNA in cell culture (Little et al. 1983), suggesting a role for this oncogene in the phenotypic conversion and malignant behavior of this SCLC phenotype. The increasing evidence that cell transformation may require two or more oncogenes acting in concert (Land et al. 1983a, b; Ruley 1983; Newbold and Overell 1983) prompted us in subsequent studies to investigate the expression of other oncogenes in human lung cancer (Griffin and Baylin 1984). We discuss the finding that c-myb is differentially expressed in human lung cancer cell lines.

---

* This work was supported by ACS Grant #PDT-108, NIH Training Grant #5 T32 CA09072-05, and a gift from the W. W. Smith Foundation.
  Abbreviations. SCLC, small cell lung cancer; *NSCLC,* non-small cell lung cancer; *SCLC-V,* small cell lung cancer, variant

Materials and Methods

Cells

All cell lines have been previously described (Goodwin et al. 1983; Gazdar et al. 1980; Quinn et al. 1949; Luk et al. 1981) except OH3, which was established in our laboratory from the pleural effusion of a patient with SCLC and which grows similarly to multiple established SCLC lines and expresses typical neuroendocrine markers such as L-dopa decarboxylase (Gazdar et al. 1980). Cells were grown in RPMI 1640 medium supplemented with 8%–15% heat-inactivated FBS, penicillin (100 units/ml), and streptomycin (100 µg/ml).

Nucleic Acids

Cytoplasmic RNA was prepared from confluent cell lines by a modification of the method of Favaloro et al. (1980). Poly(A)RNA was selected by chromotography on oligo (dT) cellulose columns (Aviv and Leder 1972). Genomic DNA was prepared from isolated cell nuclei by digestion with proteinase K 100 µg/ml in 1% SDS, followed by phenol chloroform extraction.

Probes

The 1.2-kb KpnI-XbaI fragment of the AMV genome in pBr322, containing the v-*myb* gene, was kindly provided by M. Baluda (Perbal and Baluda 1982); the 5.5-kb BamHI fragment of mouse c-*myc* genomic DNA in pBr322 was kindly provided by I. R. Kirsch (Kirsch et al. 1981); and plasmid pA1, containing a 2.0-kb actin cDNA, was kindly supplied by D. Cleveland (Cleveland et al. 1980). All probes were verified by restriction enzyme mapping and then radiolabeled using [^{32}P]dCTP by standard nick translation procedures (Rigby et al. 1977) to a specific activity of $1-3 \times 10^8$ cpm/µg. Oligolabeling (Feinburg and Vogelstein 1983) was occasionally used for the actin probe, to a specific activity of $0.5-1 \times 10^9$ cpm/µg.

Hybridization

Gel electrophoresis of RNA was performed using 1.0%–1.5% agarose containing 2.2 M formaldehyde, 2–4 µg poly A-RNA per lane being electrophoresed for 16 h at 60 V with 20 mM morpholinopropanesulfonic acid (MOPS) pH 7) as running buffer. DNA from pBr322 was cut with restriction enzymes to known sizes, made single stranded, and used as size markers.

    RNA was blotted onto nitrocellulose paper by standard methods (Thomas 1980) and then prehybridized at 43 °C in 50% formamide, 0.05 M sodium phosphate buffer pH 7, 0.8 M NaCl, 4 mM EDTA, 40 mM Tris, 5 × Denhardt's solution, and 0.15 mg/ml salmon sperm DNA for 4–6 h, according to a modification of the method of Corces et al. (1981). Fresh hybridization solution was then used along with $1-3 \times 10^7$ cpm of probe and hybridized for 24–36 h at 43 °C. Filters were then washed in 50% formamide, 0.75 M NaCl, 5 mM EDTA, 50 mM Tris, 0.5% SDS at 43 °C for 40 min, followed by 0.3 M NaCl, 2 mM

EDTA, 20 mM Tris, 0.5% SDS at 65 °C for 40 min and finally 0.07 M NaCl, 2 mM EDTA, 20 mM Tris, and 0.5% SDS at 65 °C for 30 min. Filters were then exposed to Kodak XAR-5 film at −70 °C using Lightning-Plus intensifying screens for up to 7 days.

For Southern blots, genomic cellular DNA was digested with the restriction endonucleases EcoRI, Pvu II, Xba I, Hind III, Bam HI in the manufacturers' recommended buffers at 37 °C overnight. DNA 10 µg per lane was then electrophoresed through 0.8% agarose gels at 60–80 V for 6–16 h with 0.09 M Tris-borate as electrophoresis buffer. Transfer to nitrocellulose was accomplished by standard methods (Southern 1975) and then prehybridized at 65 °C in 6 × SSC and 10 × Denhardt's solution with 200 µg/ml salmon sperm DNA for 2–4 h. Fresh hybridization solution (1 M NaCl, 10 × Denhardt's solution, 1% SDS, 2 mM EDTA, 50 mM sodium phosphate, 200 µg/ml salmon sperm DNA) with 1–5 × 10⁷ cpm radiolabeled probe was then incubated with the filters at 65 °C for 24–48 h. Filters were then washed in 2 × SSC, 0.5% SDS at room temperature for 20–30 min, followed by four washes of 30 min each in 0.25 × SSC, 0.5% SDS at 65 °C for the c-*myc* and actin probes and 0.5 × SSC, 0.5% SDS, 65 °C for the v-*myb* probe. Filters were exposed as above.

Results

In our studies of poly-A RNA from 13 established human lung cancer cell lines we found that 7 of 8 SCLC lines, including 3 of 4 variant lines (Fig. 1a), express a 3.5-kb transcript homologous to v-*myb,* as does the neuroendocrine colon cell line COLO 320. This transcript is the same size as that found in the KG1 immature myeloid cell line (Westin et al. 1982) (Fig. 1a). No *myb*-homologous RNA was detected in 5 of 5 NSCLC lines examined. The presence of intact RNA in all 13 cell lines was verified by subsequent hybridization of the same filters to an actin cDNA probe (Fig. 1b), and similar amounts of actin mRNA were found in all the samples. The amount of c-*myb* mRNA detected in the SCLC line is only about 10% of the amount found in KG1 as assayed by densitometry of the autoradiogram and normalization of the *myb* signals to those for actin mRNA (data not shown).

In our studies, we confirmed the findings of Little et al. (1983) regarding c-*myc* gene amplification and increased transcription in the variant SCLC lines; and we found an interesting potential homology between transcripts for the human c-*myb* and c-*myc* oncogenes. A second transcript of about 2.4 kb was also detected with the v-*myb* probe in three variant SCLC cell lines and the COLO 320 line. The size of the smaller *myb*-hybridizing transcript coincides with that described for c-*myc* mRNA (2.4 kb) and can only be seen in the variant SCLC lines and COLO 320 lines which are known to have *myc* gene amplification and markedly increased *myc* transcription (Griffin and Baylin 1984; Alitalo et al. 1983). Figure 2 shows amplified *myc* RNA in the lines which have the smaller *myb*-homologous band, with varying lesser amounts of *myc* RNA detected in all the other cell lines tested.

Although several larger *myb* precursor RNAs have been detected in chicken tissue (Gonda et al. 1982) and mouse plasmacytomas (Mushinski et al. 1983), only a single cytoplasmic RNA transcript has been detected in human hematopoietic cells (Westin et al. 1982). Thus, the finding of a second, smaller transcript raises questions as to its source. Besides the obvious possibility that several different-sized poly(A) RNA transcripts are present in these cell lines, the smaller band might also be due to cross-hybridization to the c-*myc* RNA species. Interestingly, the *myc* and *myb* proteins have recently been shown to have structural homology, although this homology involves conservation of amino acids rather than extensive identical nucleotide sequences (Ralston and Bishop 1983). The pos-

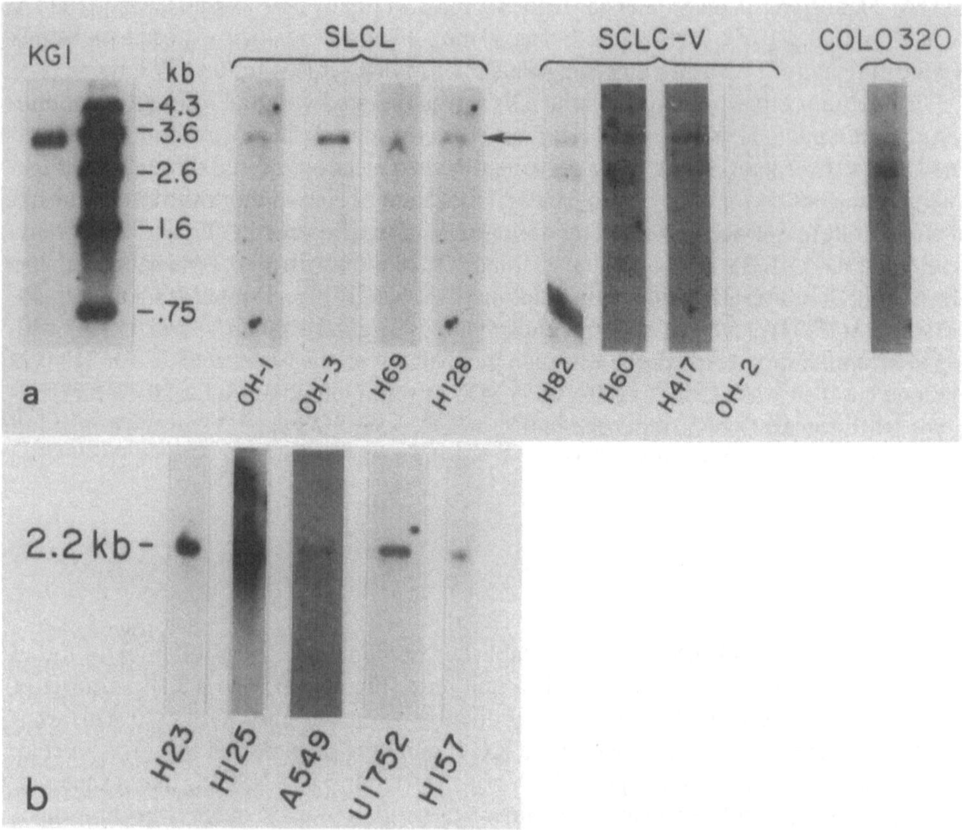

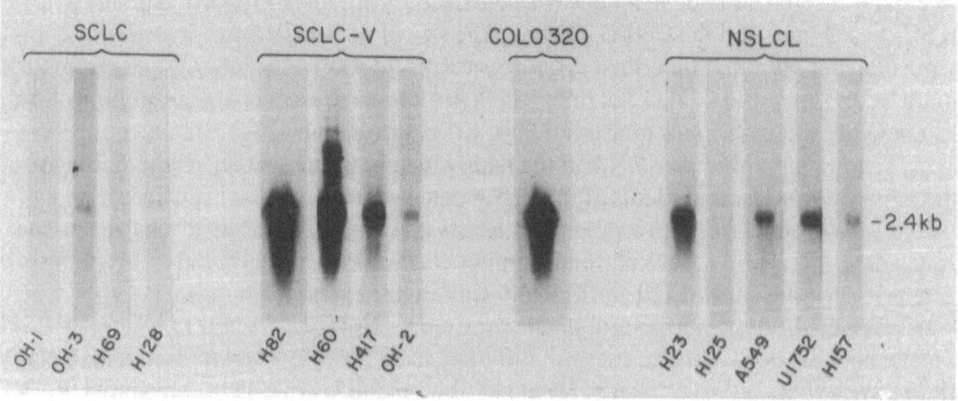

Fig. 1. a Northern blots of poly(A) RNA probed for c-*myb* in control (KG1 myeloid cell line), SCLC, SCLC-V, and Colo 320 lines show a 3.4 kb band *(arrow)* present in KG1, SCLC, SCLC-V, and Colo 320, which is not detected in the NSCLC lines. A second band at 2.4 kb can also be seen in the SCLC-V and Colo 320 lanes. **b** Northern blots probed with to verify that intact RNA in similar amounts was loaded in each lane. Only the NSCLC lanes are shown

Fig. 2. Northern blots of the same cell lines shown in Fig. 1, probed with a c-*myc* probe and showing increased amounts of c-*myc* RNA in SCLC-V and Colo 320 lines, with varying much smaller amounts *myc* RNA in SCLC and NSCLC lines

sibility that the hybridization of the v-*myb* probe to the 2.4-kb transcript might be due to weak homology between c-*myc* and c-*myb* mRNAs must be considered and further explored.

We have made a preliminary search for any alteration in c-*myb* genomic sequences which might explain activation of the gene in the SCLC cell lines. We did not detect gross rearrangments or amplification of the *myb* gene in any of these cell lines as analyzed by Southern blotting of genomic DNA restricted with EcoRI, Pvu II, Xba I, Hind III, or Bam HI (data not shown).

The mechanism of activation of the c-*myb* gene in SCLC, then, remains unclear at this time. Although rearranged *myb* sequences have been detected in mouse lymphoid tumors (Mushinski et al. 1983), we did not find rearrangements in these cell lines with the five restriction enzymes used. Other restriction fragment rearrangements or a point mutation, etc., have not been excluded. Further characterization of the c-*myb* gene present in these cells will be required to elucidate the mechanism of activation. Most genes, however, do not undergo rearrangement to trigger gene activity, so it may very well be that no easily discernable reason for activation exists.

Discussion

Our present findings (Griffin and Baylin 1984) and those of Little et al. (1983), provide some initial data concerning genes whose expression may be important for determining and/or maintaining the phenotype of some of the major forms of human lung cancer. A 3.5-kb RNA transcript homologous to v-*myb* is present in four of four lines of classic SCLC and in three of four SCLC variant lines, but not in five of five non-small cell lung cancer (NSCLC) lines tested. These results suggest a significant difference in the levels of transcription of this oncogene among the four major histologic phenotypes of lung cancer, and a possible role for c-*myb* transcription in expression of the SCLC phenotype. Similary, the findings of Little et al. (1983), confirmed in our own work (Griffin and Baylin 1984), suggest a role for increased expression of c-*myc* in the appearance and/or maintenance of the variant SCLC phenotype. This cell type, which has lost the typical neuroendocrine features of classic SCLC in culture, may represent a step in the transition from SCLC towards non-SCLC lung cancer (Baylin 1983, 1984; Goodwin et al. 1983; Gazdar et al. 1981a; Yesner 1978). Such an event in tumor progression may be important in patients, since the variant SCLC cells probably exist in the host (Goodwin et al. 1983; Yesner 1978; Matthews 1979), may be more tumorigenic (Gazdar et al. 1981a), and are relatively more radioresistant than classic SCLC cells (Goodwin and Baylin 1982; Garney et al. 1983).

In viewing the transition of SCLC in culture to the variant SCLC cells, it is important to speculate upon the timing of c-*myb* and c-*myc* gene expression. Most variant SCLC lines do not completely lose their SCLC features, especially that of growing as suspended cell aggregates in culture (Gazdar et al. 1981a; Goodwin and Baylin 1982). Their morphology shows partial but generally not total movement towards a large cell undifferentiated phenotype. The one line of variant SCLC in our studies that did not have detectable c-*myb* transcripts is thus of particular interest. This cell line, OH-2, has previously been reported in detail (Goodwin et al. 1983). Established from the pleural effusion of a patient with typical SCLC on routine histology, the line lacks the high levels of L-dopa decarboxylase activity which typically distinguish SCLC from non-SCLC in culture. The cells grew as typical large cell undifferentiated lung cancer, including formation of monolayers in cell culture and documentation of large cell histology in nude mouse heterotransplants. Chromosome analysis showed the 3p(-) deletion considered characteristic of SCLC (Whang Peng et al. 1982), and at autopsy the patient again demonstrated predominantly SCLC histology. However, foci of similar large cell undifferentiated cancer and squamous cell carcinoma were also identified (Goodwin et al. 1983). Cell surface protein analysis showed that all clones of this cell line simultaneoulsy expressed both SCLC and non-SCLC surface proteins (Goodwin et al. 1983). Other variant SCLC lines whose surface proteins have been examined do not generally exhibit non-SCLC proteins (S. B. Baylin, et al., unpublished results). Thus, this SCLC variant line contains both non-SCLC and SCLC phenotypic features simultaneously.

Unlike other variant SCLC lines (Little et al. 1983), OH-2 not only lacks transcripts for c-*myb*, but also does not have c-*myc* gene amplification. Expression of c-*myc* however (Fig. 1), is present and is shomewhat higher than in the classic SCLC lines. Thus, it may be that in a spectrum of tumor progression events involving movement of SCLC towards a non-SCLC phenotype, c-*myb* expression plus c-*myc* amplification is a transient situation associated with expression or formation of the variant SCLC cells. Further movement of the SCLC cells towards the large cell undifferentiated phenotype may involve loss of c-*myb* expression and c-*myc* amplification. Such a series of events merits further study and interpretation.

The presence of c-*myb* transcript sequences in an epithelial cell system merits special comment. Avian myeloblastosis virus, which carries v-*myb*, induces only leukemia and not carcinomas in its natural host. Because c-*myb* transcripts were initially reported only in hematopoietic cells from chickens, mice, and humans (Westin et al. 1982; Gonda et al. 1982; Mushinski et al. 1983; Rossen and Tereba 1983), and because c-*myb* transcripts are decreased in HL-60 cells chemically induced to differentiate (Westin et al. 1982), it has been postulated by other workers that the role of this oncogene is related in some way to early stages of hematopoietic cell development (Westin et al. 1982; Gallo and Wang-Staal 1982). The current detection of c-*myb* transcripts in lung cancer epithelial cell lines and in other occasional human carcinoma tissue (Slamon et al. 1984) suggests, not surprisingly, that this gene has a function not limited to cells of hematopoietic derivation.

With respect to how a gene like c-*myb* might be concerned with the specific phenotype of SCLC, our data suggest at least two possible hypotheses concerning the expression and function of c-*myb* in epithelial cells. First, c-*myb* could be associated with expression of a neuroendocrine phenotype. Among the human lung cancers, only SCLC consistently contains neurosecretory granules and expresses high levels of L-dopa decarboxylase (Gazdar et al. 1980; Baylin et al. 1978, 1980), neuron-specific enolase (Marangos et al. 1982), bb-CPK (Gazdar et al. 1981b), and bombesin (Erisman et al. 1982; Moody et al. 1981; Wood et al. 1981). Because of these properties, this tumor has been included in the APUD (amine precursor uptake and dopa decarboxylase (Baylin 1984; Pearse and Takor-Takor 1979) group of endocrine cells. Similar neuroendocrine properties are associated with endocrine cells located in intestinal epithelium (Mendelsohn and Baylin 1982). The presence of c-*myb* transcripts in SCLC cells and an intestinal tumor (Colo 320), which has been reported to express neuroendocrine properties, then suggests a possible role for this gene in cells with a neuroendocrine phenotype. However, it is important to note that the variant SCLC cells have lost the neuroendocrine phenotype (Baylin 1983, 1984; Goodwin et al. 1983; Gazdar et al. 1981a) and yet retain c-*myb* expression. Also, in our laboratory, the Colo 320 line does not contain key neuroendocrine markers such as L-dopa decarboxylase at this time. Examination of other neuroendocrine cell lines and tissues will be required for further consideration of this hypothesis.

Alternatively, c-*myb* expression may be important to normal or neoplastic bronchial epithelial cells at certain stages of differentiation along a common cell lineage from which both the SCLC and NSCLC tumors may derive. As discussed in detail elsewhere, increasing evidence suggests that each of the major forms of human lung cancer arises in cells of endodermal lineage, and that a common bronchial mucosal stem cell may have the potential for expressing the full range of differentiated cell phenotypes seen in the bronchial mucosa, including the neuroendocrine phenotype (for reviews see Baylin 1983, 1984). Descriptions have been given of how SCLC has been observed to change with time towards the NSCLC phenotype, both in culture (Goodwin et al. 1983) and in the host (Godwin et al. 1983; Gazdar et al. 1981a; Matthews 1979; Abeloft et al. 1979; Brereton et al. 1978). If, as we have recently suggested (Baylin 1984), SCLC could represent a neoplasm frozen at an earlier step in epithelial differentiation than that represented by the other types of lung cancers, then the function of the c-*myb* gene in epithelial cells may indeed correlate with early stages of cell maturation, as was initially postulated from observations of hematopoietic cells (Westin et al. 1982; Gallo and Wong-Staal 1982). In turn, *myc* amplification may be able to mediate the initiation of phenotypic conversion, represented by the SCLC variants, only in the setting of concurrent *myb* expression, which may maintain the cells in a very early stage of differentiation. Similarly, the role of other oncogenes, such as the *raf* oncogene, which has recently been reported to be transcribed in SCLC (Mark and Rapp 1984), should be considered in conjunction with *myb* and *myc*.

While the exact significance of the findings for c-*myb* and c-*myc* expression in lung cancer cells cannot be known without much further investigation, it is clear that investigations of gene expression in lung carcinoma should prove of fundamental and clinical significance. The culture lines of lung neoplasms which are becoming available for study and the increased understanding of the cellular relationships underlying both normal and neoplastic bronchial epithelial cell development make the study of lung cancer cells ideal for the investigation of genetic events which regulate cell maturation and differentiation. Also, the protein products and gene transcripts of oncogenes may become important diagnostic markers and potential therapeutic targets for lung carcinoma.

Acknowledgements. We thank M. A. Baluda, I. J. Kirsch, and D. W. Cleveland for the v-*myb*, c-*myc*, and actin probes, respectively; D. Carney, A. Gazdar, and J. Minna for multiple cell lines; and B. Nelkin and P. Pitha for helpful advice.

References

Abeloff MD, Eggleston JC, Mendelsohn G, Ettinger DS, Baylin SB (1979) Changes in morphological and biochemical characteristics of small cell carcinoma of the lung-A clinicopathologic study. Am J Med 66: 757–764

Alitalo K, Schwab M, Lin CC, Varmus HE, Bishop JM (1983) Homogeneously staining chromosomal regions contain amplified copies of an abundantly expressed cellular oncogene (c-myc) in malignant neuroendocrine cells from a human colon carcinoma. Proc Natl Acad Sci USA 80: 1707–1711

Aviv H, Leder P (1972) Purification of biologically active globin messenger RNA by chromatography on oligo thymidylic acid-cellulose. Proc Natl Acad Sci USA 69: 1408

Baylin SB (1983) Biochemical markers of human small (oat) cell lung carcinoma-Biological and clinical implications. In: Fishman WH (ed) Oncodevelopmental markers: biologic, diagnostic and monitoring aspects. Academic, New York, pp 259–277

Baylin SB (1984) The implications of differentiation relationships between endocrine and non-endocrine cells in human lung cancer. In: Hesch RD (ed) Serono Symposium. Academic, New York, (in press)

Baylin SB, Weisburger WR, Eggleston JC, Mendelsohn G, Beaven MA, Abeloff MD, Ettinger DS (1978) Variable content of histaminase, L-dopa decarboxylase, and calcitonin in small cell carcinoma of the lung. N Engl J Med 299: 105-110

Baylin SB, Abeloff MD, Goodwin G, Carney DN, Gazdar AF (1980) Activities of L-dopa decarboxylase and diamine oxidase (histaminase) in human lung cancers and decarboxylase as a marker for small (oat) cell cancer in culture. Cancer Res 40: 1990-1994

Brereton HD, Matthews MM, Costa J, Kent H, Johnson RE (1978) Mixed anaplastic small-cell and squamous-cell carcinoma of the lung. Ann Intern Med 88: 805-806

Carney DN, Mitchell JB, Kinsella TJ (1983) In vitro radiation and chemotherapy of established cell lines of human small cell lung cancer and its large cell variants. Cancer Res 43: 2806-2811

Cleveland DW, Lopata MA, MacDonald RJ, Cowan NJ, Rutter WJ, Kirschner MW (1980) Number and evolutionary conservation of - and -tubulin and cytoplasmic β- and γ-actin genes using specific cloned cDNA probes. Cell 20: 195-105

Corces V, Pellicer A, Axel R, Meselson M (1981) Integration, transcription, and control of a Drosophila heat shock gene in mouse cells. Proc Natl Acad Sci USA 78: 7038-7042

Erisman MD, Lonnoila RI, Hernandez O, DiAugustine RP, Lazarus LH (1982) Human lung small-cell carcinoma contains bombesin. Proc Natl Acad Sci USA 79: 2379-2383

Favaloro JR, Freisman R, Kamen R (1980) Transcription maps of polyoma virus-specific RNA: Analysis by two-dimensional nuclease S_1 gel mapping. Methods Enzymol 65: 718

Feinberg AF, Vogelstein B (1983) A technique for recovering and radiolabeling DNA fragments to high specific activity. Anal Biochem 132: 6-13

Gallo RC, Wong-Staal F (1982) Retroviruses as etiologic agents of some animal and human leukemias and lymphomas and as tools for elucidating the molecular mechanism of leukemogenesis. Blood 60: 545-557

Gazdar AF, Carney DN, Russell EK, Sims HL, Baylin SB, Bunn PA, Jr, Guccion JG, Minna JD (1980) Establishment of continuous, clonable cultures of small-cell carcinoma of the lung which have amine precursor uptake and decarboxylation cell properties. Cancer Res 40: 3502-3507

Gazdar AF, Carney DN, Guccion JG, Baylin SB (1981a) Small cell carcinoma of the lung: cellular origin and relationship to other pulmonary tumors. In: Greco FA, Oldham RK, Bunn PA (eds) Small cell carcinoma of the lung. Grune and Stratton, New York, pp 145-175

Gazdar AF, Zweig MH, Carney DN, van Steirteghen AC, Baylin SB, Minna JD (1981b) Levels of creatine kinase and its BB isoenzyme in lung cancer specimans and cultures. Cancer Res 41: 2773-2777

Gonda TT, Sheiness DK, Bishop JM (1982) Transcripts from the cellular homologs of retroviral oncogenes: distribution among chicken tissues. Mol Cell Biol 2: 617-624

Goodwin G, Baylin SB (1982) Relationships between neuroendocrine differentiation and sensitivity to γ-radiation in culture line O-H-1 of human small cell lung carcinoma. Cancer Res 42: 1361-1367

Goodwin G, Shaper JH, Abeloff MD, Mendelsohn G, Baylin SB (1983) Analysis of cell surface proteins delineates a differentiation pathway linking endocrine and nonendocrine human lung cancers. Proc Natl Acad Sci USA 80: 3807-3811

Griffin CA, Baylin SB (1985) Expression of the c-myb oncogene in human small cell lung carcinoma. Cancer Res 45: 272-275

Kirsch IR, Ravetch JV, Kwan SP, Max EE, Ney RL, Leder P (1981) Multiple immunoglobulin switch region homologies outside the heavy chain constant region locus. Nature 293: 585-587

Land H, Parada LF, Weinberg RA (1983a) Tumorigenic conversion of primary embryo fibroblasts requires at least two cooperative oncogenes. Nature 304: 596-602

Land H, Parada LF, Weinberg RA (1983b) Cellular oncogenes and multistep carcinogenesis. Science 222: 771-778

Little CD, Nau MM, Carney DA, Gazdar AF, Minna JD (1983) Amplification and expression of the c-myc oncogene in human lung cancer cell lines. Nautre 306: 194-196

Luk GD, Goodwin G, Marton LJ, Baylin SB (1981) Polyamines are necessary for the survival of human small-cell lung carcinoma in culture. Proc Natl Acad Sci USA 78: 2355–2358

Marangos PJ, Gazdar AF, Carney DN (1982) Neuron specific enolase in human small cell carcinoma cultures. Cancer Lett 15: 67–71

Mark GE, Rapp UR (1984) Primary structure of v-raf: relatedness to the src family of oncogenes. Science 224: 285–289

Matthews MJ (1979) Effects of therapy on the morphology and bahavior of small cell carcinoma of the lung-A clinicopathologic study. In: Muggia F, Rozencweig M (eds) Lung cancer: progress in therapeutic research. Raven, New York, pp 155–165

Mendelsohn G, Baylin SB (1982) The biological and clinical implications of polypeptide hormones. In: Sell S, Wahren B (eds) Human cancer markers. Humana, New Jersey, pp 321–358

Moody TW, Pert CB, Gazdar AF, Carney DN, Minna JD (1981) High levels of intracellular bombesin characterize human small-cell lung carcinoma. Science 214: 1246–1248

Mushinski JF, Potter M, Bauer SR, Reddy EP (1983) DNA rearrangement and altered RNA expression of the c-myb oncogene in mouse plasmacytoid lymphosarcomas. Science 220: 795–798

Newbold RF, Overell RW (1983) Fibroblast immortality is a prerequisite for transformation by EJ c-Ha-ras oncogene. Nature 304: 648–651

Pearse AGE, Takor-Takor T (1979) Embryology of the diffuse neuroendocrine system and its relationship to the common peptides. Fed Proc Fed Am Soc Exp Biol 38: 2288–2294

Perbal E, Baluda MA (1982) Avian myeloblastosis virus transforming gene is related to unique chicken DNA regions separated by at least one intervening sequence. J Virology 41: 250–257

Quinn LA, Moore GE, Morgan RT, Woods LK (1979) Cell lines from human colon carcinoma with unusual cell products, double minutes, and homogeneously staining regions. Cancer Res 39: 4914–4924

Ralston R, Bishop JM (1983) The protein products of the myc and myb oncogenes and adenonus EIA are structurally related. Nature 306: 803–806

Rigby PWJ, Dickmann M, Rhodes C, Berg P (1977) Labeling deoxyribonucleic acid to high specific activity *in vitro* by nick translation with DNA polymerase I. J Mol Biol 113: 237–251

Rossen D, Tereba A (1983) Transcription of hematopoietic-associated oncogenes in childhood leukemia. Cancer Res 43: 3912–3918

Ruley HE (1983) Adenovirus early region 1A enables viral and cellular transforming genes to transform primary cells in culture. Nature 304: 602–606

Slamon DJ, deKerion JB, Verma IM, Cline MJ (1984) Expression of cellular oncogenes in human malignancies. Science 224: 256–262

Southern E (1975) Detection of specific sequences among DNA fragments separated by gel electrophoresis. J Mol Biol 98: 503

Thomas PS (1980) Hybridization of denatured RNA and small DNA fragments transferred to nitrocellulose. Proc Natl Acad Sci USA 77: 5201–5205

Westin EH, Gallo RC, Arya SK, Eva A, Souza LM, Baluda MA, Aaronson SA, Wong-Staal F (1982) Differential expression of the AMV gene in human hematopoietic cells. Proc Natl Acad Sci USA 79: 2194–2198

Whang Peng J, Kao-Shun CS, Lee EC, Bunn PA, Carney DN, Gazdar AF, Minna JD (1982) A specific chromosome defect associated with human small cell lung cancer: deletion 3P (14–23). Science 215: 181–183

Wood SM, Wood JR, Ghatei MA, Lee YC, O'Shaugnessy D, Bloom SR (1981) Bombesin, somatostatin and neurotensin-like immunoreactivity in bronchial carcinoma. J Clin Endocrinol Metab 53: 1310–1312

Yesner R (1978) Spectrum of lung cancer and ectopic hormones. Pathol Annu 13: 217–240

Subject Index